AA003092

Particle Technology Forum

Core Programming Topic at the 2011 AIChE Annual Meeting

Minneapolis, Minnesota, USA
16-21 October 2011

ISBN: 978-1-61839-724-9

Printed from e-media with permission by:

Curran Associates, Inc.
57 Morehouse Lane
Red Hook, NY 12571

Printed by Curran Associates, Inc. (2012)

For permission requests, please contact AIChE
at the address below.

AIChE
3 Park Avenue
New York, NY 10016-5991

Phone: (203) 702-7660
Fax: (203) 775-5177

www.aiche.org

Additional copies of this publication are available from:

Curran Associates, Inc.
57 Morehouse Lane
Red Hook, NY 12571 USA
Phone: 845-758-0400
Fax: 845-758-2634
Email: curran@proceedings.com
Web: www.proceedings.com

TABLE OF CONTENTS

(25a) Filtered Models for Reacting Gas-Particle Flows

William Holloway, Chemical Engineering, Princeton University, Princeton, NJ and Sankaran Sundaresan, Department of Chemical Engineering, Princeton University, Princeton, NJ

Particle laden flows are used in a wide variety of industrially relevant processes that span applications ranging from the synthesis of specialty chemicals to the refining of petroleum products. Gas-particle flows exhibit persistent fluctuations in velocity and particle volume fraction that span a range of length and time scales. Flow structures are manifested as a consequence of the instability of the uniformly fluidized state [1], and can be predicted via continuum models that treat the particle and fluid phases as interpenetrating continua [2]. These models commonly take the form of balance equations for mass, momentum, and energy associated with fluctuating motion of particle and fluid phases. While the continuum model framework is capable of predicting these flow structures, the ability to accurately resolve all relevant length scales requires very fine grid resolution (on the order of 10 particle diameters) [3]. Such fine resolution is not affordable when simulating gas-particle flows large devices, and much coarser grid resolution is generally employed.

Recently, filtered models have been developed for the accurate simulation of non-reacting gas-particle flows on coarse numerical grids without neglecting the consequences of fine scale structure [4]. The present study is concerned with the extension of previous work to reacting gas-particle flows. Coarse-grid simulations of reacting multiphase flows require filtered species and energy balances, in addition to filtered hydrodynamic equations. In the present study we construct filtered balance equations for gaseous reactants participating in catalytic reactions on the solid surface. Such filtered equations call for filtered chemical reaction rates, which should account for modification of the reaction rates brought about by mesoscale inhomogeneous structures.

To this end we have performed highly resolved simulations of gas-particle flow in the presence of a solid catalyzed, irreversible, first order, isothermal, gas phase reaction. Our simulations reveal that the effectiveness factor associated with mesoscale structures decreases systematically with increasing filter size and reaction rate constant. This implies that coarse-grid simulations of reacting multiphase flows will overestimate conversions if proper allowances are not made for the decreased effectiveness factors. A model for the cluster-scale effectiveness factor will be presented, which was constructed by filtering high resolution reacting gas-particle flow simulations.

[1] Sangani A, Koch, DL. Particle pressure and marginal stability limits for a homoge- neous monodisperse gas-fluidized bed: kinetic theory and numerical simulations. J. Fluid. Mech. 1999; 400: 229-263.

[2] Jackson R, The Dynamics of Fluidized Particles. Cambridge University Press: Cambridge, 2000.

[3] Agrawal K, Loezos, PN, Syamlal, M, Sundaresan, S. The role of meso-scale structures in rapid gas-solid flows. J. Fluid. Mech. 2001; 445: 151-185

[4] Igci, Y, Andrews, AT, Sundaresan, S, Pannala, S, O'Brien, T. Filtered two-fluid models for fluidized gas-particle suspensions. AIChE J. 2008; 54: 1431-1448

See more of this Session: Fundamentals of Fluidization I
See more of this Group/Topical: Particle Technology Forum

(25b) Elutriation and Species Segregation Characteristics of Polydisperse Mixtures of Group B Particles In a Dilute CFB Riser

Jia Wei Chew, Drew M. Parker and **Christine M. Hrenya**, Chemical and Biological Engineering, University of Colorado at Boulder, Boulder, CO

Elutriation and Species SegregationCharacteristics of Polydisperse Mixtures of Group B particles in a dilute CFB Riser

Jia Wei Chew, Drew M. Parker, ChristineM. Hrenya*

Department of Chemical and BiologicalEngineering

University of Colorado at Boulder

Boulder, CO 80309

Elutriationhas long been recognized as an issue in gas-solid fluidized beds, with alargely empirical understanding focused on binary mixtures. In this work, experiments in agas-solids circulating fluidized bed (CFB) have been carried out using twocategories of polydispersity: binary mixtures with varying compositions andcontinuous particle size distributions (PSDs) with varying widths. Measurementsinclude total and species elutriation flux, and local mass flux and speciessegregation. Qualitative differences between the binary mixtures and continuousPSDs were observed. First, whilean increase of the total elutriation flux is linked to an increase in the mass% of fine particles (with terminal velocity, U_t, less than the superficial gas velocity, U_s) for the binary mixtures,the trends are opposite for the continuous PSDs, the latter of which is counterto the plethora of composition-based elutriation correlations. Second, whilethe elutriation flux of coarse particles ($U_t > U_s$) increasesnon-monotonically with mass % of fines for binary mixtures, that of continuousPSDs is monotonic. Both of these qualitative differences can be explained bythe increasing disparity of size differences with an increase in the distributionwidth for the continuous PSDs, while the size disparity remains constant inbinary mixtures of varying compositions. Third, while the mass % of coarse particles decreases non-monotonicallywith riser height for binary mixtures near the wall, the decrease is monotonicfor continuous PSDs. These observations underscore the inherent differencesbetween binary mixtures and continuous size distributions, and point to theneed for independent experiments and model validation for each type of polydispersity.

*Corresponding author: hrenya@colorado.edu

See more of this Session: Fundamentals of Fluidization I
See more of this Group/Topical: Particle Technology Forum

(25c) Quantification of GasPhase Velocity Fluctuations In Statistically Homogeneous GasSolid Flow Using ParticleResolved Direct Numerical Simulation

Sudheer Tenneti[1], Rahul Garg[2] and Shankar Subramaniam[1], (1)Mechanical Engineering, Iowa State University, Ames, IA, (2)National Energy Technology Laboratory, Morgantown, WV

Gas–phase velocity fluctuations in statistically homogeneous gas–solids flow are quantified using particle–resolved direct numerical simulation (DNS). The kinetic energy associated with the gas–phase velocity fluctuations k_f in steady flow past fixed random assemblies of monodisperse spheres is characterized as a function of solid volume fraction ϕ and the Reynolds number based on the mean slip velocity Re_m. The DNS approach is based on a new formulation we refer to as the Particle–resolved Uncontaminated–fluid Reconcilable Immersed Boundary Method (PUReIBM). A simple scaling analysis is used to explain the dependence of k_f on ϕ and Re_m. The steady value of k_f results from the balance between the source of k_f due to interphase transfer of kinetic energy, and the dissipation of k_f in the gas–phase. It is found that it is appropriate to model the dissipation of k_f in gas–solid flows using a length scale that is analogous to the Taylor microscale used in single–phase turbulence.

See more of this Session: Fundamentals of Fluidization I
See more of this Group/Topical: Particle Technology Forum

(25d) Interpretation of Specularity Coefficient Used In Boundary Conditions for Rapid Granular Flows

Tingwen Li, URS Corp. National Energy Technology Laboratory, Morgantown, WV and Sofiane Benyahia, National Energy Technology Laboratory, Department of Energy, Morgantown, WV

Interpretation of Specularity Coefficient Used in Boundary Conditions for Rapid Granular Flows

Tingwen Li[1,2], Sofiane Benyahia[1]

1. National Energy Technology Laboratory, Morgantown, WV 26505, U.S.A.

2. URS Corporation, Morgantown, WV 26505, U.S.A. Abstract

In this presentation, we revisit Johnson and Jackson boundary conditions for rapid granular flows (Johnson PC, Jackson R. Frictional-collisional constitutive relations for granular materials, with application to plane shearing. Journal of Fluid Mechanics. 1987; 176: 67-93). The oblique collision between a particle and a flat wall is analyzed by adopting the classic rigid-body theory and a more realistic semi-analytical model. Based on the kinetic granular theory, the input parameter for the partial-slip boundary conditions, specularity coefficient, which is not measureable in experiments, is then interpreted as a function of the particle-wall restitution coefficient, the frictional coefficient, and the normalized slip velocity at the wall. An analytical expression for the specularity coefficient is suggested for a flat, frictional surface with a low frictional coefficient. The procedure for determining the specularity coefficient for a more general problem is outlined, and a working approximation is provided.

See more of this Session: Fundamentals of Fluidization I
See more of this Group/Topical: Particle Technology Forum

(25e) Experimental Study of the Fluidization of Geldart A Type Catalyst In a Rotating Fluidized Bed In a Static Geometry

Waldo Rosales Trujillo, Dept. Materials and Process Eng., Universite catholique de Louvain (UCL), Louvain-la-Neuve, Belgium and Juray De Wilde, Materials and Process Engineering, UCL, Louvain-la-Neuve, Belgium

The fluidization of Geldart A type FCC catalyst in a rotating fluidized bed in a static geometry (RFB-SG) was experimentally studied. In RFB-SGs, the particle bed rotational motion is introduced by the tangential injection of the fluidization gas in the fluidization chamber, via multiple gas inlet slots. Three different gas inlet designs were tested, essentially varying the number of gas inlet slots and their size. This allowed evaluating the effect of the uniformity of gas injection and of the amount of energy and tangential momentum injected with the gas. Two different systems for feeding the solids were also considered. In the first, the solids are fed axially via one of the end plates of the reactor, in the second they are fed via one of the gas inlet slots.

Measurements over a range of gas flow rates and solids feeding rates were carried out. The pressure drop over the fluidization chamber and over the particle bed, the solids loading in the fluidization chamber, and the particle bed height, average density, uniformity and rotational speed were determined by means of pressure sensors, high speed cameras, and a rotating flag inserted in the particle bed.

Using the experimental data, the hydrodynamic behavior in a RFB-SG of Geldart A type particles is investigated. The presence of waves in the system and the importance of a free vortex in the particle bed freeboard region are analyzed. The optimization of the fluidization chamber design for the fluidization of Geldart A type and smaller particles is discussed. The potential of the technology for application in Fluid Catalytic Cracking (FCC) and in other processes is evaluated.

See more of this Session: Fundamentals of Fluidization I
See more of this Group/Topical: Particle Technology Forum

(25f) Investigation On Electrostatics In Fluidized Bed

Yongpan Cheng[1], Eldin Wee Chuan Lim[1], Chi-Hwa Wang[1], Jun Yao[2], Guoqing Guan[3], Chihiro Fushimi[4] and Atsushi Tsutsumi[4], (1)Chemical and Biomolecular Engineering, National University of Singapore, Singapore, Singapore, (2)School of Energy Research, Xiamen University, Xiamen, China, (3)North Japan Research Institute for Sustainable Energy (NJRISE), Hirosaki University, Hirosaki, Japan, (4)Institute of Industrial Science, The University of Tokyo, Tokyo, Japan

Electrostatics is unavoidable in fluidized bed due to the frequent collisions among solid particles, as well as between solid particles and the wall. It can cause serious problems for fluidized bed, altering hydrodynamics, causing agglomeration, interfering with instrumentation, generating nuisance discharges and even creating the danger of explosions. The accumulated electrostatic charges on the solid particles or walls can greatly affect the performance of fluidized bed. Despite these negative effects of electrostatics, current study is far from complete. Our study will focus on charge accumulation in the solid particles and wall of fluidized bed. The influence of solid particle properties, fluidizing velocities on the electrostatics will be studied through experiments, thus we can have better understanding into the mechanism of electrostatics generation and transport. The influence of electrostatics on the image reconstruction in Electrical Capacitance Tomography will also be addressed.

*Corresponding author. Tel.: +65 6516 5079; Fax: +65 6779 1936; E-mail address: chewch@nus.edu.sg (C. H. Wang)

See more of this Session: Fundamentals of Fluidization I
See more of this Group/Topical: Particle Technology Forum

(25g) Pressure Fluctuations In Fluidized Beds of Different Diameters and the Effect of Baffles and Operating Conditions

Allan Issangya[1], S. B. Reddy Karri[1], Ted Knowlton[1] and Ray Cocco[2], (1) Particulate Solid Research, Inc., Chicago, IL, (2)PSRI, Chicago, IL

This paper discusses differential pressure fluctuations measured in 0.3, 0.6, 0.9 and 1.52 m diameter units for FCC catalyst particles. The influence of fines content, baffles, pressure, bed height and gas velocity are examined.

See more of this Session: Fundamentals of Fluidization I
See more of this Group/Topical: Particle Technology Forum

(27a) Adsorption and Desorption of Polymeric Nanoparticulate Drug Delivery Systems On Cellulose Surfaces

Ming Zhang and **Mustafa Akbulut**, Chemical Engineering, Texas A&M University, College Station, TX

With the increasing consumption of nanomedicine and a lack of regulation governing their disposal, the occurrence and fate of medicines in the environment and the potential consequences for human health have been increasingly recognized as issues warranting consideration. To properly assess the distribution and fate of nanomedicine and develop successful strategies for minimizing any side effects on the environment, one has to understand how nanomedicine interacts and adsorbs on environmental interfaces. To this end, we investigated the adsorption and desorption behavior of polymeric nanoparticulate drug delivery systems (PNDDS), which represent one of the most commonly used forms of drug delivery, on cellulose surfaces. In this study, cellulose is selected because it is the most abundant component of plant roots and important in the context of determining how PNDDS may accumulate on the plant roots, which can be consumed by animals or affect symbiotic relationships with fungi and bacteria. It was found that the adsorption was partially reversible, and strongly influenced by PNDDS size.

See more of this Session: Health and Environmental Effect of Nanoparticles
See more of this Group/Topical: Particle Technology Forum

(27b) Effects of Charged Polymeric Nanoparticles On Pulmonary Surfactant Function

Amir Mohammad Farnoud, Chemical and Biochemical Engineering, University of Iowa, Iowa City, IA and Jennifer Fiegel, Chemical and Biochemical Engineering and Pharmaceutics and Translational Therapeutics, University of Iowa, Iowa City, IA

Pulmonary surfactant (PS) is vital for maintaining the normal function of the lungs during respiratory maneuvers. This surfactant layer reduces the surface tension of the alveolar fluid, thereby reducing the energy required to inflate the lungs and stabilizing the alveoli to avoid collapse. Disrupted function of the surfactant layer can lead to impaired gas exchange and decreased lung compliance. *In vitro* studies have shown that certain nanoparticles have the ability to alter surfactant function. However, it remains unclear which physicochemical properties of the particles cause surfactant inhibition. The present study aims to elucidate the effects of nanoparticle charge on the function of two pulmonary surfactant model systems: a monolayer containing the primary lipid of PS, dipalmitoylphosphatidylcholine (DPPC), and a complex surfactant model derived from calf lungs, Infasurf.

Aqueous solutions containing charged nanoparticles (200 nm carboxyl- or amine-modified polystyrene) were aerosolized as droplets onto monolayers of DPPC or Infasurf. The dynamic surface tension of the surfactant films was measured during compression and expansion cycles using a computer-controlled Langmuir trough equipped with a Wilhelmy plate balance. Surfactant microstructure was studied at various surface pressures using fluorescence microscopy, after addition of the fluorescent probe Texas Red-DHPE to the surfactant film. To aid interpretation of the data, particle properties (size, charge and surface elemental composition) were characterized before and after surfactant interaction.

Nanoparticle-induced changes to surfactant function were based on the composition of the surfactant film, particle dose and charge. On the DPPC films, nanoparticles led to a dose dependent increase in the liquid expanded-liquid condensed coexistence region, independent of particle charge. This led to early gelation of lipid condensed domains and, in some cases, a torn monolayer. Nanoparticle aggregation was observed upon interaction with DPPC, but did not induce a change in particle charge. These results suggest a stabilizing effect of charged nanoparticles on DPPC films due to an electrostatic screening effect. This effect reduces electrostatic repulsion between the DPPC head groups, leading to the formation of smaller condensed phases and an excess of liquid lipids during the transition phase.

See more of this Session: Health and Environmental Effect of Nanoparticles
See more of this Group/Topical: Particle Technology Forum

(27c) No Evidence of Cerium Oxide Nanoparticle Translocation In Maize Plants

Karin Birbaum[1], **Robert N. Grass**[2], Detlef Gnther[1], Ludwig K. Limbach[3] and Wendelin J. Stark[3], (1)Departement of Chemistry and Applied Biosciences, ETH Zurich, Zurich, Switzerland, (2)Institute for Chemical and Bioengineering, ETH Zurich, Zurich, Switzerland, (3)ETH Zurich, Zurich, Switzerland

The rapidly increasing production of engineered nanoparticles has raised questions regarding their environmental impact and their mobility to overcome biological important barriers. Nanoparticles were found to cross different mammalian barriers, which is summarized under the term translocation. The present work investigates the uptake and translocation of cerium dioxide nanoparticles in maize plants as one of the major agricultural crops.

In order to simulate the gas-phase exposure of maize plants to nanoparticles, flame spray pyrolysis (FSP) was utilized to directly synthesize a cerium oxide comprising aerosol. (Birbaum et al. 2010) The aerosol was produced in an enclosed cabinet (glove box, 2 $m^{3,}$ see Figure 1) in which the maize plants were cultivated. After digestion of the maize leaves, the cerium concentrations were measured by a sector field inductively coupled mass spectrometer (ICP-SF-MS), allowing a detection limit of less than 1 ng of cerium per gram of leaf. (< 1 ppb mass/mass). The plant leaves were analysed for the cerium content directly after exposure and 12 weeks post exposure in order to measure the cerium translocation within the plant.

Element analysis results showed that cerium oxide nanoparticles were present on/ in the maize leaves after the aerosol exposure. While it was not possible to discriminate between particles taken up by the plant cells and particles adsorbed to the surface of the leaves, the deposited nanoparticles could not be washed off with water (simulating rainfall). Examination of the concentrations in the leaves 3 months after the exposure experiments revealed that the cerium oxide nanoparticles did not translocate into newly formed leaves. No cerium oxide (detection limit of 0.4 ng cerium / g leaf) could be measured in the leaves grown during the three months post exposure period.

The presentation will further show the influence of light (exposure under light and dark for open and closed stomata) and results on the suspension exposure of maize plants (via the soil). The results of the maize plant exposure will be discussed and compared with *in vitro* lung cell exposure experiments (Rothen-Ruthishauser et al. 2010, Raemy et al. 2011) performed under the same exposure conditions.

Refs:

Birbaum, K., Brogioli, R. Schellenberg, M., Martinoia, E., Stark, W.J., Günther, D. and Limbach, L.K. (2010) *Environ. Sci. Technol.* **44**, 8718-8723.

Raemy, D.O., Limbach, L.K., Rothen-Rutishauser, B., Grass, R.N., Gehr, P., Birbaum, K., Brandenberger, C., Günther, D. and Stark, W.J. (2011) *Eur. J. Pharm. Biopharm.* doi: 10.1016/ j.ejpb.2010.11.017.

Rothen-Rutishauser, B., Grass, R.N., Blank, F., Limbach, L.K., Muehlfeld, C., Brandenberger, C., Raemy, D.O., Gehr, P. and Stark W.J. (2009) *Environ. Sci. Technol.* **43**, 2634-2640.

See more of this Session: Health and Environmental Effect of Nanoparticles
See more of this Group/Topical: Particle Technology Forum

(47a) A Discrete Population Balance Model for Particle Breakage In Dense-Phase Particulate Systems

Ecevit Bilgili[1], Maxx Capece[1] and Rajesh Dave[2], (1)Chemical, Biological and Pharmaceutical Engineering, New Jersey Institute of Technology, Newark, NJ, (2) Department of Chemical, Biological and Pharmaceutical Engineering, New Jersey Center for Engineered Particulates, Newark, NJ

Population balance models (PBMs) provide a quantitative understanding at the process length scale for comminution processes. Not only can PBMs serve as a quantitative tool for simulation, design, and optimization [1,2], but they can also elucidate the breakage mechanism(s) such as massive fracture, cleavage, and/or attrition [3]. Time-discrete or space-discrete PBMs are commonly used for comminution processes in which the average residence (retention) time of the particles is relatively short and the process can be described by considering a series of elementary breakage events each with a small duration of Dt *or with a small length DL*, respectively [1].

Traditional discrete *linear* PBMs (DL-PBMs) assume the independence of particle breakage from the surrounding particle population. However, at the particle ensemble scale, particles of all sizes mechanically interact with each other. Particles, which have a dynamically varying size distribution and inter-particle contacts, deform and transmit forces among themselves. The distribution of contact forces (stresses) evolves with the continuously changing population density and thus leads to a population-dependent breakage probability. These multi-particle interactions become dominant especially in dense-phase comminution processes. Toward accounting for these multi-particle interactions explicitly, a non-linear kinetics framework for rate-based comminution processes has been proposed [4], which decomposes the specific breakage rate into an apparent first-order breakage rate and a population-dependent functional. The functional describes different types of non-first order breakage kinetics.

In this study, using the aforementioned framework, we formulate a fully discrete, *non-linear* population balance model (DNL-PBM) in an attempt to analyze and model the multi-particle interactions for dense-phase comminution processes. Numerical simulations considering the breakage of binary- and poly-dispersed particle mixtures were performed to assess the predictive capability of the DNL-

PBM over the traditional DL-PBM. They provided significant insight into the retardation effect of fines on the breakage probability of the coarser particles experimentally observed in particle bed compression tests. With a flexible mathematical structure able to account for multi-particle interactions explicitly, the DNL-PBM can serve as a framework to model dense-phase comminution processes. The practical applicability, accuracy, and stability aspects of the model are also discussed.

References

[1] L.G. Austin, A review: introduction to the mathematical description of grinding as a rate process, Powder Technol. 5 (1971) 1–17.

[2] A.D. Randolph, M.A. Larson, Theory of Particulate Processes, Academic Press, San Diego, 1988.

[3] E. Bilgili, R. Hamey, B. Scarlett, Nano-milling of pigment agglomerates using a wet stirred media mill: elucidation of the kinetics and breakage mechanisms, Chem. Eng. Sci. 61 (2006) 149–157.

[4] E. Bilgili, J. Yepes, B. Scarlett, Formulation of a non-linear framework for population balance modeling of batch grinding: beyond first-order kinetics, Chem. Eng. Sci. 61 (2006) 33–44.

See more of this Session: Population Balance Modeling for Particle Formation Processes: Nucleation, Aggregation and Breakage Kernels
See more of this Group/Topical: Particle Technology Forum

(47b) On a Class of Analytical Solutions for the Population Balance Equation with Aggregation and Breakage

Tony Saad, Sean T. Smith, Alex Abboud and Terry A. Ring, Chemical Engineering, Institute for Clean and Secure Energy, University of Utah, Salt Lake City, UT

The analysis of reactive, dispersed-phase systems relies almost entirely on the accurate representation of the behavior of particles, their chemistry, and hydrodynamic interaction. This type of analysis is generally approached by solving the population balance equation (PBE) coupled with species transport to track the progress of chemical reactions.

Population balances are ubiquitous in the many branches of scientific and engineering applications. Albeit the theoretical foundations of PBE models have long been established, the literature appears to be lagging in analytical solutions except for a handful of studies. Analytical solutions provide an avenue for closed form, simple evaluations that help in identifying the key complexities of a particular model as well as benchmarks for experimental and numerical investigations.

In this talk, we discuss two approaches for deriving exact and approximate solutions for the population balance equation. We first present a novel transformation that maps the non-homogeneous PBE into a simple advection equation with the internal coordinate growth rate acting as the advecting velocity. This applies to a special class of population balances, namely, those that track one internal coordinate r with birth and death rates dependent only on r. Our transformation allows us to derive exact solutions for an arbitrary initial population. These are documented and applied to a plug flow reactor model for a mineral carbonation process where the formation and precipition of calcium carbonate is tracked via the population balance equation.

We then extend our analysis to more complex, nonlinear systems where exact, closed form solutions cannot be found. We apply the Homotopy Analysis Method (HAM) to an expanded version of the plug flow reactor example where the supersaturation ratio is made to be time dependent. Our results indicate that the required simplicity of exact analytical solutions for PBE models can be relaxed by using novel analytical techniques that aid in linearizing the complex birth and death rates.

See more of this Session: Population Balance Modeling for Particle Formation Processes: Nucleation, Aggregation and Breakage Kernels
See more of this Group/Topical: Particle Technology Forum

(47c) Solution of a Four-Component Population Balance Model for Wet Granulation Via Constant-Number Monte Carlo

Carl L. Marshall Jr.[1], Themis Matsoukas[1] and Pavol Rajniak[2], (1)Chemical Engineering, The Pennsylvania State University, University Park, PA, (2) Pharmaceutical Commercialization Technology, Merck & Co., Inc., West Point, PA

In the past, granulation population balance models (PBMs) typically were one-dimensional with granule size as the characteristic of interest. However, various attributes such as particle morphology, size, porosity, wettability, and binder viscosity can play key roles in determining the overall size and composition of the granules that are produced. Our previous research considered the impact of composition on two-component granulation [1,2]. In this work, we present a stochastic model for multicomponent granulation in which binder, moisture and two solid powders (e.g., pharmaceutical drug and excipient) are treated as explicit components that affect granulation. A modified KTGF (kinetic theory of granular flow) kernel that incorporates a critical Stokes number analysis for evaluating collision efficiency [3] is used in the model. Specifically, we combine the physical description for surface roughness given by Stepanek et al. [4] with a consideration for the degree of moisture evaporation from binder due to heat effects in order to evaluate granulation growth profiles and granule composition. Because we expect the smoothest powder to be assimilated into a granule faster than the second powder, the compositional distribution across various size classes ("sieve cuts") and at various times in the granulation is studied. In addition, higher temperatures are expected to decrease the coagulation constant contained within the kernel due to an increased rate of moisture removal and binder solidification. Therefore, heterogeneities within granule structures due to a local binder "age" are examined. All simulations are conducted via constant-number Monte Carlo and results compared with experimental data (obtained from a fluidized bed granulator with continuous addition of binder). Upon solving the PBM, other comparison points include: granule growth profile, loss on drying (LOD), binder distribution, particle size distribution. The status of this research is presented.

References:

[1] C. L. Marshall Jr., P. Rajniak, T. Matsoukas, Numerical simulations of two-component granulation: Comparison of three methods, Chemical Engineering Research and Design In Press, Corrected Proof (2010) doi:10.1016/j.cherd.2010.06.003.

[2] C. L. Marshall, Jr., T. Matsoukas, P. Rajniak, Multicomponent Population Balance Modeling of Granulation with Continuous Addition of Binder (2011--in preparation)

[3] L. Liu, J. Litster, S. Iveson, B. Ennis, Coalescence of deformable granules in wet granulation processes, AIChE Journal 46 (2000) 529-539.

[4] F. Stepanek, P. Rajniak, C. Mancinelli, R. Chern, R. Ramachandran, Distribution and accessibility of binder in wet granules, Powder Technology 189 (2009) 376-384.

See more of this Session: Population Balance Modeling for Particle Formation Processes: Nucleation, Aggregation and Breakage Kernels
See more of this Group/Topical: Particle Technology Forum

(47d) A Multi-Dimensional Population Balance Model Validation Approach to High-Shear Wet Granulation (HSWG) Processes

Anwesha Chaudhury, Chemical and Biochemical Engineering, Rutgers, The State University of New Jersey, Piscataway, NJ, Preetanshu Pandey, Bristol-Myers Sqibb, New Brunswick, NJ and Rohit Ramachandran, Department of Chemical and Biochemical Engineering, Rutgers, The State University of New Jersey, Piscataway, NJ

Process understanding and control, by the implementation of Quality by Design (QbD) is critical to the pharmaceutical industry, to ensure high quality products. An important unit operation in the manufacture of final dosage forms is high shear wet granulation (HSWG). High shear granulation results in improved flow properties of the resulting granules for better weight control during tablet compression. HSWG can also provide effective control over granule porosity and density, thus maintaining content uniformity of the drug substance. These are some of the critical quality attributes that need to be tightly regulated due to their impact on dissolution, disintegration and overall bioavailability.

This study focuses on the development of a multi-dimensional population balance model that incorporates mechanistic descriptions of the main granulation rate processes. (e.g. nucleation, aggregation, breakage). The rate processes incorporate important material properties of solid and liquid as well as process and design parameters. The multi-dimensional model framework is able to track the evolutions and distributions of key granule properties such as size, liquid content and porosity. Lab-scale experimental data were used for model calibration. An objective function that minimizes the maximum likelihood function was formulated to estimate optimal parameters. Experimental data were further analyzed to improve model physics. The validated model was then used to predict the effect of key processing parameters (such as impeller speed, water amount etc) on granule characteristics. The model results were compared to experimentally obtained results.

Results show the potential of a model-based approach to HSWG processes by providing the industrial practitioner with more science-based approach that will then result in reducing the number of experiments needed during process optimization and scale-up studies.

See more of this Session: Population Balance Modeling for Particle Formation Processes: Nucleation, Aggregation and Breakage Kernels
See more of this Group/Topical: Particle Technology Forum

(47e) Efficient Simulation of Population Balance Models Via Parallel and Distributed Computing

Anuj Varghese, Chemical and Biochemical Engineering, Rutgers, The State University of New Jersey, Piscataway, NJ, Atul Dubey, Chemical and Biochemical Engineering, Rutgers University, Piscataway, NJ, Rutgers University, Piscataway, NJ and **Rohit Ramachandran**, Department of Chemical and Biochemical Engineering, Rutgers, The State University of New Jersey, Piscataway, NJ

Computer-aided modelling and simulation is a crucial step in developing, integrating and optimizing unit operations and subsequently, entire processes, in the Chemical/pharmaceutical industry. Industries actively involved in the modelling and simulation of unit operations like granulation, blending or crystallization, are increasingly recognizing the need for minimizing simulation run-times of models that are computationally intensive. These simulations typically involve multiple iterations over several increasing or different data sets for prolonged time periods. MATLAB Parallel and Distributed Computing [1] and the Graphics Processing Unit (GPU) Toolboxes affords a convenient way of speeding up these calculations by allowing the programmer to fully exploit the parallel processing capabilities of today's multi-core multi-processor desktops.

This study details two methods of reducing the computational time to solve complex process models, namely the population balance model which given the source terms can be very computationally intensive. Population balance models are also widely used to describe the time evolutions and distributions of many particulate processes and its efficient and quick simulation would be very beneficial. The first method illustrates utilization of the parallel/distributed computing toolbox and the second method makes use of GPU computing to speed up simulations. Results show significant reduction in computational time for the same accuracy. This lends credence to the use of high-fidelity models (in place of reduced order models) for control and optimization of particulate processes.

References

1. Piotr Luszczek: Parallel Programming in MATLAB. IJHPCA 23(3): 277-283 (2009)

See more of this Session: Population Balance Modeling for Particle Formation Processes: Nucleation, Aggregation and Breakage Kernels
See more of this Group/Topical: Particle Technology Forum

(47f) A Study of Population Balance Modeling In a Large-Eddy Simulation with Carbonate Precipitation

Alex W. Abboud, Sean T. Smith, Tony Saad and Terry A. Ring, Chemical Engineering, Institute for Clean and Secure Energy, University of Utah, Salt Lake City, UT

For precipitation in a liquid turbulent pipe flow a simulation of a population balance in large-eddy simulation will be presented. The moments of the population balance will be used to track the size distribution of the particles as it evolves in the flow field. The flow field will be simulated in a computational fluid dynamics code using large-eddy simulation, which resolves large-scale components and models sub-grid scale effects. Specifically, the research will be based on the reaction of calcium chloride and sodium carbonate to produce calcium carbonate, in an effort to reproduce experimental results from the literature. The population balance is coupled with the near instantaneous acid-base chemistry equilibria. A thermodynamic program is used to deal with the complexities in the interacting activity coefficients of the possible species present. This data is tabulated over a large variation of mixture fraction and reaction progress so that it can be read into large-eddy simulation. The computational code will be able to calculate the evolution of the population balance.

See more of this Session: Population Balance Modeling for Particle Formation Processes: Nucleation, Aggregation and Breakage Kernels
See more of this Group/Topical: Particle Technology Forum

(47g) Incorporating Surfactant Coverage and Drop Coalescence Into Population Balance Models of High Pressure Emulsification

Shashank N. Maindarkar, Chemical Engineering, University of Massachusetts, Amherst, MA, Peter Bongers, Process Science Deparment, Unilever R&D Vlaardingen, Vlaardingen, Netherlands and Michael H. Henson, Department of Chemical Engineering, University of Massachusetts Amherst, Amherst, MA

Populationbalance equation (PBE) models of emulsification processes allow the predictionof the drop size distribution, a critical determinant of emulsion properties. ManyPBE models that account only for drop breakage have been developed for modelemulsion systems with relatively low oil-to-surfactant ratios. However,industrial practice is to reduce manufacturing costs by minimizing surfactantuse and establish process conditions under which drop coalescence isappreciable. In this study, we incorporated surfactant coverage and drop coalescenceinto our previously developed breakage-only PBE model of high pressurehomogenization (Raikar et al., 2009, Raikar et al., 2010) to allow the prediction of drop sizedistributions for high oil-to-surfactant ratios used industrially.

Dropbreakage under turbulent homogenization conditions was modeled with twodistinct breakage rate functions and a distribution function that accounted forthe formation of multiple daughter drops from a single breakage event. Dropcoalescence was incorporated through the addition of two functions for the dropcollision rate and the coalescence efficiency of binary collision events. The coalescenceefficiency was assumed to be dependent on the surfactant surface coverage ofeach drop, with no coalescence possible when both drops have maximum coverageand the efficiency increasing as both drop become less covered. We investigatedboth equilibrium and dynamic equations for predicting the surface coverage fromthe drop size distribution and the free surfactant concentration. The system ofequations was closed by adding a dynamic balance on the free surfactant, whichalso allowed a time varying interfacial tension to be computed.

Byutilizing nonlinear optimization to estimate six adjustable parameters in the breakageand coalescence functions from measured drop distributions, the combinedbreakage-coalescence model was shown to provide superior predictions ascompared to the breakage-only model for Pluronic F-68 surfactant and emulsionswith high oil-to-surfactant ratios. Because mechanistic breakage andcoalescence functions that included emulsion properties and homogenizationconditions were used, the model was able to satisfactorily predict drop sizedistributions at other surfactant concentrations and operating pressureswithout re-estimation of the parameters. The model was able to generateacceptable predictions for other non-ionic surfactants if the appropriatesurfactant coverage model was used and model parameters were re-estimated usingdata for the particular surfactant.

References

RaikarN.B., Bhatia S.R., Malone M.F. and Henson M.A, Experimental Studies andPopulation Balance Equation Models for Breakage Prediction of Emulsion DropSize Distributions, *Chemical EngineeringScience*, 64 (2009), 2433-2447.

Raikar N.B., BhatiaS.R., Malone M.F., McClements D.J., Almeida-Rivera C., Bongers P. and HensonM.A., Prediction of emulsion drop size distribution with population balanceequation models of multiple drop breakage,*Colloidsand surfaces A: Physicochemical and engineering Aspects, doi:*10.1016/j. colsurfa.2010.03.020

See more of this Session: Population Balance Modeling for Particle Formation Processes: Nucleation, Aggregation and Breakage Kernels
See more of this Group/Topical: Particle Technology Forum

(55a) A Multi-Scale Modeling Approach to Continuous Blending Processes

Maitraye Sen and Rohit Ramachandran, Department of Chemical and Biochemical Engineering, Rutgers, The State University of New Jersey, Piscataway, NJ

Continuous processing is attracting increasing attention within the pharmaceutical industry today because it could lead to significant decreases in production costs while improving product quality. Regulatory authorities have also recognized the potential for improved QbD brought about by continuous processing. Within this, continuous blending is becoming an increasingly common process in industries that handle particulate material with the purpose of producing a homogenous blend with desired properties. Critical quality attributes such as API composition, relative standard deviation (RSD) of the composition and even particle size distribution (PSD) have to be tightly controlled due to the stringent quality control measures imposed.

This study focuses on the development of a hybrid population balance model and discrete element model (PBM-DEM) that addresses the multi-scale nature of the blending dynamics. The DEM model simulates the micro-scale particle-level behaviour and the resulting axial and radial fluxes are incorporated into the meso-scale particle cluster-level to simulate the key particle/blend properties. Results show the potential for the overall model to qualitatively predict trends in the literature (residence time distribution, RSD, API composition) and even particle size distribution changes (for a case of cohesive particles) at a fraction of the time taken for a full DEM simulation.

See more of this Session: Solids Handling and Processing
See more of this Group/Topical: Particle Technology Forum

(55b) Model-Based Scale-up of Impact Milling

Brian T. Gettelfinger[1], Stephen R. Glassmeyer[2], Mark A. Pinto[3] and Sean K. Bermingham[3], (1)Chemical Systems Modeling Section, Modeling and Simulation, Corporate R&D, Procter & Gamble, West Chester, OH, (2)Particle Processing Section, Process Technologies, Corporate Engineering, Procter & Gamble, West Chester, OH, (3)Process Systems Enterprise, London W6 7HA, United Kingdom

The Vogel& Peukert model uses separate material properties and mill parametersdetermined from bench top experiments to predict the particle size distributionof the output of impact mills. This turns mill modeling into a tool that can beused every day at P&G. This talk covers our implementation of this modeland use within the gSOLIDS environment (Process Systems Enterprise, UK). We determined the material properties for an absorbent gelling material used indiapers from sieving and single impact milling experiments. Wesuccessfully deployed the population balance model of Vogel and Peukert thatpredicts the output of our bench top pin mill in gSOLIDS. The gSOLIDStool allowed us to perform parameter estimation of this highly nonlinear modelwhich gave us distinct material and mill properties. We then madesuccessful model predictions of mill scale-up using these same parameters. Thismethod could potentially save millions annually in experimental costs as we cangeneralize this method to any powder that is broken in an impact mill. We have thus developed a model-based work process for impact mill scale-up thatuses gSOLIDS at its core.

References:

L. Vogel, W.Peukert (2005) From single impact behaviour to modelling of impact mills. *ChemicalEngineering Science* **60**, 5164-5176.

See more of this Session: Solids Handling and Processing
See more of this Group/Topical: Particle Technology Forum

(55c) Jamming In Granular Systems

Shrikant Swaminathan and Carl R. Wassgren, Mechanical Engineering, Purdue University, West Lafayette, IN

This talk focuses on jamming of particles in a bladed mixer geometry. Simple analyses and discrete element method (DEM) computer simulations are used to investigate the conditions that result in particles jamming between the mixer blade and container. Preliminary findings have shown that the gap-to-particle diameter ratio and friction coefficient have a significant influence on the likelihood that particles will jam. Limiting values of these properties are observed. The blade speed and bed depth are found to play only a secondary role.

See more of this Session: Solids Handling and Processing
See more of this Group/Topical: Particle Technology Forum

(55d) Tablet Press Feed Frames: The Influence of Design and Operation On Powder Flow and Tablet Quality

William R. Ketterhagen, Pfizer Inc., Groton, CT

Tablet presses are used in the manufacture of pharmaceutical tablets to compress a powder blend containing an active pharmaceutical ingredient (API) into dosage forms suitable for use by patients. There are several powder handling challenges within this process where powder must transferred without significant segregation from a large bin blender to a die where it is ultimately compressed into tablets. The tablet press feed frame is a key component used to transfer the powder from the feed chute to the empty dies. In this work, the discrete element method (DEM) is used to model powder flow within the feed frame of a lab scale tablet press. Varying paddle wheel designs, rotation speeds, and rotation direction are considered in conjunction with varying levels of powder cohesion. The effects of these variables on the powder flow pattern, powder residence time, and tablet quality (e.g. tablet weight variability) are assessed.

See more of this Session: Solids Handling and Processing
See more of this Group/Topical: Particle Technology Forum

(55e) Use of a Couette Shearing Cell As a "Powder Rheometer" to Predict Stress and Flow Fields of Dense Granular Matter

Mehrdad Kheiripour Langroudi, Chemical Engineering, City College of the CUNY, New York, NY, Mehrdad Shahnam, Computational Science Division, National Energy technology Laboratory, Morgantown, WV, James N. Michaels, Merck and Co. Inc., West Point, PA, Paul R. Mort, Procter & Gamble co., Cincinnati, OH and Gabriel I. Tardos, Department of Chemical Engineering, The City College of the City University of New York, New York, NY

The basic motivation for this work was to develop a reliable continuum model that relates stresses inside the system to shear rates and is valid for both quasi-static and intermediate regimes of powder flow. To achieve the goal, a traditional Couette shearing cell was modified by providing continuous, low flow-rate axial powder feed and discharge, which allows for the bed dilation and provides space for powder particles to collide. A visco-plastic constitutive equation was developed from the Couette data, and this was then generalized by considering the effects of size, shape, surface texture, compressibility, deformability, rigidity and elasticity of particles on the rheological index. As a second step, this constitutive equation was validated in a more complicated geometry that of the spheronizer: a device widely used in the pharmaceutical industry. We used a computational fluid dynamics (CFD) approach to model the free surface centripetal flow of granular matter in the spheronizer. We constructed a lab-scale spheronizer and measured all relevant parameters. Experimental results, such as surface height and shape, solid volume fraction and wall normal stress as a function of rotational speed were compared with those obtained from computations and showed a good agreement.

The picture that emerges from all the experimental, theoretical and computational work is the fact that one can use a continuous flow Couette cell as a "Powder Rheometer" and find powder flow characteristics to enable the prediction of flow and stress fields in complex systems such as spheronizer. This is a significant first step since flows in high shear granulators and mixers are somewhat similar and, in general, pose similar problems.

See more of this Session: Solids Handling and Processing

Investigation of Pressure and Energy Consumption in Bypass Pneumatic Conveying Systems

Bin Chen, Kenneth Charles Williams, Mark Glynne Jones, Ying Wang
Centre of Bulk Solids & Particulate Technologies, The University of Newcastle, Callaghan, NSW, Australia

Abstract: Bypass pneumatic conveying systems provide the capacity of transporting some materials that are not naturally suited to dense phase flow in a low velocity, dense phase flow regime. Bypass pneumatic conveying systems also provide a passive capability to reduce minimum particulate transport velocities. Therefore, particle degradation and pipe line wear can be much reduced.

In this paper, the operation of internal bypass system was investigated by both experiments and modelling. An entire bypass system was numerically modelled based on the mass conservation. An integrated version of the Ideal Gas Equation was applied to evaluate pressures at the central point of each node for air flow in the bypass pipe and the main pipe. The bypass pneumatic experimental system was built with a main pipe of 79mm in diameter and an internal bypass pipe with orifice plate flute arrangement. Fly ash and alumina were used in the tests. High speed video camera visualization and differential pressure transmitters were employed to investigate the operation of dense phase bypass pneumatic transport systems and the mechanism of material blockage inhibition provided by this system.

The bypass system was found to consume more energy than conventional system when using the same air mass flow rate due to the increase of friction. The conveying velocity of alumina in bypass system was much lower than that of conventional pipelines, which resulted in much reduced specific energy consumption. In this system, particulate material blockages were inhibited in bypass systems due to the air penetration into the particulate volume, as was reflected in differential pressure transmitter measurement data and flow visualization.

Key words: Pneumatic conveying; bypass system; modelling; pressure; specific energy consumption.

1. INTRODUCTION

Bypass pneumatic conveying systems provide a passive capability to reduce conveying velocity and therefore reduce attrition and abrasion in the process of conveying many fragile and erosive particulate solids. Barton [1] studied overall alumina conveying performance of bypass pneumatic conveying systems. Mathematical equations were developed to determine the bypass pipe diameter and flute spacing. Some bypass system modelling work was established by Solomon [2] to simulate the bypass systems using a finite element approach. However, in this model an appropriate initialisation pressure at the pipe inlet was not provided, which undoubtedly increases the inaccuracy of the results.

Recently, the non-invasive flow visualization techniques (e.g. electrical capacitance tomography (ECT) and high speed video camera (HSVC)) have been employed to obtain detailed information of the solid phase during pneumatic conveying process by researchers. The flow patterns were observed in the internal bypass pneumatic system by Xu et al. [3] using ECT. Most recently, Chen et al [4-6] employed high speed camera to visualize the flow patterns in an internal bypass system.

The aim of this study is to theoretically and experimentally investigate the differential pressure between bypass pipe and main pipe as well as the pressure drop in bypass pneumatic conveying system. In particularly, the focus is to be placed on the study of the differential pressure whereby the mechanism of material blockages inhibition was exposed. The influences of bypass configurations on specific energy consumption were also discussed.

2. MODELLING OF BYPASS PNEUMATIC CONVEYING SYSTEMS

Chen et al. [7] developed a bypass system model with the application of the mass conservation principle and the integrated version of ideal gas equation. The pressure in an element after a short time period would simply be the initial pressure plus a change in pressure. The pressure at node N_i after a short time interval [2] is:

$$P_{N_i(t+\Delta t)} = P_{N_i(t)} + \frac{P_0}{\rho_0} \cdot \frac{\sum \dot{m}}{V_{N_i}} \Delta t \tag{1}$$

where $P_{N_i(t+\Delta t)}$ is the pressure of node N_i at time $t + \Delta t$, $P_{N_i(t)}$ is the pressure of node N_i at time t, P_0 is the atmosphere pressure, ρ_0 is air density at atmosphere pressure, $\sum \dot{m}$ is the total instantaneous air mass flow, Δt is the short time interval, and V_{N_i} is the volume of the element.

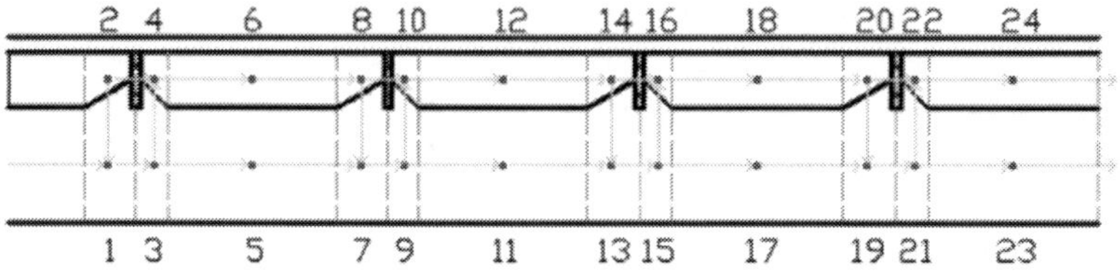

Fig. 1 Bypass pneumatic conveying analysis

The air flow rate [2] in the bypass pipe was determined by rearranging the Darcy Equation.

$$\dot{m}_{i,i+2} = \frac{\pi d^2}{4} \sqrt{\frac{\left(P^2_{N_i(t)} - P^2_{N_{i+2}(t)}\right) d}{\lambda_f RTL}} \tag{2}$$

where d is bypass diameter; λ_f is the friction factor due to the gas; R is constant (287 J/kg/K); L is the distance between node N_i and node N_{i+2}.

When the air flows through an orifice plate located in the flute of the bypass pipe, the air flow rate of the bypass pipe was obtained by orifice plate theory.

$$\dot{m}_{i,i+2} = K_0 \frac{\pi \theta^2}{4} \sqrt{\left(P^2_{N_i(t)} - P^2_{N_{i+2}(t)}\right) \frac{\rho_0}{P_0}} \tag{3}$$

where K_0 is orifice discharge coefficient, and θ is the orifice diameter.

The air flow rate in the main pipe can be expressed as follows.

$$\dot{m}_{i,i+2} = \frac{\pi (D^2 - d^2)}{4} \sqrt{\frac{\left(P^2_{N_i(t)} - P^2_{N_{i+2}(t)}\right)(D-d)}{(\lambda_f + \mu \lambda_s) RTL}} \tag{4}$$

where D is main bypass diameter; μ is solids loading ratio; and λ_s is solids friction factor, which was determined using the model developed by Jones and Williams [8].

The air mass flow between bypass pipe and main pipe can be derived as Eq. (5).

$$\dot{m}_{i+1,i} = A_o \frac{\left|P^2_{N_{i+1}(t)} - P^2_{N_i(t)}\right|}{P^2_{N_{i+1}(t)} - P^2_{N_i(t)}} \sqrt{\frac{2 d_o \left|P^2_{N_{i+1}(t)} - P^2_{N_i(t)}\right|}{(\lambda_f + \mu \lambda_s)(D-d) RT}} \tag{5}$$

where A_o is the area of bypass pipe opening, d_o is equivalent diameter of bypass pipe opening.

An algorithm was developed using the C++ program to implement the bypass model to obtain the pressure in both main pipe and bypass pipe.

3. EXPERIMENTAL SET-UP AND PROCEDURES

The pneumatic conveying tests were conducted in a 6.5 m system to obtain experimental pressure data as shown in Fig. 2. The pipeline consists of a main pipe with an internal diameter (ID) of 79 mm and a bypass pipe ID of 27mm. The orifice plate approaches were employed to design the internal bypass pipe, and the bypass configurations, e.g. bypass flute sizes and flute spacing, are shown in Tab. 1. Alumina and fly ash were used in this study, and their properties are listed in Tab. 2. Pressure transmitters were utilised in the system to measure the gauge pressure while differential pressure transmitters were employed to measure the pressure difference between the bypass pipe and main pipe. A Lab-view program was used to monitor and record the data at an acquisition rate of 50Hz. HSVC Phantom 5 was used to obtain detailed information of solid phase flow behavior in pneumatic conveying pipelines.

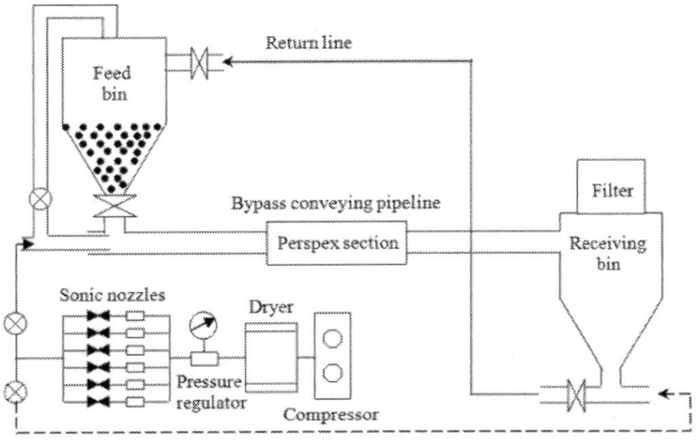

Fig. 2 Schematic diagram of the bypass system

Tab. 1 Bypass configurations

Bypass pipe ID, m	Bypass flute spacing, m	Orifice plate diameter, m	Angle of bypass opening, deg
0.027	0.4	0.007	45

Tab. 2 Physical properties of the materials

Material	d_p , μm	ρ_p, kg/m3	ρ_b, kg/m3	Permeability 10^{-7} $m^2/(Pa\cdot s)$	u_{mf} mm/s
Alumina	76.7	4088	1003	3.6	17.7
Fly ash	14.7	2093	775	6.6	84.7

4. RESULT AND DISCUSSION

4. 1 Pressure and differential pressure (DP) results

The pressure and differential pressure of conveying alumina were measured under the operation condition of air mass flow rate of 0.0417 kg/s and solid mass flow rate of 0.57 kg/s, and the measurements are plotted as a function of time in Fig. 3 and Fig. 4, respectively. Also, the model, Eq. (1), was used to calculate the pressure and differential pressure under this condition, and the model results are also shown in Fig. 3 and Fig. 4, respectively. The experimental results showed that the pressure signals at the pipe inlet were relatively stable. From Fig. 3 and Fig.4, it can be seen that the calculation results are reasonably close to the measurements. From the differential pressure data, it can be derived that air came into main pipe and penetrated into the particulate volume continuously before the orifice plate thus material blockages were inhibited. The differential pressure also illustrated that materials may go into the bypass pipe as the pressure in the bypass pipe after orifice plate is lower than that in main pipe. The differential pressure after the orifice plate showed more fluctuation than before the orifice plate because turbulence was generated by the acceleration of gas and material mixture thorough orifice plate.

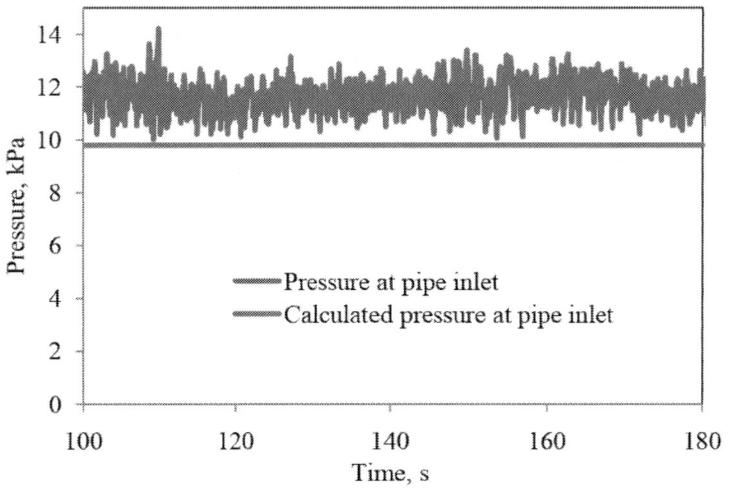

Fig. 3 Comparison between measured and calculated pressure at pipe inlet

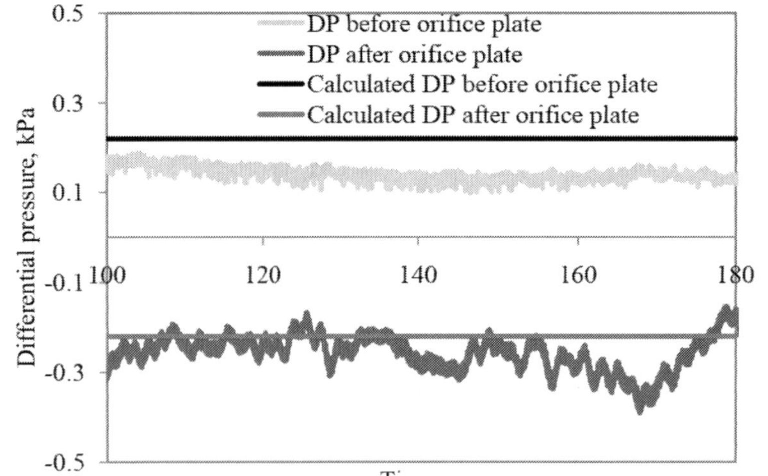

Fig. 4 DP results between bypass pipe and main pipe

4.2 Visualization of unblocking the pipeline

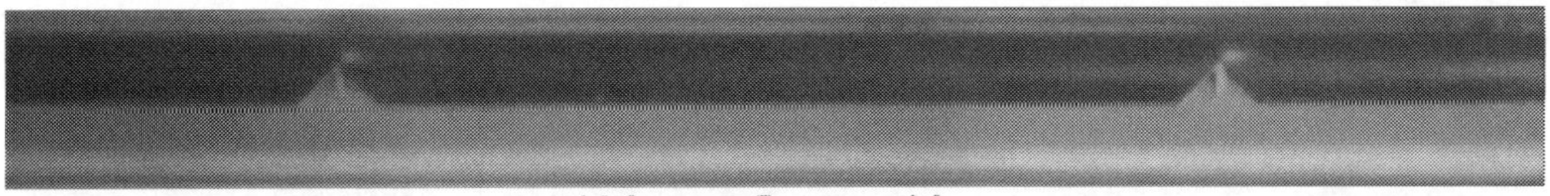
(a) bypass flutes position

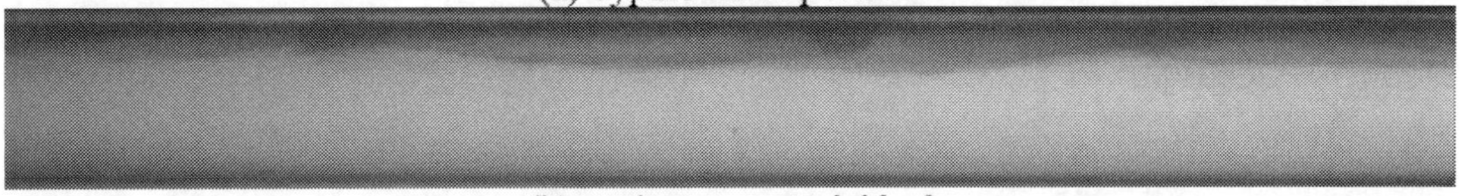
(b) stationary material bed

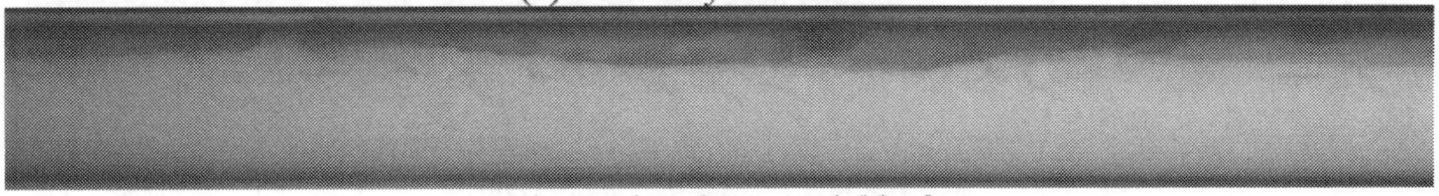
(c) aerating the material bed

Fig. 5 Visualization of aerating the alumina bed

(a) bypass flutes position

(b) stationary material bed

(c) aerating the material bed

Fig. 6 Visualization of aerating the fly ash bed

Chen et al [4-6] investigated the alumina and fly ash blockage clearance mechanisms using HSVC visualization. The results are presented in Fig. 5 and Fig. 6, respectively. The stationary alumina bed almost filled the whole cross-section of the main pipe as shown in Fig. 5 (b). As well as the motionless fly ash bed was shown in Fig. 6 (b). When the aerating process within the material bed commenced, cracks in the material bed structure were observed, as shown in Fig. 5 (c). This "crack structure" phenomena was similarly observed in the fly ash experiments, as shown in Fig. 6(c). In Fig. 6(c), the fly ash layer height increased and the material bed expanded, with the cracks close to the bypass flutes larger than those away from the flute. For alumina, the phenomenon was not very obvious as it has a slightly larger particle size and higher density than fly ash. The materials were then fluidized slowly and blockage was cleared from the blocked location.

4.3 Specific Energy

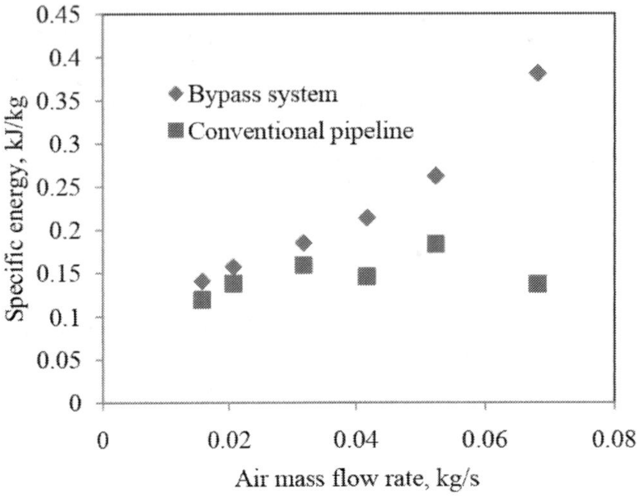

(a) fly ash

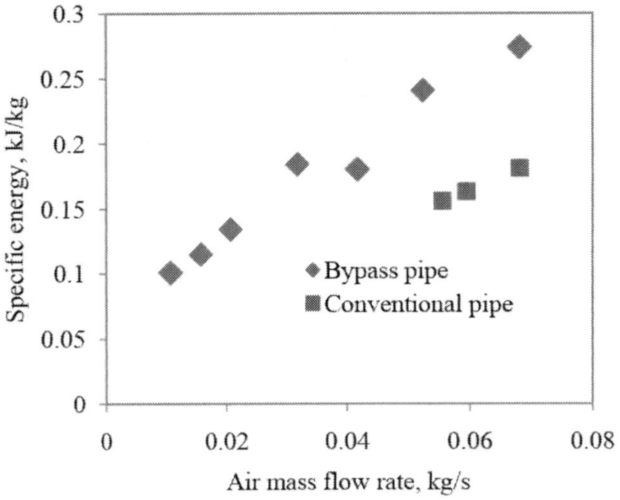

(b) alumina

Fig. 7 Comparison of specific energy consumption between bypass system and conventional pipeline

The specific energy which represents the energy consumption required to convey certain amount solids was given by the Eq. (6) [9].

$$Specific\ Energy = 2RT\frac{M_a}{M_s}\ln\left(\frac{p_1}{p_2}\right) \qquad (6)$$

where M_a is air mass flow rate (kg/s); M_s is solid mass flow rate (kg/s); R is universal gas constant (kJ/kg/K); T is absolute temperature (K); p_1 is conveying line inlet absolute pressure, p_2 is conveying line outlet absolute pressure.

The results of specific energy in both bypass system and conventional pipeline are compared as shown in Fig. 7. For both bypass pipe and conventional pipe, the specific energy increased with the increase of mass flow rate of conveying air. From the experimental results using our 6.5 m length test rig, it can be seen that the bypass pipeline had higher specific energy consumption than conventional system when using the same air mass flow rate and solids mass flow rate. Compared to conventional pipeline, the existence of the bypass pipe caused the increase of wall friction and flow resistance of a number of orifice plates when the bypass pipe was put inside of the main pipe. The difference in specific energy consumption between two conveying systems decreased with the decrease of conveying velocity due to the reduced flow resistance.

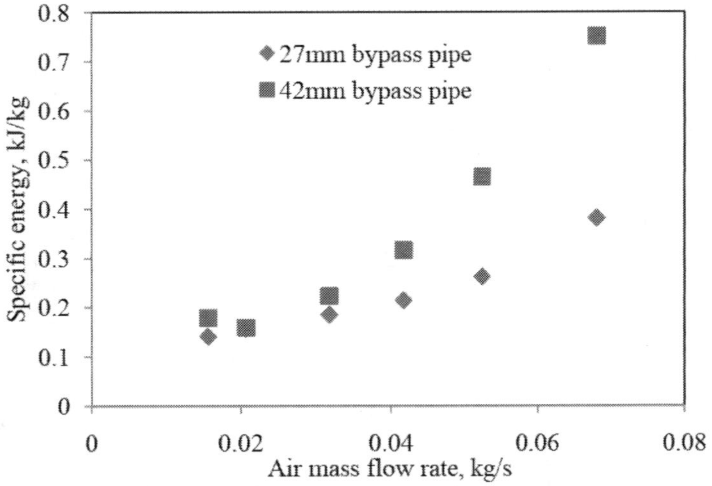

Fig. 8 Comparison of the specific energy consumption between two bypass configurations

The specific energy consumption of the fly ash conveying bypass system with different bypass pipe diameters is plotted as a function of the air mass flow rate in Fig. 8. It has been found that the specific energy consumption of the bypass system with a 42 mm bypass pipe was higher than that with a 27 mm bypass pipe for the same solids and air flow rates, and the difference became smaller at a lower air flow rate. The main reason is that a 42 mm bypass pipe took up 2.4 times the circulation area as a 27 mm bypass pipe did, thus dramatically reducing the circulation area of the main pipe. Furthermore, the wall friction also increased due to the larger wall surface area. Therefore, it is important to choose the appropriate bypass pipe diameter as a suitable diameter of bypass pipe can ensure good performance of bypass systems with lower energy consumption.

4. CONCLUSION

In this study, the operation of internal bypass system was investigated by both experiments and modelling. Based on the measurements and modelling analysis of the differential pressure, it has been derived that the mechanism of blockage inhibition in bypass system is that air flows into main pipe and aerates the material volume as a result of orifice plate airflow resistance. The pipeline pressure drop results showed that pressure drop in the bypass systems was higher than that in the conventional system when using the same operating parameters due to the increase of friction. Compared to the bypass system with a 27 mm bypass pipe, energy consumption in the system with a 42 mm bypass pipe was higher because of the dramatically reducing circulation area of the main pipe and the increasing wall friction due to larger wall surface area.

REFERENCES

1 Barton S. (1997) The Effect of Pipeline Flow Conditioning on Dense Phase Pneumatic Conveying Performance, PhD Thesis, Glasgow Caledonian University

2 Solomon M. (2002) Bypass pneumatic conveying systems. Final project report. School of Engineering, University of Newcastle, Australia.

3 Xu H, Liu S, Wang H, and Jiang F. (2002) Experimental Study on Wavy-flow Pneumatic Conveying in Horizontal Pipe. J. of Thermal Science, 11(2): 114-120

4 Chen B, Williams KC and Jones MG. (2010) Experimental Investigation of Low Velocity Pneumatic Transport of Fly Ash in Bypass System. The Fourth Baosteel Biennial Academic Conference, Shanghai, China

5 Chen B, Cenna AA, Williams KC, Jones MG and Wang Y. (2011) Investigation of Energy consumption and wear in Bypass Pneumatic Transport of Alumina, 6th World Congress on Engineering Asset Management, Cincinnati, USA

6 Jones MG, Chen B, Williams KC, Cenna AA and Wang Y. (2011) High Speed Visualization of Pneumatic Conveying of Materials in Bypass System, The 9th International Conference on Measurement and Control of Granular Materials. Shanghai, China

7 Chen B, Jones MG, Williams KC, Tan S and Wang Y. (2010) Investigation of Pressure and Gas Velocity in Bypass Pneumatic Conveying Systems, The World Congress on Particle Technology 6, Nuremberg, Germany

8 Jones MG and Williams KC. (2003) Solids friction factors for fluidized dense-phase conveying, Particulate Science and Technology 21:45–56

9 David D. (2004) Pneumatic conveying design guide, Elsevier Butterworth-Heinemann, 247-252

(55g) Controlling Granular Segregation In Rotating Cylinders

Suman K. Hajra, TBA, Pittsburgh, PA and Joseph J. McCarthy, Chemical Engineering, University of Pittsburgh, Pittsburgh, PA

The segregation of flowing granular materials is a commonly observed phenomenon in industrial processes and in nature. Segregation causes problems in maintaining uniformity of the processed material and leads to deviations in the quality of the final product. Granular materials are widely used in process industries, such as coal gasification, iron & steel, chemical, pharmaceutical, cement, food, detergents, mining, ceramic, and fertilizers. In order to design equipment such as rotary kilns, bins, silos, combustors, hoppers, chutes, hydro-cyclones, etc., in an effective and economical way, a thorough understanding of the various factors governing the flow characteristics of granular materials must be obtained. These needs have already motivated extensive analytical and experimental investigations of the flow of granular materials. Mixing, blending, heating, and granulation of these materials are frequently carried out in rotating cylinders or tumblers; however, when mixtures of two different types of particles are rotated in tumbler, they tend to segregate. Mixing and segregation in two-dimensional mixers depends on many factors including the type of components involved, the speed of rotation, the filling amount, and the mixer shape. All these parameters play an important role in the quality, degree, and the speed of segregation. We study the granular flow and segregation in rotating cylinders for different size and different density systems. We have found that periodic flow inversions of free-flowing materials may be used to effectively control both size and density segregation in free-surface flows. Parameters for the periodic flow inversion technique will be discussed. In a rotating cylinder, these inversions may be induced by placement of an axially located baffle within the flow. Flow induced granular mixing will be discussed. Segregation model will be presented.

See more of this Session: Solids Handling and Processing
See more of this Group/Topical: Particle Technology Forum

(67a) Modeling of Solid Deformation During Roller Compaction Using Air Pocket Deformation

Leaelaf M. Hailemariam, Dow Chemical Company, Midland, MI

Solid deformation during roller compaction has been the topic of major research interest over the years [1]. In addition to solid particle deformation, the air trapped between the solid particles is compressed, leading to internal voids or fractures when the roller compaction pressure is released [2]. This issue is frequently resolved in industry using vacuum pumps or special solid feeding techniques to remove the air. This work looks at the deformation of the air pockets inside the powder during roller compaction making possible a better understanding of the ribbon structure. This understanding can be used to develop better products and processes.

[1] Cunningham, J.C., Winstead, D., Zavaliangos, A. (2010) Understanding variation in roller compaction through finite element-based process modeling Computers and Chemical Engineering 34 1058-1071

[2] Teng, Y., Qiu, Z., Wen, H. (2009) Systematical approach of formulation and process development using roller compaction European Journal of Pharmaceutics and Biopharmaceutics 73 219–229

See more of this Session: Agglomeration and Granulation Processes
See more of this Group/Topical: Particle Technology Forum

(67b) Monitoring of Fluidized Bed Dryers and Granulators Using Pressure Fluctuation Signals

Lilian de Martn and J. Ruud van Ommen, Department of Chemical Engineering, Delft University of Technology, Delft, Netherlands

Monitoring of fluidized bed dryers and granulatorsusing pressure fluctuation signals

Lilian deMartn and J. Ruud van Ommen

Delft University of Technology, Department of Chemical Engineering,Product & Process Engineering, Julianalaan 136, 2628 BL Delft, TheNetherlands

Themonitoring of the particle moisture content and the particle size is of greatimportance in fluidized bed drying and granulation. Due to the implementationof the Process Analytical Technology initiative in the pharmaceutical industry[1], there is an increasing interest to apply on-line monitoring tools forprocess control. In spite of the amount of the existing techniques developed tomonitor the beds, there are only a few techniques available to do monitorgranule properties on-line. Near infrared microscopy and microwave absorptionare the most used. Nevertheless, these techniques present several drawbacksthat compromise their applicability. Consequently, the challenge of developinga suitable technique for the on-line monitoring of fluidized bed dryers andgranulators is still open.

Given theadvantages of the pressure fluctuation measurement over other techniques (e.g.low cost and easy implementation), three signal processing approaches (dominantfrequency, standard deviation of a narrow frequency band and attractorcomparison [2]) have been applied to pressure data to get information about theparticle moisture and the granule size inside a fluidized bed dryer/ granulator.

Theanalysis of the pressure signals show that some frequency bands of thesesignals are more sensitive to the changes in the bed dynamic related with theparticle moisture and the granule size than others. By processing the signalproperly it is possible to have clear relationships between the particlemoisture or granule size and the standard deviation of the pretreated signal. Moreover, in the case of the granulator, the relationship between the granulesize and the standard deviation is linear. The results show that the pressurefluctuation measurements are a promising tool to monitor on-line the fluidizedbed drying and granulation process.

[1] Guidancefor Industry PAT - A Framework for Innovative Pharmaceutical Development,Manufacturing and Quality Assurance, U.S. department of health and humanservices. Food and Drug Administration, 2004.

[2] J. R. vanOmmen, M.-O. Coppens, C. M. van den Bleek, J. C. Schouten, Early warning ofagglomeration in fluidized beds by attractor comparison, AIChE J. 46 (2000)2183 - 2197.

See more of this Session: Agglomeration and Granulation Processes
See more of this Group/Topical: Particle Technology Forum

(67c) Distribution Nucleation: Quantifying the Liquid Distribution On a Particles Surface Using the Dimensionless Particle Coating Number

Ben Freireich[1], Waithira I. J. Kariuki[2], Rachel M. Smith[2], Martin Rhodes[3] and Karen P. Hapgood[2], (1)Core R&D - Solids Processing, The Dow Chemical Company, Midland, MI, (2)Department of Chemical Engineering, Monash University, Clayton, Australia, (3)Department of Chemical Engineering, Monash University, Victoria, Australia

Nucleation is a very important first step in wet granulation, as it affects the appearance and function of the final granules. Over the last decade, researchers have focused on immersion nucleation, where the drops are much larger than the particles. Far less is known about distribution nucleation, where the drops are much smaller than the particles, as tends to be the case in fluidized beds.

In this paper a newly derived dimensionless parameter, the particle coating number, has been derived with the objective of understanding liquid distribution of small drops over a relatively large particle. When aided with a simple probabilistic analysis the surface coverage of an individual particle as a function the particle coating number.

Experiments were conducted to validate the parameter by adding drops of PEG1000 (10 μ L or 20 μ L) randomly over the surface of five different sized particles (20-50 mm diameter). The surface coverage of the particles was measured using image analysis. The results demonstrated the capability of the particle coating number to predict surface coverage and allow the surface coverage of a particle to be predicted *a priori*. The effects of drop size, particle size, and particle surface area were investigated and agree qualitatively with independent modeling results.

The results are expected to be valuable in a variety of particle wetting and coating applications where small drops are distributed over larger particles. The new particle coating number parameter avoids the need for sophisticated modeling and fitted coefficients and will be able to be used in DEM modeling of fluid bed granulation, particularly for predicting the probability of particle collision at a wetted section. It is also expected to form part of a new set of design rules for fluid bed granulation.

See more of this Session: Agglomeration and Granulation Processes
See more of this Group/Topical: Particle Technology Forum

(67d) Mapping of Nucleation Mechanisms for Regime Separated Granulation

Heather N. Emady[1], Defne Kayrak-Talay[1], William C. Schwerin[2] and James D. Litster[3], (1)School of Chemical Engineering, Purdue University, West Lafayette, IN, (2)Honeywell, Des Plaines, IL, (3)School of Chemical Engineering, Department of Industrial and Physical Pharmacy, Purdue University, West Lafayette, IN

Wet granulation is the process of adding a liquid binder to a fine powder in order to get larger granules for improved particle properties. This process is used in a variety of industries, including pharmaceuticals, food, agricultural chemicals, and detergents. The typical industrial equipment used for this process includes high shear mixers, drums, pans, and fluid beds. Currently, many of the granulation rate processes occur simultaneously in industrial applications, making it difficult to control product properties, such as size and density. These three rate processes include wetting and nucleation, consolidation and growth, and breakage and attrition. The aim of this research is to physically separate the different rate processes to get better control of the product granule attributes.

The first step of granulation is nucleation, and drop controlled nucleation is the most desirable form. In this case, one drop forms one granule. Single drop granule experiments with a syringe and a dish filled with powder are used to simulate drop controlled nucleation. Many model powders were used in single drop granule experiments, with a variety of binders and process conditions. The resulting nuclei granules were excavated, and their size and density were measured using a novel prism technique with image analysis. The nuclei granule properties were then mapped to powder properties, binder properties, and process variables. Different nuclei formation mechanisms occur for different combinations of these variables, explaining the formation of nuclei granules with varying properties. The desired outcome is to be able to predict the nucleus formation mechanism and subsequent nuclei granule shape based on the knowledge of powder and binder properties, as well as process variables.

(67e) Rotating Doublets of Wetted, Solid Particles

Carly M. Donahue, William M. Brewer, Robert H. Davis and Christine M. Hrenya, Chemical and Biological Engineering, University of Colorado, Boulder, CO

Collisions between wetted particles appear in nature in examples such as pollen capture, landslides, and avalanche flow. Furthermore, wetted collisions have industrial applications in granulation, coagulation, and water filtration. Unlike previous investigations that focused on head-on collisions between multiple, wetted particles or oblique collisions between a particle and a wetted wall, oblique particle-particle collisions form a doublet that rotates. Therefore, the centrifugal force plays an essential role as to whether the particles will stick or rebound. Using both experiments and theory, we have explored the possible outcomes associated with oblique collisions between wetted particles for the case of Stokes flow in the liquid layer. The wet particles are observed to stick for low Stokes numbers and low impact angles (i.e., nearly head-on collisions). However, at low Stokes numbers and high impact angles, particles initially stick, rotate for a long time (relative to the collision time of a head-on collision), but then rebound due to centrifugal forces. Bouncingcollisionswith short collision times have been observed for high Stokes numbers;in this case, both centrifugal forces and elastohydrodynamics play a significant role in the dynamics of the collision. The model predictions of the critical Stokes number, the normal restitution coefficient, and the rotation angle are consistent with experiments in which parameters such as oil viscosity and ball materials are varied.

See more of this Session: Agglomeration and Granulation Processes
See more of this Group/Topical: Particle Technology Forum

KINETICS OF DRY NEUTRALIZATION OF DODECYL-BENZENESULFONIC ACID WITH RESPECT TO DETERGENT GRANULATION

Marek Schongut, Zdenek Grof, Frantisek Stepanek
Institute of Chemical Technology, Prague, Czech Republic

1. INTRODUCTION

Sodium dodecyl-benzenesulfonate [1-6] is one of the most common anionic surfactant used in powder detergents. It is a product of reactive granulation of dodecyl-benzenesulfonic acid, which also functions as a binder, with a solid base (in this case sodium carbonate powder). Since the original acid is not being produced as a pure single component but as a mixture of C_{10}-C_{13} isomers, in the literature it is often denoted as HLAS, where LAS stands for Linear Alkylbenzene Sulfonate (the sodium salt is then denoted as NaLAS). The stoichiometric equation of the dry neutralization chemical reaction can be written as:

$$2HLAS + Na_2CO_3 \rightarrow 2NaLAS + H_2O + CO_2$$
$$(2A + B \rightarrow 2C + D + E)$$

Fully neutralized NaLAS is a soft-solid material which at room temperature can form three different mezophases (liquid crystal phase between liquid and crystalline phases), all of them probably of lamellar type [7] (Fig. 1).

Figure 1. Layer of a liquid-crystalline phase of NaLAS/H_2O present between HLAS (dark bottom layer) and a Na_2CO_3/H_2O solution (top clear layer).

NaLAS has earned its popularity through high washing efficiency, favorable cost/performance ratio and environmental friendliness (fast and easy biodegradation). Only a small amount of non-degraded surfactant will reach water treatment plants, with no harmful intermediates or side-products being produced during the degradation [9]. Despite its unquestionable technical importance, there is only a limited amount of information about the physical properties and phase changes of HLAS and NaLAS during reactive granulation.

In a typical granulator the acid is fed in to the vessel and sprayed on to the fluidized or mechanically agitated powder. With on-going neutralization, the stickiness of the original acid grows and makes the inter-particular bridges strong enough to bind primary particles of unreacted sodium carbonate together and create agglomerates larger than the original particles. Final granule properties such as size distribution, or porosity, as well as the overall degree of neutralization are controlled by process parameters such as impeller speed, acid/base ratio, type of feeding nozzles, temperature and moisture, as well as the primary particle size distribution [1, 3, 4]. A challenge is to correlate the properties of final product with changes in process conditions or feed material properties. What makes this particular process even more complex to describe is the fact that apart from agglomeration also a chemical reaction occurs simultaneously [10]. Properties of the liquid bridges (e.g. viscosity) are changing continuously with time and the actual process of wetting is directly dependent on a chemical reaction between the binder and particles. An accurate analysis of micro-scale wetting [11] and reaction-diffusion processes is therefore crucial for understanding and rational design of reactive granulation.

Wetting kinetics can be evaluated from an observation of a contact angle between the solid substrate and the liquid during wetting and the velocity of the three-phase contact line [12, 13]. For a constant volume and homogeneous substrates, the Tanner's law can be used for describing the wetting kinetics [14-16]. This law also describes the relationship between the capillary and viscous forces. Although the Tanner's law is strictly valid just for Newtonian fluids, it could be adjusted for liquids with their viscosity dependable on shear stress. Wettability does not only influence the degree and rate of binder spreading over particles but also the shape and strength of liquid bridges [17, 18].

The aim of this study is to measure the wetting and reaction kinetics of dry neutralization as function of temperature, in order to determine the rate-limiting step. Data obtained during this work will be used as input into a numerical model which would be able to predict structural properties of granules (porosity, spatial distribution of components) depending on the operating conditions of the granulation process (i.e. temperature, liquid-solid ratio, primary particle size etc.).

2. EXPERIMENTAL SECTION

To simulate actual processes that take place during reactive granulation, the following approach has been taken for reaction and wetting kinetics estimation: a small drop of HLAS (volume approx. 4 micro liters, purchased from Acros Organics, mixture of $C_{10} - C_{13}$ isomers) is placed on a substrate pellet (Fig. 2) which has been prepared from anhydrous sodium carbonate fine powder (purchased from Lach-ner, mean particle size 14 μm) and compacted on a CARVER pellet press at 7000 psi resulting in a microscopically smooth surface. An SEM image of a freshly prepared unreacted tablet is shown in Figure 3.

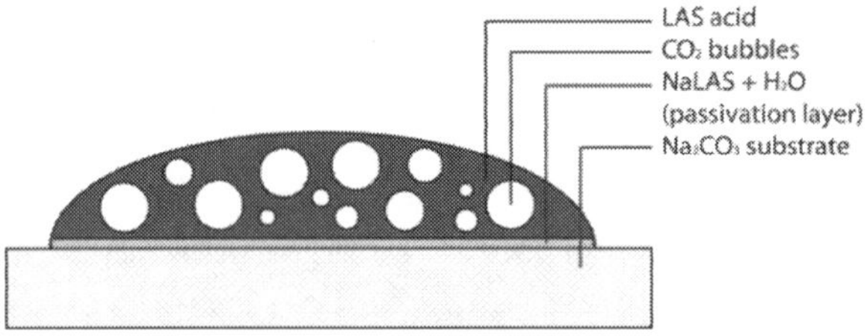

Figure 2. Scheme of an acid drop spreading over a substrate pellet.

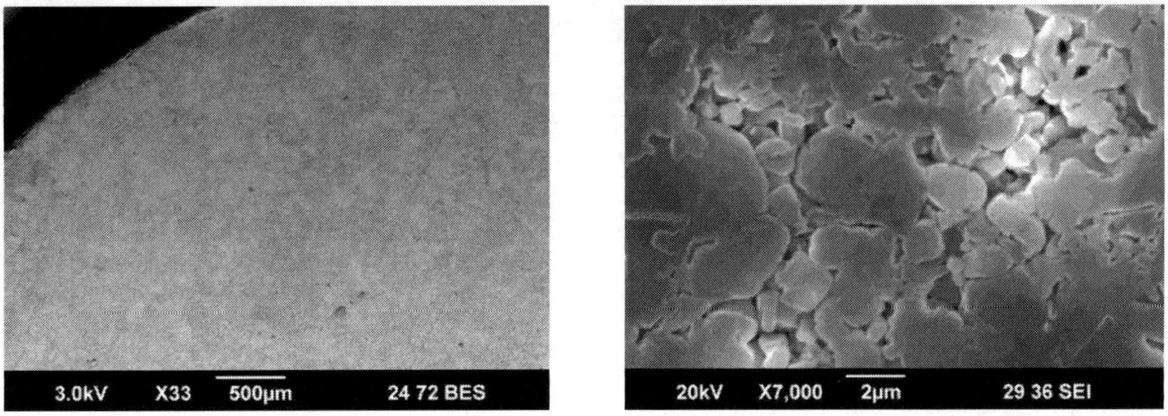

Figure 3. SEM image of the surface (left) of a freshly prepared Na_2CO_3 pellet with a closer look at the pellet's pores (right).

The acid was dropped using a standard medical syringe with a 0.8 mm diameter needle. The distance between the needle and the pellet's surface is 15 mm which is small enough to neglect the effect of gravity on the drop and its spreading on the substrate, leaving only capillary forces to be taken into account (low Weber number deposition). With these conditions, a spherical cap approximation can be used for a drop fitting during wetting. The spreading experiment itself takes approximately 30 seconds, which was found to be long enough for the contact line between the drop and the pellet to reach a quasi-stationary state (contact line pinning). The tabled was placed into a measuring cell with controlled temperature (Fig. 4) and droplet images were recorded using a EOS 40D camera with Sigma EX 105mm macro-lens, at a frequency of approximately 6 frames per second.

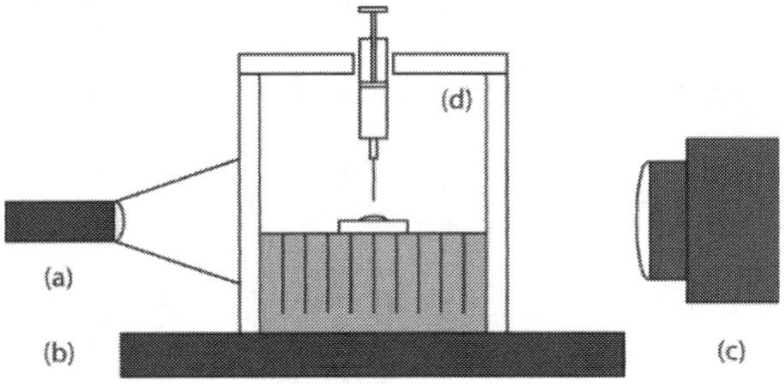

Figure 4. Experimental setup (a) light source, (b) heater, (c) Canon camera (d) heated cell.

The images were subsequently cropped and adjusted in the image processing software ImageJ and then merged in to a video file, making them suitable for further image analysis. The droplet shape was evaluated by the Sessile Drop Fit method (SDF) using the Drop Shape Analysis software (Kruss, Inc.). The desired parameters, namely: drop base diameter, drop height, drop volume and contact angle, were thus obtained for each point in time. The velocity of the three-phase contact line was then calculated from the time change of the droplet base diameter. The relationship between the contact line velocity and the dynamic contact angle is often used for a description of wetting kinetics. Reaction kinetics and passivation layer growth were calculated from the volume of evolved CO_2 (increase in the overall droplet volume with time). During the reaction rate estimation, the following simplifying assumptions were used: (i) ideal gas equation of state, (ii) the volume of CO_2 evolved equals the difference between the current drop volume and drop volume at contact line arrest (i.e., no CO_2 is being dissolved during the chemical reaction), (iii) all evolved CO_2 forms bubbles with atmospheric pressure that do not leave the droplet interior. Under these assumptions, the reaction kinetics can be estimated from the knowledge of droplet base radius (contact area) and the change in droplet volume.

3. RESULTS AND DISCUSSION

The evolution of droplet shape in the course of one experiment is shown in Figure 5. The following parameters were evaluated from each experiment: drop base diameter, drop height drop volume and contact angle. The dependence of contact line velocity and dynamic contact angle on time is used for the wetting kinetics characterization (Fig. 6), whereas the time dependence of drop volume is the most important parameter for the evaluation of reaction kinetics (Fig. 7).

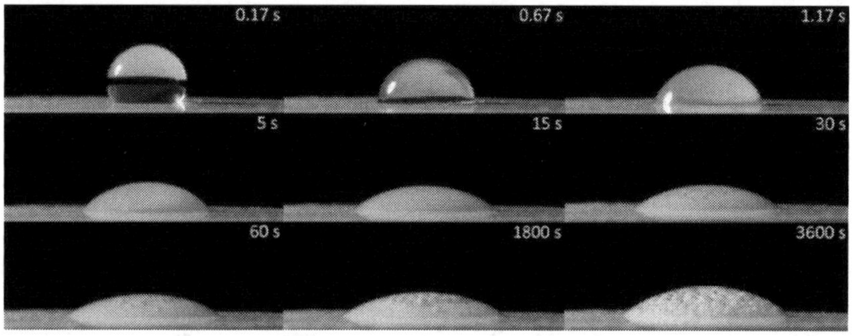

Figure 5. An example of droplet shape evolution during reactive wetting at 20°C. Note that the first CO_2 bubbles only start to be apparent at approx. 30 s, while at 3600 s the droplet is already significantly foamed.

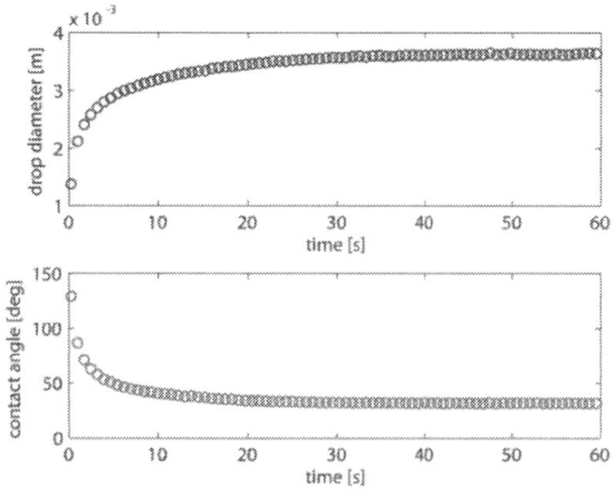

Figure 6. Change of base diameter and instantaneous contact angle evaluated from a wetting experiment at 20°C.

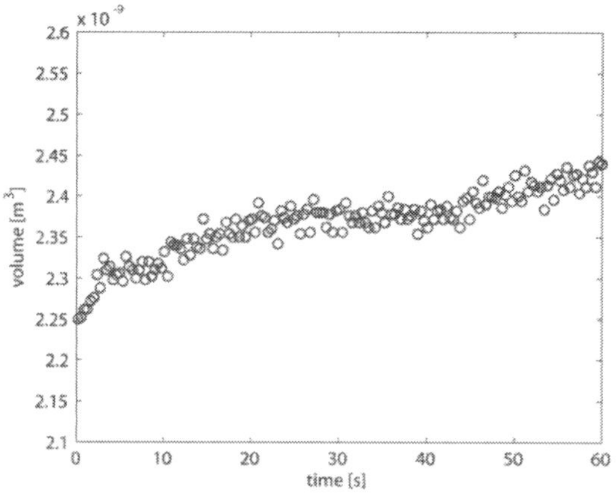

Figure 7. Change of drop volume with time with visible two phases of reactive wetting at 20°C.

3.1. Wetting kinetics

During the first phase of wetting (which starts with the drop touching the substrate and ends with contact line arrest at a quasi-stationary state) chemical reaction is being neglected because there is no visible evolution of carbon dioxide. A decrease of volume is ascribed to absorption of acid into the substrate's surface, which is a common phenomenon during reactive wetting [13]. After reaching the quasi-stationary state, the chemical reaction starts to be noticeable, indicated by the evolution of carbon dioxide bubbles leading to foaming of the drop and increase of its volume. The contact line velocity is calculated as a time derivative of the drop base radius using the finite difference formula (Eq. 1):

$$U(t) = \frac{dr}{dt} = \frac{r(t_{i+1}) - r(t_i)}{t_{i+1} - t_i} \quad \text{Eq. (1)}$$

Where U is the contact line velocity, R drop radius and t is time. The wetting kinetics equation is mostly defined as a relationship between the contact line velocity and dynamic contact angle and it can be fitted by a linear function (Eq. 2, Fig. 8). To get a relation with an asymptotic contact angle (which describes what would the equilibrium value of the contact angle be if there were no chemical reaction).

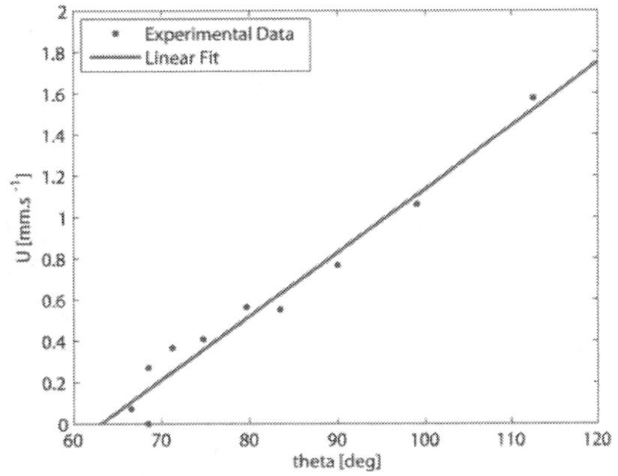

Figure 8. Estimation of wetting kinetics parameters for experimental data measured at 20°C.

$$U(t) = e(\theta(t) - \theta_0) \quad \text{Eq. (2)}$$

Where e is a fitting parameters, $\theta(t)$ is the dynamic contact angle and θ_0 is the asymptotic contact angle. The regression parameters e and θ_0 and their standard deviations for 20, 40 and 60°C are reported in Table 1.

3.2. Reaction kinetics

To estimate the temperature effect on wetting and reaction kinetics several experiments at different temperatures (namely 20, 40 and 60°C) have been carried out using a closed cell with a passive heat sink (Fig. 4). Unfortunately, method introduced in this work is inaccurate to use for experiments at temperatures above 80°C because HLAS viscosity drops so significantly, the carbon dioxide bubbles start to leave the drop.

Passivation layer thickness has been evaluated from reaction stoichiometry (Eq. 3) while using simplifications denoted in section 3 (Eq. 4).

$$\frac{dL}{dt} = \frac{3}{A}\left(\frac{2}{3}\frac{M_C}{\rho_C} + \frac{1}{3}\frac{M_D}{\rho_D}\right)\frac{d\xi}{dt} \quad \text{Eq. (3)}$$

$$\frac{d\xi_{exp}}{dt} = \frac{dn_{CO2}}{dt} = \frac{p}{RT}\frac{dV}{dt} = \frac{p}{RT}\frac{V(t_i) - V(t_{i-1})}{t_i - t_{i-1}} \quad \text{Eq. (4)}$$

Where L is passivation layer thickness, A wetted area, M_C, M_D, ρ_C, ρ_D are molar masses and densities of NaLAS and water respectively, p is atmospheric pressure, R ideal gas constant, T thermodynamic temperature, V drop volume, t time of experiment, ξ extend of reaction.

Experimental values of the passivation layer thickness at different temperatures are plotted in Figure 9. The time is taken as a relative time from the point of contact line arrest, at which stage the droplet volume starts to be measured. Figure 10 is an example of fitting the experimental data (in this case at 60°C) with a numerical model (cf. Section 3.3). An overview of the final values of the kinetic parameters is provided in Table 1.

Table 1. Reaction and wetting kinetics overview

T	k	D_A	e	θ_0
[°C]	[$m^3.mol^{-1}.s^{-1}$]	[$m^2.s^{-1}$]	[$m.s^{-1}.deg^{-1}$]	[deg]
20	(6.05 ±1.13) x 10^{-5}	(1.88 ± 0.74) x 10^{-14}	(2.83 ± 0.41) x 10^{-5}	67.4 ± 5.4
40	(4.73 ± 3.47) x 10^{-4}	(3.04 ± 0.83) x 10^{-14}	(3.31 ± 0.70) x 10^{-5}	40.2 ± 5.5
60	(1.78 ± 0.41) x 10^{-3}	(1.48 ± 0.37) x 10^{-14}	(4.42 ± 0.95) x 10^{-5}	48.8 ± 5.7

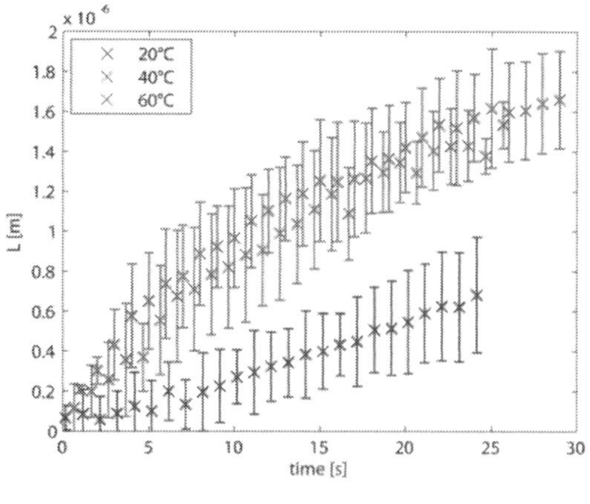

Figure 9. Experimental data on passivation layer growth at different temperatures.

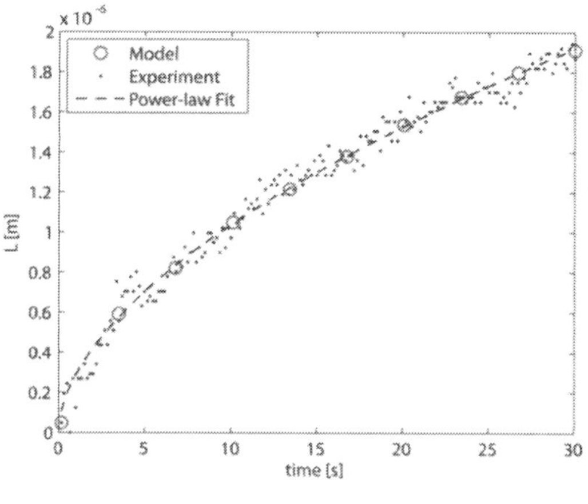

Figure 10. Experimental data (60°C) fitted with model.

From these experimental data it is clear that rate constant *k* of dry neutralization increases with increasing temperature in line with the Arrhenius law (Fig. 11). The reason why the estimated diffusion coefficient D_A does not follow a monotonous pattern is not completely clear at the moment. One possible explanation can be the fact that although diffusion coefficient on its own should increase with temperature, the phase behaviour the passivation layer made from sodium salt liquid crystals is more complex and leads to an overall decrease of the diffusion coefficient. From a study of wetting kinetics came up an obvious influence of temperature on the primary phase of wetting. With increasing temperature the viscosity of the acid decreases which leads to a faster spreading over the substrate and faster convergence towards the quasi-equilibrium state. The asymptotic contact angle decreases when temperature is increased from 20 to 40 °C but then stays more or less constant, within the experimental error.

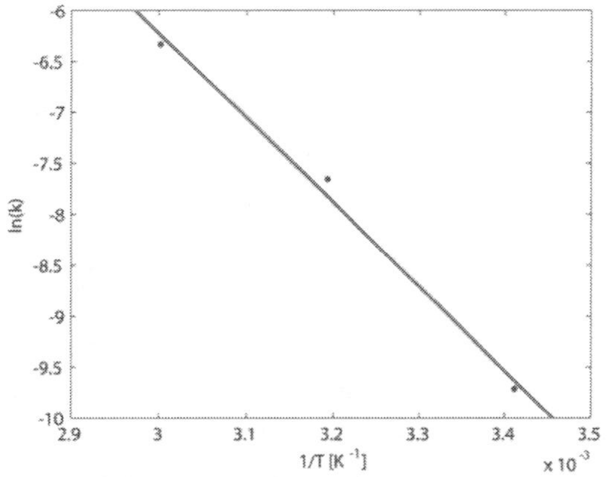

Figure 11. Change of logarithmic value of the kinetic constant according to Arrhenius law.

3.3. Pseudo-steady state model

A reaction-diffusion model based on the solution of a system of differential equations has been developed and solved using MATLAB, considering the mass balance of the two reacting species (acid and base, denoted A and B, Eq. 5) By solving this system of ODEs concentration profiles for both reactants can be obtained and then overall reaction rate within passivation layer (Eq. 6):

$$\frac{dC_i}{dt} = D_i \frac{d^2 C_i}{dx^2} - |v_i| k C_A C_B \quad \text{Eq. (5)}$$

$$\frac{d\xi}{dt} \equiv R = A \int_0^L k C_A(x) C_B(x) \quad \text{Eq. (6)}$$

where the *i* index is for A (acid) and B (base), *k* is the rate constant.

Reaction-diffusion equations need to be solved in order to obtain concentrations of HLAS and Na_2CO_3, C_A and C_B, respectively. If the characteristic time of the layer growth is sufficiently long compared to the characteristic time of concentration profile evolution then pseudo-steady state assumption can be used and PDEs reduce to ODEs boundary value problem with the following boundary conditions applied (Fig. 12) where C_{A0} is a ratio between density and molar weight of acid and c_{B0} is the concentration of saturated Na_2CO_3 solution.

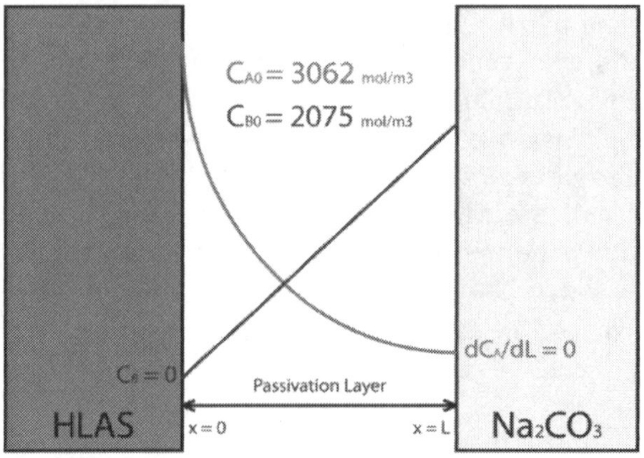

Figure 12. Model of the passivation layer with concentration profiles and boundary conditions.

4. CONCLUSION

From the experiments it appears that kinetics of reactive wetting of sodium carbonate by HLAS has two distinct phases. The primary phase is fast and the effect of the chemical reaction can be neglected because there is no volume growth. Rather a small decrease can be ascribed to absorption on the pellet surface. The primary phase is associated with rapid spreading; meaning a fast decrease of the drop height and contact angle, and an increase of the drop base diameter. The secondary phase starts after reaching a quasi-stationary state (contact line arrest). During this phase the drop volume change becomes clearly visible because of carbon dioxide evolution which also changes the drop height (and therefore contact angle). From this experimentally observed behavior the reaction rate of dry neutralization at different temperatures was evaluated. Increase of temperature positively influences the rate of wetting and reaction. Faster wetting and reaction lead also to shorter primary phase and to a lower value of the quasi-stationary contact angle. Reaction and wetting kinetics obtained by this work will be important input parameters for models leading to quantitative description of reactive granulation in the field of laundry detergents manufacture.

Acknowledgements: This work has been supported by the European Research Council (200580 – Chobotix) and by the Specific University Research grant (MSMT 21/2011)

5. REFERENCES

1. Appel, P.W., *Modern methods of detergent manufacture.* Journal of Surfactants and Detergents, 2000. **3**(3): p. 395-405.
2. Masters, K., *Spray drying handbook.* 5th ed. 1991, Burnt Mill, Harlow, Essex, England, New York: Longman Scientific & Technical; Wiley. xiv, 725 p.
3. Pietsch, W., *Size enlargement by agglomeration.* 1991, Chichester, West Sussex, England; New York, Aarau: Wiley; Salle & Sauerländer. xii, 532 p.
4. Smulders, E., W. Rähse, and G. Jakobi, *Laundry detergents.* 2002, Weinheim: Wiley-VCH. ix, 270 p.
5. Scheibel, J.J., *The evolution of anionic surfactant technology to meet the requirements of the laundry detergent industry.* Journal of Surfactants and Detergents, 2004. **7**(4): p. 319-328.
6. Yu, Y.X., J. Zhao, and A.E. Bayly, *Development of surfactants and builders in detergent formulations.* Chinese Journal of Chemical Engineering, 2008. **16**(4): p. 517-527.
7. Ramaraju, S.M., et al., *The liquid crystalline phases formed by linear-dodecylbenzene sulphonic acid during neutralisation with sodium carbonate.* Colloids and Surfaces a-Physicochemical and Engineering Aspects, 2006. **288**(1-3): p. 77-85.
8. Jönsson, B., *Surfactants and polymers in aqueous solution.* 1998, Chichester; New York: John Wiley & Sons. xii, 438 p.
9. Mungray, A.K. and P. Kumar, *Fate of linear alkylbenzene sulfonates in the environment: A review.* International Biodeterioration & Biodegradation, 2009. **63**(8): p. 981-987.
10. Simons, S., et al., *LAS acid reactive binder: Wettability and adhesion behaviour in detergent granulation.* Powder Technology, 2009. **189**(2): p. 385-393.
11. Iveson, S.M., et al., *Nucleation, growth and breakage phenomena in agitated wet granulation processes: a review.* Powder Technology, 2001. **117**(1-2): p. 3-39.
12. Leger, L. and P. Silberzan, *The Spreading of Drops on Solid-Surfaces.* Journal of Physics-Condensed Matter, 1990. **2**: p. Sa421-Sa425.
13. Prabhu, K.N. and G. Kumar, *Review of non-reactive and reactive wetting of liquids on surfaces.* Advances in Colloid and Interface Science, 2007. **133**(2): p. 61-89.
14. Rowe, R.C., *Correlation between Predicted Binder Spreading Coefficients and Measured Granule and Tablet Properties in the Granulation of Paracetamol.* International Journal of Pharmaceutics, 1990. **58**(3): p. 209-213.
15. Sprackling, M.T., *Liquids and solids.* Student physics series. 1985, London ; Boston: Routledge & K. Paul. ix, 237 p.
16. Lopez, V.H. and A.R. Kennedy, *Flux-assisted wetting and spreading of Al on TiC.* Journal of Colloid and Interface Science, 2006. **298**(1): p. 356-362.
17. Orr, F.M., L.E. Scriven, and A.P. Rivas, *Pendular Rings between Solids - Meniscus Properties and Capillary Force.* Journal of Fluid Mechanics, 1975. **67**(Feb25): p. 723-742.
18. Mason, G. and W.C. Clark, *Liquid Bridges between Spheres.* Chemical Engineering Science, 1965. **20**(10): p. 859-866.

(67g) High Shear Wet Granulation of a Lubricated Pharmaceutical Formulation

David A. Mota Aguilar, Chemical Engineering, University of Puerto Rico at Mayaguez, Mayaguez, PR and Carlos Velzquez, Department of Chemical Engineering, University of Puerto Rico at Mayaguez, Mayaguez, PR

Granulation plays an important role in pharmaceutical manufacturing as a common process in a solid dosage production line. Usually, excipients and active ingredients undergo this process to enhance the flow properties of the material in the compaction stage. Among the excipients, lubricants are widely used to enhance the compression efficiency, however, most of them modify the hydrophobicity of pharmaceutical formulations, which can alter significantly the process and the final product.

In this work, the effect of the concentration of lubricant in a high shear granulation process was evaluated experimentally. A lactose-based placebo formulation was prepared using magnesium stearate as lubricant at two different concentrations (1% and 3% w/w) and the granulation performance was compared with a non-lubricated formulation. A diluted p*olyvinylpyrrolidone (PVP) solution was used as binder.*

Lubricant dispersion at different times was estimated by a hydrophobicity measurement based on the Washburn equation according to the procedure described by Llusa et al. in 2010 *[Powder Technology, 198 (2010) 101-107]*. The evolution of the particle size of the material in the massing stage was monitored off line by means of standard laser diffraction for fine particles (< 1mm) and standard sieving test for coarse granules (>1mm). Liquid content distributions were also determined by loss on drying method, while geometric and physical characteristics of the granules can be observed in scanning electron microscope images.

Results indicate quantitatively the effect of the concentration of magnesium stearate on the growth kinetics of the granulation and the overall development of the material wettability.

See more of this Session: Agglomeration and Granulation Processes
See more of this Group/Topical: Particle Technology Forum

(71a) An Integrated Approach to Separations Process Research

Kyle P. Kostroski, Refining & Logistics Technology, BP Products North America Inc., Naperville, IL

The importance of separations processes is demonstrated by their ubiquity in the chemical and petrochemical industries. Many separations processes are also technologically mature, however, and it is often tempting for companies to forego rigorous separations research by limiting capital spending on benchtop- and pilot-scale programs. For obvious economic reasons, experimental work has been replaced by the use of process simulators. While this is understandable, it comes with a caveat: simulators are only as good as the data they use to make predictions. To avoid the so-called "garbage in, garbage out" problem associated with simulation, an integrated approach to separation process research might be necessary.

This talk will focus on how bench-scale and pilot-scale experimental facilities can be leveraged cost-effectively to do the key experiments required for enhancing process simulations, informing process economic evaluations, and enabling process scale-up. Development of a solvent extraction process for hydrocarbon separation will serve as a case study. Three efforts were undertaken in parallel: bench-scale solvent screening, pilot-plant validation and determination of scale-up parameters, and integration of experimental data into process simulations. The results of this study indicated that neither simulation results nor experimental data alone were adequate to assess the overall performance of the extraction process. However, by integrating key bench and pilot plant data into the simulation, a more accurate prediction of separation performance was achieved.

See more of this Session: Best Practices In Pilot Plant Separation
See more of this Group/Topical: Process Development Division

(71b) Drying Aqueous Biphenol Salt for Polyetherimide Monomer Synthesis

James M. Silva[1], David Woodruff[1], Thomas Guggenheim[2] and Albert Stella[1], (1) General Electric Global Research Center, Niskayuna, NY, (2)SABIC Innovative Plastics, Mt. Vernon, IN

Polyetherimide synthesis via displacement polymerization is important for process simplification and product flexibility. A key step in the synthesis of biphenol-based polyetherimides is the production of biphenol (BP) disodium salt with minimal moisture content. To achieve sufficient control over the salt composition, the BP disodium salt is synthesized as an aqueous solution. Salt drying is accomplished by introducing an aqueous salt solution into a vessel containing a hot organic solvent. This removes most of the water from the aqueous BP salt solution as an overhead product. The process of adding the aqueous BP salt solution into hot solvent is called "sprayover".

During the laboratory and semi-pilot development phases of this program, we found that the sprayover vessel must be maintained at or above a certain minimum target temperature in order to avoid agglomeration of the BP salt and caking on the vessel walls. We utilized the sprayover vessel target temperature at the 50-liter and 50-gallon scales. However, as aqueous BP salt is sprayed into the hot solvent a large amount of solvent is lost to the overhead stream. Such a high solvent evaporation rate causes serious solvent inventory control problems for a sprayover vessel with a short-path (non-reflux) condenser configuration.

By utilizing a partial reflux condenser in the overheads system, we avoided agglomeration and caking in the sprayover vessel and achieved stable operation at the target temperature. We were initially concerned that the refluxing solvent, which is saturated with water, would yield a higher water concentration in the sprayover vessel (compared with the short-path condenser configuration). However, application of the phase rule and ASPEN Plus® simulations of the solvent-H_2O system showed that the equilibrium water concentration in the sprayover vessel depends only on the sprayover vessel temperature and pressure, and is independent of the condenser configuration. Measured solvent moisture levels were comparable for both condenser configurations.

The calculated heat duty to operate a sprayover vessel at the target temperature is

6% higher with a partial reflux condenser than with a short-path condenser. However, the calculated amount of solvent in the distillate at the target temperature is 33X lower with a partial reflux condenser than with a short-path condenser.

Sprayover tests in 50-liter and 50-gallon sprayover vessels demonstrated that with a partial reflux condenser, the solvent can be contained in the system at high target temperatures. In these tests, solvent losses were minimal, the vessel walls were clean after sprayover at both scales, and the salt product polymerized normally.

See more of this Session: Best Practices In Pilot Plant Separation
See more of this Group/Topical: Process Development Division

(71c) Design of Reactive Distillation Column for Non-Parameterized Components - Which Experiments Are Essential

Oliver Litzmann[1], Thomas Grtzner[2], Gnter Wozny[1] and Jens-Uwe Repke[3], (1) Chair of Process Dynamics and Operation, Berlin Institute of Technology, Berlin, Germany, (2)Lcdrn, Lonza AG, Visp, Switzerland, (3)Institut fr Thermische Verfahrenstechnik, Umwelt- und Naturstoffverfahrenstechnik, TU Bergakademie Freiberg, Freiberg, Germany

Design of reactive distillation column fornon-parameterized components - which experiments are essential

Litzmann, O., TU Berlin, Berlin; Lorenz, H.-M., TU Berlin, Berlin; Wozny, G., TU Berlin, Berlin; Grtzner,T., Lonza AG, Visp/CH; Repke, J.-U., TU Freiberg, Freiberg

Designing reactive distillation processesrequires a great number of fitted parameters depending on the model used. Theseparameters are known for test-systems like MTBE or TAME but not for systemswhich have never before used in reactive distillation. As fitting of parametersrequire a lot of time and money for trials it is crucial to know the impact ofan improved model to the prediction accuracy. If this is known it is possibleto carry out fewer experiments with and fit a set of sufficient parameters inorder to describe the RD process as good as needed to strivethe goal.

Application to a new industrial substancesystem

The system analysed consists of an organic acid, analcohol, the corresponding ester and water whereupon some components cannot bedescribed by standard UNIFAC methods. The esterficationtakes place in a reactive distillation column designed for this system. The reactionis catalysed by the heterogeneous catalyst Amberlyst 36.

In order to describe the results of thecolumn, models are used which take different phenomena into account and usedifferent sets of parameters which have been measured independently.

Fig. SEQ Fig. * ARABIC 1: Comparison of the composition forecast by pseudo homogeneous and adsorption based model

For instance the reaction can be described either byequilibrium or by kinetics which can be corrected by adsorption effects in addition.This kinetic have been measured in a batch reactor and is described with significanthigher accuracy by an adsorption based kinetic model (see Fig.1) Theportability of these model improvement and similar effects to the reactivecolumn simulation will be presented and discussed by means of comparison withthe experimental results from the lab scale plant.

See more of this Session: Best Practices In Pilot Plant Separation
See more of this Group/Topical: Process Development Division

(71d) Design and Fabrication of Pilot Scale Gas/Liquid Separators

Oliver Orrell, Zeton Inc, Burlington, ON, Canada

Separation processes are an important component of chemical and petrochemical plants. In large facilities, gas/liquid, liquid/liquid, and gas/liquid/liquid separators can benefit from engineered internals such as coalescing plate packs, weirs, and baffles. In smaller pilot facilities, these internals can be difficult to employ due to size constraints. This session will discuss various approaches of designing and fabricating pilot scale separators that are effective without the benefit of internals available to their full-scale counterparts. The main focus will be on a recent redesign of a small scale three phase separator that has been proven to have less liquid carryover than earlier designs.

See more of this Session: Best Practices In Pilot Plant Separation
See more of this Group/Topical: Process Development Division

(71e) Pilot Plant Development of Molecular Separations by Organic Solvent Nanofiltration

Andrew G. Livingston, Andrew T. Boam, Andrea Meniconi, Fui-Wen Lim and Marcin Makowski, Evonik Membrane Extraction Technology Limited, London, United Kingdom

Organic Solvent Nanofiltration (OSN) separates molecules present in solutions of organic solvents. It is the same in principle as the well known RO/nanofiltration used in water treatments, except that the separations are based in organic solvents. This opens up a whole range of separations applications in organic processes – concentration, purification, catalyst retention and solvent recovery are just a few of these.

In OSN, a pressure gradient is applied across a solvent stable nanoporous film, inducing transport of liquid across the film. Depending on relative permeation rates, molecular species in a feed liquid stream can be concentrated and separated. Typically OSN will use a small fraction (less than 10%) of the energy of thermal methods such as evaporation and distillation, and can work at temperatures around ambient. Challenges for application of OSN technology in the refining, chemicals and pharmaceutical industry have been: (i) the lack of membranes with wide solvent resistance, and which retain their nanostructure with exposure to aggressive liquids such as dipolar aprotic solvents (DMF, DMSO, THF); and (ii) acceptance of the introduction of a new process technology to large scale chemical processes.

With respect to the first challenge, this presentation will describe the recent development of two new classes of commercially available OSN membranes and modules, and their applications to processes typically used in chemical manufacturing routes. The DuraMem™ series OSN membranes are made through post-formation processing including crosslinking of polymer membranes, which makes them stable in polar solvents such as acetone, DMF and THF. The PuraMem™ S series are thin film composite membranes which show exceptional performance in apolar solvents. Both these membranes are available for lab testing as flat sheets, and as spiral wound cartridges for scale up.

To meet the second challenge, the development of organic processes where OSN is used as a molecular separation step proceeds from lab bench scale to kilo/pilot scale, and then to process scale. The laboratory bench testing programme must yield key performance data to show that OSN is a viable technical solution, and that the technology is economically viable. Suitable process models are employed to predict the performance of the process, and to decide the size of the appropriate pilot scale tests that will provide convincing data for full scale plant design. Next, it is essential to move to a pilot plant testing programme utilising spiral wound membrane modules. The spiral modules must remain stable in the organic solution being processed, and so it is very important that the construction of these modules is the same for the pilot programme as will be used for the scaled up process. Membrane lifetime is a key factor that must be estimated from the data obtained in the pilot testing programme. Typically, at Evonik we have used spiral wound modules of 1.8" x 12" (0.2m2) through to 2.5" x 40" (2 m2) for the testing programmes; sometimes we also employ process size 4" x 40" modules for advanced tests. Where the application of OSN is in the pharmaceutical industry, compliance with GMP requirements and the need for demonstrating that leachable materials from the spiral element will not affect the overall process, is also essential.

The paper will describe examples of pilot testing programmes that have been carried out and which have led to process scale installations of OSN technology. OSN membrane modules are being successfully applied in cGMP manufacturing processes in the pharmaceutical industry (GSK, Johnson and Johnson), and for larger scale chemicals processing including homogenous catalyst retention at Evonik, and these applications will be described.

See more of this Session: Best Practices In Pilot Plant Separation
See more of this Group/Topical: Process Development Division

(71f) Pilot Scale Liquid-Liquid Extraction Process Development for Emulsifying Viscous Organic-Aqueous Mixtures

William Hollar[1], Babatunde Oguntade[1], Pankaj S. Gautam[2] and Zeljko Kuzelevic[1], (1)Sabic Innovative Plastics, Mount Vernon, IN, (2)Process Technology, SABIC Innovative Plastics, Mt. Vernon, IN

The tendency of specific two-phase liquid-liquid systems to emulsify makes proper pilot scale process development extremely important for successful process scale-up. This paper will focus on a case study of a polymer solution – water mixture that exhibited strong emulsification tendencies and the challenges associated with piloting and scale-up. Its behaviors and unique features will be compared to a system with low emulsification tendency. Efforts to measure liquid-liquid mixture properties and correlate them to input process conditions and equipment design will be detailed. A brief description of ongoing process improvement efforts will be included.

See more of this Session: Best Practices In Pilot Plant Separation
See more of this Group/Topical: Process Development Division

(85a) Discrete Approximation of a Continuous Size Distribution for Use In Kinetic-Theory Predictions of Solids Flows

J. Aaron Murray, Chemical Engineering, University of Colorado at Boulder, Boulder, CO and Christine M. Hrenya, University of Colorado, Boulder, CO

Discrete Approximation of a Continuous Size Distribution for use in Kinetic-theory Predictions of Solids Flows

J. Aaron Murray, Chemical and Biological Engineering, University of

Colorado at Boulder, Boulder, CO

Christine M. Hrenya, Chemical and Biological Engineering, University of

Colorado at Boulder, Boulder, CO

It is common in nature and industry for the distribution of particle sizes within a granular mixture to be essentially continuous. The focus of this work pertains to kinetic-theory-based models for rapid flows of continuous particle size distributions (PSDs). Such models are restricted to a finite number (s) of particle species. Accordingly, a continuous PSD must be approximated as a discretization of s particle diameters and associated volume (or number) fractions. The kinetic-theory-based model used as a basis in this study was proposed by Garzó, Hrenya, and Dufty (2007; hereafter referred to as GHD). The overall aim of the current work is to determine the suitability of the GHD theory to predict the flow behavior of a continuous size distribution of solids. In particular, the objective of this effort is twofold: (i) to determine the number of *discrete* species required to accurately approximate a *continuous* PSD, and (ii) to validate these results via a comparison with molecular dynamics (MD) simulations for continuous PSDs. In both cases, the discrete approximation is determined by matching an increasing number of moments of the distributions as the value of s increases. Regarding (i), the number of discrete species needed for a desired level of accuracy is found via an evaluation of all scalar transport coefficients for increasing values of s. This process was carried out for a variety of Gaussian and lognormal PSDs, as well as an experimental bimodal PSD supplied by Department of Energy (DOE NETL). With regards to (ii), a comparison between MD simulations (Dahl, Clelland and Hrenya, 2003) and GHD predictions of granular pressure and shear viscosity for a simple shear flow is conducted. The results indicate that wider continuous PSDs require a larger number of particle species for a given level of accuracy, whereas the effect of overall solids fraction and restitution coefficient is less pronounced. Also, the GHD predictions match the MD data quite well (on par with monodisperse predictions) for the values of *s* determined above.

References:

Dahl, S., R. Clelland and C. Hrenya (2003). "Three-dimensional, rapid shear flow of particles with continuous size distributions." Powder Technology **138**(1): 7-12.

Garzó, V., J. W. Dufty and C. M. Hrenya (2007). "Enskog theory for polydisperse granular mixtures. I. Navier-Stokes order transport." Physical Review E **76**(3): 031303.

Garzó, V., C. M. Hrenya and J. W. Dufty (2007). "Enskog theory for polydisperse granular mixtures. II. Sonine polynomial approximation." Physical Review E **76** (3): 031304.

See more of this Session: Dynamics and Modeling of Particulate Systems I
See more of this Group/Topical: Particle Technology Forum

(85b) DEM Validation Using An Annular Shear Cell

Adel F. Alenzi[1], Martin C. Marinack Jr.[2], C.F. Higgs III[2] and J. J. McCarthy[1], (1) Department of Chemical and Petroleum Engineering, University of Pittsburgh, Pittsburgh, PA, (2)Mechanical Engineering, Carnegie Mellon University, Pittsburgh, PA

Studying the flow of granular materials is important for its impact on both industrial applications and natural phenomena. A fundamental understanding of the flow of granular materials is lacking, and this results in difficulties in modeling and predicting their flow behavior. The discrete element method (DEM) is often used as the "golden standard" for comparison to continuum-level theories of granular material flows due to its derivation from first-principal constructs, like contact mechanics. In this paper, we continue our work on quantitative validation of DEM simulations using detailed measurements of simple, well-characterized flows that allow us to examine the effect of rough surfaces and rotational rates on granular flow using an annular shear cell. Experimentally, we use digital particle tracking velocimetry (DPTV) to obtain velocity, solids fractions, and granular temperatures profiles. Computationally, we compare the results obtained using different contact mechanics force laws to those of experimental measurements as well as perform a sensitivity analysis on device and particle geometry and different material properties employed. In previous work, we have found that an elasto-plastic normal force model is critical to obtaining accurate results as is detailed matching of both particle and system geometry, while friction model choice is found to be unimportant. Here, we examine the robustness of these observations to both particle materials properties as well as systemic variables (such as total system solids fraction).

See more of this Session: Dynamics and Modeling of Particulate Systems I
See more of this Group/Topical: Particle Technology Forum

(85c) A New Algorithm to Solve Efficiently the Boltzmann Equation for Inelastic Particles

Matteo Strumendo, Department of Chemical Engineering (DIPIC), University of Padua, Padua, Italy

Granular flows are described by the Enskog-Boltzmann equation for inelastic particles (except in the regions where frictional forces are dominating); in dilute systems, the Enskog-Boltzmann equation reduces to the Boltzmann equation. Several methods are available to solve the Boltzmann equation, i.e. generalizations of the Chapman-Enskog expansion, method of moments, Monte Carlo simulations, discretization of the Boltzmann equation in the velocity space, Lagrangian methods; in this work, the focus is on the solution of the Boltzmann equation by the method of moments [1,2,3].

The Boltzmann equation can be interpreted as a multi-variate population balance equation, in which the internal variables are the components of the particle velocities. In this perspective, concepts and techniques developed in the solution of population balance equations can be used for the solution of the Boltzmann equation, and vice versa.

The FCMOM (Finite size domain Complete set of trial functions Method Of Moments) was developed by Strumendo and Arastoopour ([4]-[6]) and by Strumendo ([7]) to solve mono-variate and bi-variate population balance equations. The fundamental idea is to construct a method of moments on a finite domain of the internal variables (typically the particle size in mono-variate population balance equations, or the particle velocities in the Boltzmann equation). An advantage of the FCMOM is that it provides both the moments *and* the reconstructed distribution of the particle internal variables. Further, the domain of the internal variables is always well defined (a property which is relevant in multi-variate applications). In the applications developed, the numerical solutions provided by the FCMOM converged rapidly to the exact solutions.

In principle, the FCMOM technique could be extended for solving the Boltzmann equation for inelastic particles. However, some difficulties arise when applying the FCMOM algorithm to the Boltzmann equation. In fact, when inelastic particles are considered the numerical solution converges slowly and the algorithm performance is poor. This appears clearly in the numerical solution of the Homogeneous Cooling State problem.

Additionally, several different regimes need to be considered when testing if a numerical technique can solve efficiently the Boltzmann equation for granular systems, namely low/high Knudsen number flows and low/high Mach number flows (low/high Stokes number flows need to be considered as well if the fluid-particle interaction is included in the kinetic equation). Therefore, it is crucial to demonstrate the capability of solving simultaneously the Boltzmann equation in different regimes, and thus the transition between different regimes.

In this work, a new technique, which modifies the FCMOM algorithm, is proposed for solving efficiently the Boltzmann equation for inelastic particles. The proposed method is illustrated through homogeneous and in-homogeneous applications (relaxation of elastic particles to the Maxwellian state, Homogeneous Cooling State solution for inelastic particles, impulsive start-up problem).

[1] Grad, H., "On the Kinetic Theory of Rarified Gases", Communications on Pure and Applied Mathematics, 2, 331-407, (1949).

[2] Strumendo, M., Canu P., "Method of Moments for the Dilute Granular flow of Inelastic Spheres", Physical Review E 66, 041304/1-041304/20, (2002).

[3] Fox, R.O., "A quadrature-based third-order moment method for dilute gas-particle flows",
Journal of Computational Physics, Volume 227, Issue 12, 6313-6350, (2008).
[4] Strumendo, M., Arastoopour, H., "Solution of PBE by MOM in Finite Size Domains", Chemical Engineering Science 63, 2624-2640, (2008).

[5] Strumendo, M., Arastoopour, H., "Solution of Bivariate Population Balance Equations Using the FCMOM", Industrial and Engineering Chemistry Research 48 (1), 262-273, (2009).

[6] Strumendo, M., Arastoopour, H., "Solution of population balance equations by the FCMOM for in-homogeneous systems", *Industrial and Engineering Chemistry Research*, 49(11), 5222-5230, 2010.

[7] Strumendo, M., "Solution of population balance equations using the FCMOM", PBM 2010. Proceedings of the Fourth International Conference on Population Balance Modelling, Berlin, Germany, September 15-17, 2010, 293-300.

(85d) Granular Mixing In the Flowing Layer In Rotating Cylinders

Suman K. Hajra, TBA, Pittsburgh, PA and Devang V. Khakhar, Department of Chemical Engineering, Indian Institute of Technology, Bombay, Mumbai - 400076, India

Granular materials are widely used in industries such as iron & steel, mining, cement, construction, lime, fertilizers and pharmaceuticals. In order to design equipment such as rotary kilns, bins, silos, combustors, hoppers, chutes, hydro-cyclones, etc., in an effective and economical way, a thorough understanding of the various factors governing the flow characteristics of granular materials must be obtained. These needs have already motivated extensive analytical and experimental investigations of the flow of granular materials. Mixing is a central feature of many processes such as in the reduction of iron ore, cement clinkering, coal gasification, food processing, and tablet manufacturing. Ineffective mixing, or the inability to control particle segregation, is always costly in terms of rejected materials, extra blending time, and defective end products. Physical parameters that control the extent of mixing are not well-studied even in a simple geometry such as rotating cylinder. We study granular mixing in a rotating cylinder. A thin layer of particles in the free surface moves at very high velocity, whereas rests of the bed materials are rotated as a solid body rotation. Particles in the flowing layer experience a random motion due to inter particle collisions and this random motion causes the particles to diffuse in the transverse direction with respect to flow direction. Results of various parameters affecting the mixing will be presented. Role of particle size and cylinder speed of rotation on granular mixing will be discussed. Mixing is well-predicted with Lagrangian simulation based theory. Mixing in both radial and azimuthal directions will be discussed.

See more of this Session: Dynamics and Modeling of Particulate Systems I
See more of this Group/Topical: Particle Technology Forum

(85e) A Combined Experimental and Computational Approach to the Scale-up of High-Shear Wet Granulation

Preetanshu Pandey[1], Jing Tao[1], Julia Z. Gao[1], Dilbir Bindra[1], Ajit Narang[1], Rohit Ramachandran[2] and Anwesha Chaudhury[2], (1)Bristol-Myers Squibb, New Brunswick, NJ, (2)Department of Chemical and Biochemical Engineering, Rutgers, The State University of New Jersey, Piscataway, NJ

High-shear wet granulation is widely used in the pharmaceutical industry to achieve particle size enlargement, which leads to improved properties such as flowability and bulk density. Scale-up of wet granulation is always a challenge due to a wide range of geometrical designs of granulators and the varying amounts of shear rates in them, which in turn dictates the dynamics of granule formation and granule characteristics.

The objective of the current study is to find the smallest appropriate scale for conducting formulation and process optimization studies that can then be considered scalable. The ability to conduct key experiments at a small scale can lead to significant reduction in time and resources. As a first step, a full factorial design of experiments was conducted on three key process parameters to gain understanding of their effects on granule characteristics at 1-L scale. There were four center-point repeats and each parameter (water amount, impeller speed, wet massing time) was studied at two levels excluding the centers. The formulation used consisted primarily of microcrystalline cellulose, lactose, hydroxypropyl cellulose (added dry), and croscarmellose sodium. The main responses studied included granule porosity (mercury intrusion), granule particle size (sieve analysis), flow (Erweka® flow tester), bulk density, compactability (Stylcam® tablet press simulator), and dissolution. It was found that granule particle size was relatively insensitive to changes in process conditions (other than water amount), but those changes were well-captured by granule porosity and compactability. Statistical analysis was conducted on the responses and an optimized 1-L process was selected based on granulation flow and compactability.

Scale-up experiments were conducted using different impeller speed power laws, such as constant tip speed, constant Froude number, and constant shear stress, as scale-up rules. Three sizes of granulators were used for this study: 1-L, 10-L and 65-L. Water amount and water addition time was kept constant. Scale-up criteria included achieving similar granule characteristics (porosity and compactability) and a similar tablet dissolution profile across scales. It was found that in addition to impeller speed, wet massing time played a key role when comparing granule characteristics (especially pore size distribution) between scales.

A multi-dimensional population balance model that incorporates kernels based on fundamental physics and chemistry has been developed. The model will be validated and thereafter used to predict evolutions and distributions of granule particle size and porosity, thus alleviating the need for potentially labor and capital intensive experiments.

See more of this Session: Dynamics and Modeling of Particulate Systems I
See more of this Group/Topical: Particle Technology Forum

(85f) A Three-Dimensional Cellular Automata Framework to Model Granular Shear Flows

Martin C. Marinack Jr. and C. Fred Higgs III, Mechanical Engineering, Carnegie Mellon University, Pittsburgh, PA

Granular flows continue to be a complex problem in nature and industrial sectors where solid particles exhibit solid, liquid, and gaseous behavior in a manner which is often difficult to predict both locally and globally. A popular method in the modeling and prediction of granular flows is the discrete element method (DEM), which provides a rigorous physical treatment of particle interactions, and has been shown to be the "gold standard" for simulating many granular systems. One preliminary supplement to a DEM-based design or engineering analysis framework is cellular automata (CA). The CA modeling framework provides a platform for obtaining fast first-order approximations of simple and complex granular systems in the kinetic regime, and recently, in two-phase or transitional regimes (where both the frictional and kinetic regime are present) as well. It has the flexibility to employ rule-based mathematics, first-principle physics, or both to rapidly model physical processes, such as granular flows. Prior physics-based two-dimensional (2D) CA models were developed to predict granular flow in a 2D annular Couette shear cell. In this work, these previous 2D models are extended into the third-dimension to model a three-dimensional (3D) Couette shear cell. This specifically involves the expansion of a CA particle's interaction neighborhood to now include out-of-plane neighbor cells. The predictive capabilities of the newly developed 3D model are tested and validated against experimental results. Expanding the CA framework for granular shear flows into 3D, expands the scope and applicability of the CA framework. More specifically, it allows for the framework to be more easily extended and applied to more realistic and complicated engineering applications.

See more of this Session: Dynamics and Modeling of Particulate Systems I
See more of this Group/Topical: Particle Technology Forum

(85g) Aerosol Brownian Coagulation In a Strong Temperature Gradient

Daniel E. Rosner, Chemical Engineering, Yale University, New Haven, CT and Manuel Arias-Zugasti, Vis. Asst Prof.-ChE Dept., Yale University, New Haven, CT

For particle populations in a non-isothermal carrier gas, size-dependent particle *thermophoresis* provides an often-overlooked coagulation mechanism[1] But strong temperature gradients can either induce the coagulation of non-Brownian particles, or, more generally, alter the collision/coagulation frequency for Brownian particles. Here we consider a rapidly coalescing aerosol of uniform composition, Brownian spherical particles evolving under the combined influence of both mechanisms: ie., well-known Brownian coagulation (Smoluchowski, Friedlander, *et.al.*) and thermophoresis-induced coagulation (Rosner and Arias-Zugasti [2, 3]).

We first derive a combined coagulation frequency based on a thermophoresis-modified continuum-limit coagulation frequency. The Brownian + thermophoresis combined coagulation frequency is modeled in a way similar to the Brownian + sedimentation coagulation frequency [4] (with the gravitational sedimentation velocity replaced by the size-dependent near-continuum thermophoretic velocity).

The relative intensity of each coagulation mechanism can be characterized by a dimensionless Peclet number, defined by the corresponding ratio of characteristic coagulation frequencies. Introducing the combined coagulation frequency as a kernel into a Smoluchowski-type population-balance integro-PDE, we perform a systematic parametric study of the time evolution of a coagulation-aged, initially log-normal population, as a function of the aforementioned Peclet number (Pe) and particle/gas Fourier thermal conductivity ratio, kp/kg. We have also investigated the errors associated with the frequently made "additive kernel" approximation.

Our numerical results show that in the long-time limit quasi-self-preserving populations (QSPPs) are reached. When the reference Pe is much smaller than, say, 0.1, previously well-studied Brownian self-preserving population results are recovered. For Pe-values larger than about 10 our QSPP become indistinguishable from our recently reported TP-"dominated" results [1]. However, for intermediate Pe-values characteristic distortions set in, corresponding to increased population spread and skewness, and slightly smaller departures from log-normality.

Extensions of this research of current interest [3] include accounting for a) unequal "particle" thermal conductivities, b) thermo-capillary fluid motion, c) non-ideal carrier vapor effects at higher pressures, and d) finite-rate particle "coalescence" rates.

a. For presentation at AIChE 2011Mtg (October), Minneapolis MN

b. Chemical & Environmental Engineering Dept., Yale University, New Haven, CT 06520-8286, USA

.c. Dept. de Física Matemática y de Fluidos, UNED, Apdo: 60141, 28080 Madrid, Spain

Supported by NSF under Grant: CBET-1037733 at Yale University and Ministerio de Ciencia e Innovación (#ENE2008-06515-C04-03) and Comunidad de Madrid (#S2009/ENE-1597) at UNED.

[1] Rosner DE and Arias-Zugasti M (2011) *Phys. Rev. Lett.* **106** 015502 (January 7, 2011)

[2] Arias-Zugasti M and Rosner DE (2011) *Phys. Rev. E* (submitted)

[3] Rosner DE and Arias-Zugasti M (2011) *Industrial & Engineering Chemistry Research* [submitted (for Churchill 90 issue)]

[4] Simons S, Williams MMR, and Cassell JS (1986) *Journal of Aerosol Science* **17** 789

See more of this Session: Dynamics and Modeling of Particulate Systems I
See more of this Group/Topical: Particle Technology Forum

(91a) Flow Regime Identification In Fluidized Beds Based On Statistical Analysis of Both Pressure Fluctuations and Optical Probe Data

Faraj M. Zaid, Stoyan Nedeltchev and Muthanna Al Dahhan, Department of Chemical and Biological Engineering, Missouri University of Science and Technology, Rolla, MO

The fluidized bed reactors can operate in different flow regimes. The reactor performance can change significantly as a result of flow regime change. It is very important to know how to identify the boundaries of the different flow regimes since reactor volume productivity, mixing, mass and heat transfer are affected by the prevailing flow regime. In this work pressure transducer and optical probe were used to identify the main flow regime transitions in two different plexiglas fluidized beds (0.14 m and 0.45 m in ID). By means of statistical and autocorrelation analyses applied to gauge pressure fluctuations (pressure readings above the atmospheric pressure) and optical probe signals the boundaries of the main hydrodynamic regimes were derived. Such information is needed for preparing a flow regime map.

The gauge pressure fluctuations were measured at a sampling frequency of 67 Hz and the optical probe readings were recorded at 5 kHz. The measurements were performed at three different axial positions. Air was used as a gas phase and glass beads (solid density ρ_s=2500 kg/m^3, mean particle size d_p=150-210 μm) were used as a solid phase in both columns. The same type of gas distributor (made of a porous polyethylene sheet with a pore size of 15-40 mm) was used in both columns.

In the case of small fluidized bed the autocorrelation function showed that the minimum fluidization velocity occurred at 0.12 m/s and the transition velocity to bubbling fluidization regime occurred at 0.38 m/s. The profile of variance was capable of identifying the onsets of both turbulent fluidization regime (0.91 m/s) and fast fluidization regime (1.6 m/s). In the case of bigger fluidized bed the profile showed that the minimum fluidization velocity occurred at 0.1 m/s and the transition velocity to bubbling fluidization regime occurred at 0.29 m/s. We confirmed these results by applying the same statistical methods to the optical probe data.

REFERENCES

[1] A. Chehbouni, J. Chaouki, C. Guy, D. Klvana, "Characterization of the Flow Transition between Bubbling and Turbulent Fluidization," *Ind. Eng. Chem. Res.* **33**, 1889 (1994).

[2] D. Bai, A. S. Issangya, J. R. Grace, "Characteristics of Gas-Fluidized Beds in Different Flow Regimes," *Ind. Eng. Chem. Res.* **38**, 803 (1999).

[3] S. Nedeltchev, S. Aradhya, F. Zaid, M. Al-Dahhan, "Flow Regime Identification in Both Spouted and Fluidized Beds Based On Different Entropies Derived From Gauge Pressure Fluctuations," *Proc. 1st Intl. Symp. on Multiscale Multiphase Process Engineering*, Kanazawa city, Japan, October 4-7 2011, Japanese Society of Chemical Engineers (2011) (CD-ROM).

See more of this Session: Fundamentals of Fluidization II
See more of this Group/Topical: Particle Technology Forum

(91b) Effects of Fine Powder Accumulation and Interstitial Gas Flow Rate On the Formation of Particle Bridging In a Moving Bed

Dawei Wang, William G. Lowrie Department of Chemical and Biomolecular Engineering, Ohio State University, Columbus, OH, Qiang Zhou, Chemical and Biomolecular Engineering, Ohio State University, Columbus, OH, Andrew Tong, William G. Lowrie Department of Chemical and Biomolecular Engineering, The Ohio State University, Columbus, OH and L. S. Fan, Chemical and Biomolecular Engineering, The Ohio State University, Columbus, OH

Bridging (or arching) of non-fluidized solids particles is a particularly dangerous hazard in the operation of particle handling moving bed system. The analysis of arching mechanism lies on the particulate media mechanics which is originally developed for the discharge characteristics of powders from hoppers with particle layer stress analysis. However, bridging may occur even when the moving bed system is originally designed correctly to allow desired continuous mass flow of solids particles. Experimental study reveals the appearance of bridging of solids particles normally accompanies with appearance and accumulation of fine powders among the moving coarse particles and the change of interstitial gas flow. The counter-current interstitial gas flow provides an opposite resistant in terms of interphase drag to prevent continuous solids flow driven by gravity. Its change may account for stress redistribution inside the layered particles, which yields an interlock among particles and forms a bridge. The accumulation of fine powders may alter the hydrodynamic characteristics of solids particles as well as greatly decrease the local voidage inside the moving bed. The decrease of voidage may significantly increase the inter-phase drag while the change of hydrodynamic characteristics may change the tendency of bridging of solids particles. In the proposed study, a theoretical stress analysis on layered particles in a vertical moving bed with counter-current interstitial gas flow, which takes into account of the effects of characteristics of fine powder, such as size distribution and volume fraction, and the gas flow rate, will be developed to predict the appearance of bridging phenomenon. Experiments will be performed to validate the theoretical predictions for different combinations of fine powder distribution, percentage and gas flow rates.

(91c) Studies of Solid-Solid Mixing Behaviors In a Downer Reactor

Yongpan Cheng[1], Eldin Wee Chuan Lim[1], Chi-Hwa Wang[1], Guoqing Guan[2], Yu Nakamura[3], Masanori Ishizuka[3], Chihiro Fushimi[4] and Atsushi Tsutsumi[4], (1) Chemical and Biomolecular Engineering, National University of Singapore, Singapore, Singapore, (2)North Japan Research Institute for Sustainable Energy (NJRISE), Hirosaki University, Hirosaki, Japan, (3)Institute of Industrial Science, University of Tokyo, Tokyo, Japan, (4)Institute of Industrial Science, The University of Tokyo, Tokyo, Japan

Downer reactor has the very short residence times, narrow residence time distributions and uniform solids holdup distributions with the aid of gravitational acceleration. As a result, it has become an attractive alternative to its counterpart, the riser reactor, for certain classes of chemical reactions. These include gasification reactions, fluid catalytic cracking, Fischer-Tropsch reactions and others. In particular, in downer reactor, high reaction efficiencies and low tar formation can be achieved in fast pyrolysis of coal or biomass. The flow behavior of solids is a key to the successful operation of the reactor within the downer, because it determines the contacting pattern as well as the extents of mixing between different phases of materials. Despite the research efforts that have been invested towards both experimental and numerical studies of hydrodynamics and flow behaviors in downer reactor systems, current understanding of such systems is far from complete. In particular, relatively little attention has been given towards the studies of mixing behaviors between two or more types of solids within a downer reactor. Our work will focus on the characterization of the mixing between two types of solids through both experiment and numerical studies.

*Corresponding author. Tel.: +65 6516 5079; Fax: +65 6779 1936; E-mail address: chewch@nus.edu.sg (C. H. Wang)

See more of this Session: Fundamentals of Fluidization II
See more of this Group/Topical: Particle Technology Forum

Kinetic Theory Based CFD Simulation and Experiments of Dynamic Characteristics in Gas-Solid Fluidized Bed

A two-dimensional Eulerian-Eulerian model integrating the kinetic theory of granular flow is used to simulate the bubble and particle dynamic characteristics in a gas-solid fluidized bed. Experiments of pressure fluctuation and acoustic energy (AE) signals in the fluidized bed are measured to validate the CFD simulation. The simulated results show that the Syamlal-O'Brien drag model predicts more reasonable results than the Gidaspow drag model. The power spectrum of pressure and voidage fluctuation is analyzed to associate with bubble motion. As the bed height increases, the simulated and experimental main frequency of pressure fluctuation first stays constant and then rises sharply, while the simulated main frequency and amplitude of voidage fluctuation both increase. The bubblelike granular temperature calculated from turbulent energy of solid phase has the same variation tendency compared to the experimental AE energy generated by particle impact on the wall. According to this variation tendency, the fluidized bed can be divided into stagnant zone, main circulation zone and freeboard. The computed vertical turbulent energy of particles is stronger in wall region and dense region of the fluidized bed. Based on the cascade theory of turbulence, the vertical turbulent energy spectrum of particles is divided into energy-containing range, inertial range and dissipation range. In inertial range at high frequency, the Levy-Kolmogorov law is obeyed. The flatness factors of particle fluctuation velocity and 8 scales wavelet decomposition coefficients of particle fluctuation velocity are calculated to investigate the intensity of intermittence caused by coherent structures in the flow field. Coherent structures are worm-like vortexes such as filamentary structures and contain most of the flow energy. They exist in turbulent flow and are responsible for most of the momentum transfer. The intermittence in dissipation range is much stronger than that in energy-containing and inertial range, and reinforces rapidly as the radial distance and bed height increase.

To further investigate the intermittence of the gas-solid flow field, the continuous wavelet transform technique is for the first time introduced to analyze the particle vortex behavior. The results show that the time scale of a particle vortex has a linear relationship with the wavelet scale factor. Some particle vortexes perform periodic motion and merge with or separate from others until the end of the vortex period. As the superficial gas velocity and bed height increase, the time scale decreases, while the particle vortexes are stretched and their boundaries become unclear. Meanwhile, the probability density function (PDF) of wavelet coefficients is computed. It is found that the wavelet coefficients and the probabilities of intermittent events both improve as the wavelet scale, superficial gas velocity and bed height increase. Since intermittency is caused by coherent structures in the flow field, the effects of coherent structures on the intermittency are studied through a coherent structure extracting method and the extended self-similarity (ESS) scaling law. Before extraction, wavelet coefficients at the same scale increase as the superficial gas velocity and bed height increase, while the scaling law curve at Ug=0.75 m/s deviates from those at other superficial gas velocities, but shows no obvious change with the bed height. After extraction, the discriminations of scaling law curves under different operating conditions are eliminated and their slopes become identical.

A CFD SIMULATOR for Multiphase FLOW IN RESERVOIRS and Pipes

Dimitri Gidaspow[1], Jing Huang[2] and Fang Li[1], (1)Chemical and Biological Engineering, Illinois Institute of Technology, Chicago, IL, (2)Energy System Division, Argonne National Laboratory, Argonne, IL

Introduction

We have developed a computational fluid mechanics (CFD) computer code for flow of oil, gas and sand in reservoirs and pipes that will help to understand flow in wild wells that are drilled for offshore oil production .We have simulated the flow of mud, gas and oil, as described in the Sunday, December 26,2010 New York Times article, " Deepwater Horizon's Final Hours". With a low level of mud in the pipe and some trapped gas in the well, the mud gets blown out, followed by gas and then oil. The diving force for flow is the buoyancy between the oil, gas and mud in the pipe and an equivalent height of water.

We have used a modification of our CFD code described in the 2009 book, " COMPUTATIONAL TECHIQUES " by Dimitri Gidaspow and Veeraya Jiradilok, Nova Science Publishers Conventional state of the art multiphase flow simulations in the reservoirs are done using Darcy's law (Economides and Nikolaou, 2011). Our simulations using CFD show that near the well the flow is highly turbulent. Our measure of turbulence is the computed granular temperature of the sand near the well.

In the tall well we are computing a core-annular flow in dilute oil-gas flow, in agreement with literature. We have also computed the friction coefficients of gas and oil flow in the well at the high gas and oil flow rates. In the one dimensional models, such as those used for licensing nuclear reactors, the fully developed friction coefficients are an input into the codes (Solbrig et. al., 1978).

Case I: Mud, Gas and Oil Flow (BP Oil Case)

We have simulated the flow of mud, gas and oil, as described in the Sunday, December 26, 2010 New York Times article, "Deepwater Horizon's Final Hours". Figure 1 shows the initial pressure distribution in the well and in the reservoir. In the BP oil spill, the pipe broke at the bottom of the ocean at the depth of 5,000 feet. The pressure is 1.5×10^7 Pa. The height of the well is 13,000 feet. We modeled the reservoir as shown in figure 2 with an inflow of gas, oil and sand at velocities of 4.04, 3.37 and 3.37 cm/s. The initial pressure at this boundary was estimated to be equivalent to that given by a height of water of 13,000 feet. Initially, we assumed the reservoir to contain a mixture of 50% sand, 25% oil and 25%gas. We assumed the sand to have a diameter of 10 microns and a density of 2.94 g/cm^3. We assumed the oil to have a droplet diameter of 100 microns and a density of 0.85 g/cm^3.

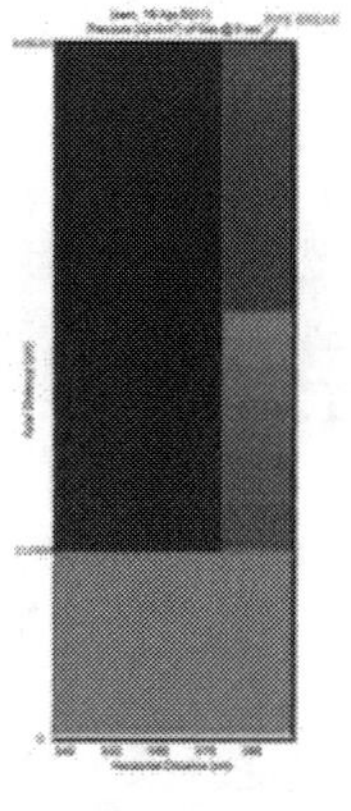

Figure 1

Figure 2 shows the mud volume fraction in the well due to BP oil gulf spill as a function of time. The figure shows that after 18 seconds, the mud got blown out. Figure 3 shows the gas volume fraction in the well. It shows that following the mud flow, gas is discharged through the pipe break. Figure 4 shows the computed oil volume fractions. In the accident, the oil was discharged in the form of a mist. The next simulation shows this behavior more clearly.

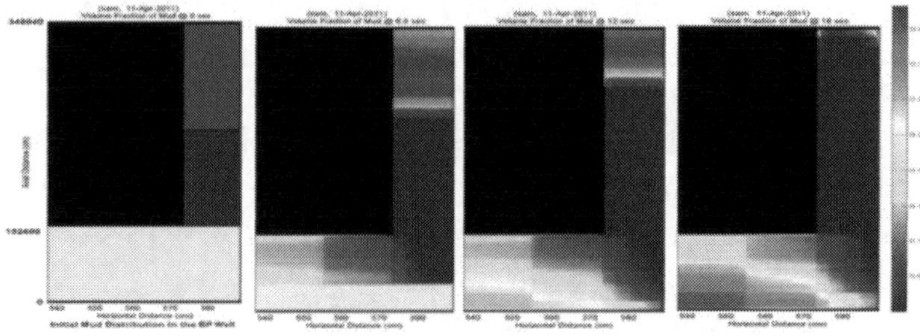

Figure 2

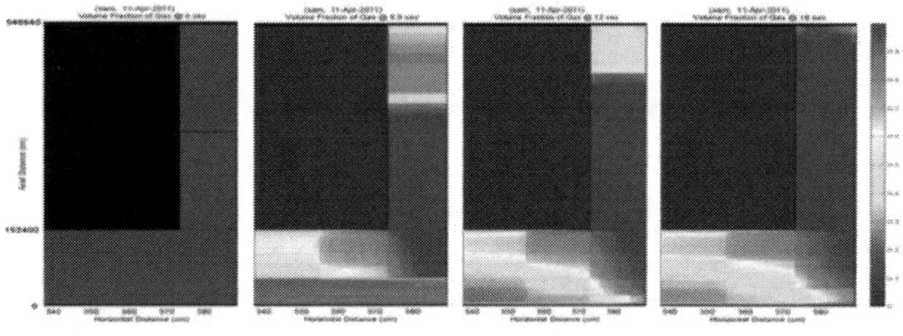

Figure 3

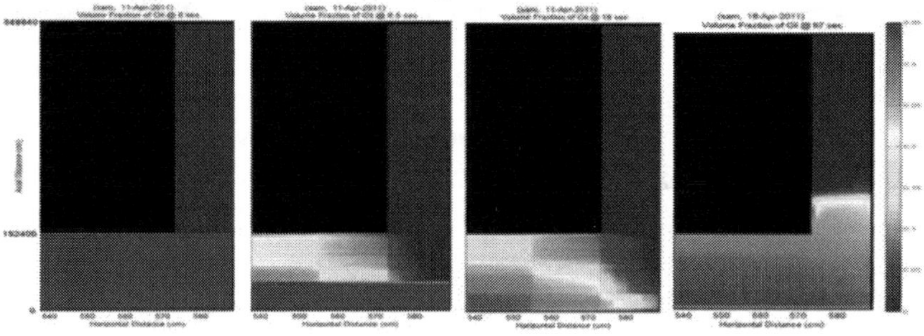

Figure 4

Case II Gas-filled well

In the second simulation, we assumed the well is filled with gas only. Initially the reservoir still contained a mixture of 50% sand, 25% oil and 25%gas as shown in figure 5. Figure 6 shows the oil volume fraction as a function of time. For dilute gas-oil flow, we obtain core annular flow for oil, in agreement with the flow regime map for vertical flow in the map as shown in the figure 7. Figure 8 shows the methane volume fraction as a function of time. Figure 9 shows the sand volume fraction distribution. We again see the core-annular region for sand. In our simulation, we do not have a grid to prevent the sand from coming into the pipe.

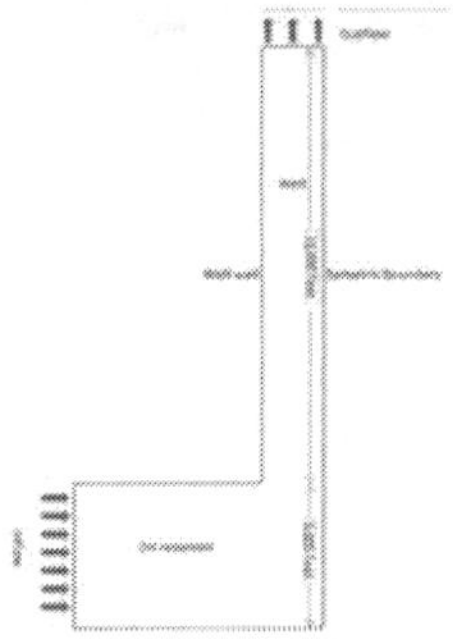

Figure 5

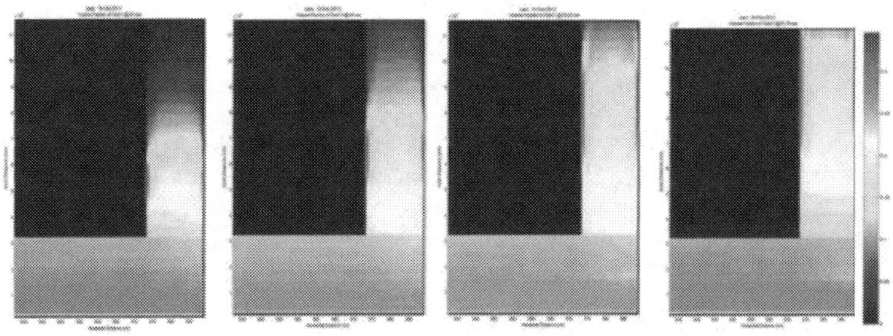

Figure 6

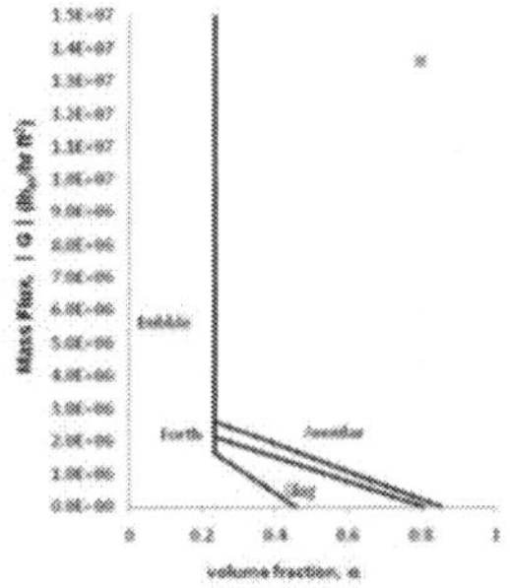

Figure 7

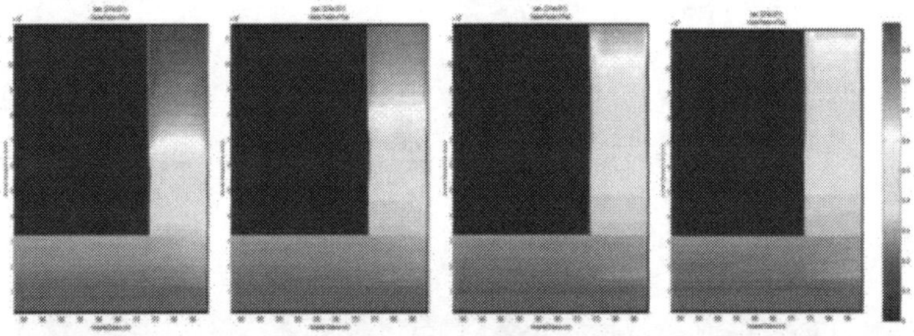

Figure 8

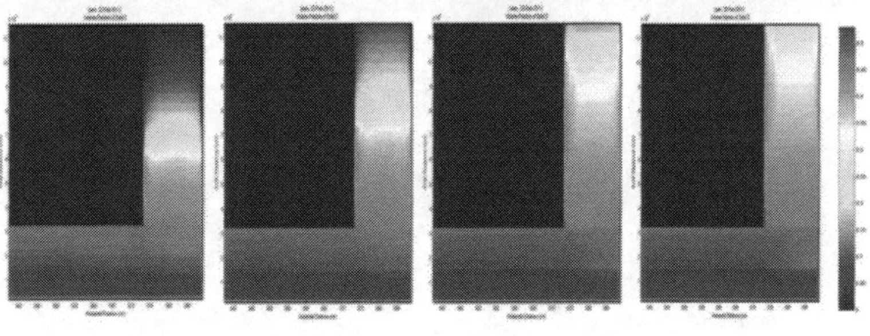

Figure 9

Computation of Turbulence in the Reservoir

Figure 10 shows the axial and radial velocities and the spectral distribution the center of the reservoir. There are large velocity oscillations which cannot be predicted using Darcy's Law. Figure 11 shows the granular temperature in the reservoir as a function dimensionless radial distance from the left boundary for various times. The computed granular temperature at the dimensionless radial distance of 1 is very large. Such a large turbulence produces a large entrance effect near the pipe, which cannot be predicted using the conventional Darcy's Law model. Figure 12 shows the axial velocity and its spectral decomposition. The computed frequency is 0.024 s^{-1}. This computation explains the origin of slow gas bursts in the wild wells.

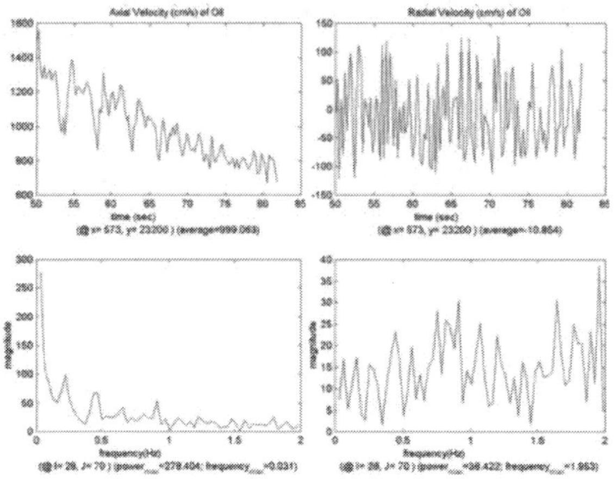

Figure 10

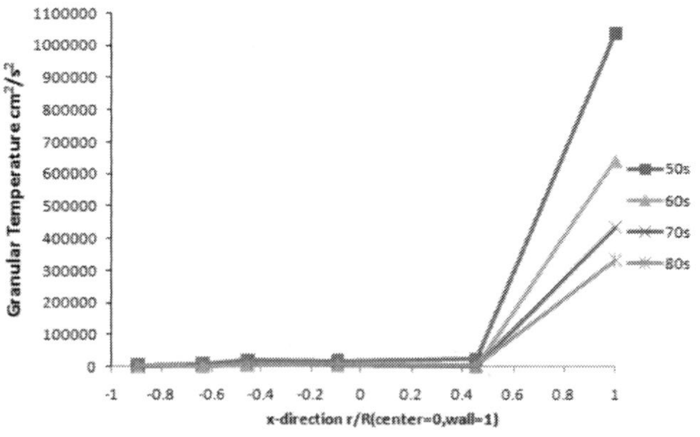

Figure 11

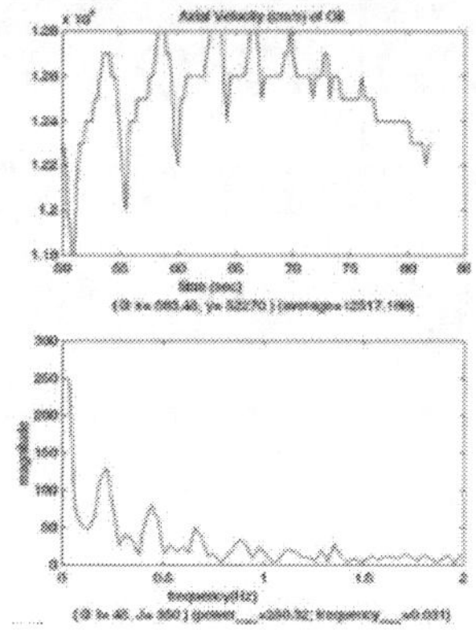

Figure 12

References

Michael J. Economides and Michael Nikolaou, Technologies for Oil and Gas Production: Present and Future, AICHE Journal Vol. 57, 1974-1982, 2011

Solbrig, C. W., J. H. McFadden, R. W. Lyczkowski and E. D. Hughes, Hear Transfer and Friction Correlations Required to Describe System-water Behavior in Nuclear Safety Studies, AICHE Symposium Series, Vol. 74, 100-128, 1978.

(91f) Effect of Emulsion Fluidization State On the Fluctuations In Gas and Particle Velocities Inside the Plume of a Gas Jet Penetrating a Fluidized Bed

Deepak Rangarajan[1], Alexander G. Mychkovsky[2], Jennifer Sinclair Curtis[3] and Steven L. Ceccio[2], (1)Chemical Engineering, University of Florida, Gainesville, FL, (2)Mechanical Engineering, University of Michigan, Ann Arbor, MI, (3) University of Florida, Gainesville, FL

The enhanced mass, momentum and energy transfer between the emulsion phase and the gas jet that occurs when a jet impinges a fluidized bed is exploited favorably in many industrial applications, such as coal and biomass gasification. In order to characterize such flows, earlier works have focused on the measurement and analysis of the resulting plume size and shape as well as mean velocity and concentration profiles within the plume. Though the jet plume is a region of turbulent mixing, no experimental data concerning fluctuations in the velocities has been reported, primarily due to the limitations in the measurement techniques employed so far.

This talk presents variation in the fluctuations of both gas and particle phase velocities inside the jet plume region with change in the fluidized state of the emulsion. The measurements are achieved using a Laser Doppler Velocimetry technique. It is seen that below minimum fluidization there is a decrease in the fluctuating intensity in both phases upon increasing the fluidization rate, but increasing the fluidization rate beyond minimum fluidization significantly increases the fluctuations. These fluctuation measurements would help in the development of more fundamentally sound two-phase models which are based on the principles of turbulent flows to explain the complex transport that occurs in a bubbling bed.

See more of this Session: Fundamentals of Fluidization II
See more of this Group/Topical: Particle Technology Forum

CFD Simulations for Prediction of Scaling Effects in Pharmaceutical Fluidized Bed Processors at Three Scales

James Parker[1], Keirnan LaMarche[2], **Wei Chen**[2], Ken Williams[1], Howard Stamato[2] and **Scott Thibault**[1]

(1) CPFD Software, LLC, Albuquerque, NM,
(2) Bristol-Myers Squibb Company, New Brunswick, NJ

Effective operation of fluidized bed processors requires an even gas distribution and good mixing of particles. Scaling up a fluidized bed process from lab scale to production scale presents unique challenges, such as preventing granule attrition, fines loss, or segregation. In addition, current methods make it difficult to optimize processor performance without large scale experimentation due to the scale dependence of influences on the fluidization and related effects on the fluidized particle (e.g. attrition). Fluidization behavior is dependent on the reactor geometry and internals and well as the particle size distribution and mechanical properties of the powder. To accurately simulate fluidized bed behavior, it is necessary to choose a numerical method capable of accounting not only for the particle-fluid effects (e.g., drag) but also for particle-wall impacts and particle-particle interactions in three dimensions.

The objective of this work was to perform a "blind test" validation of the Barracuda® commercial computational fluid dynamics (CFD) software package from CPFD Software, LLC (Albuquerque, NM) for the modeling of fluidized bed processors. Barracuda rigorously simulates fluid-particle interactions in three-dimensions using the CPFD® numerical method (Snider, 2001; Snider, 2007; O'Rourke et al, 2009) and is an established commercial simulation software package for the study of research-scale and industrial-scale fluidized bed reactors (Zhao et al, 2006; Snider et al, 2010; Parker, 2011). No experimental results were shared with the CFD analyst until the simulations had been completed.

Experimental Operation

Experimental data were collected for the operation of three different fluidized bed processors:

- Vector Lab Micro: 45 – 140 mm diameter, 710 mm height
- Glatt GPCG-1: 140 – 300 mm diameter, 900 mm height (including filters)
- Niro MP4: 560 – 910 mm diameter, 2800 mm height

Each processor was operated in the conditions listed in Table 1. Particles were dry, ranged in size from 2 micron – 1500 micron diameter, and had a material density (skeletal) of 1370 kg/m^3 and a bulk density of 450 kg/m^3.

Table 1 - Experimental cases studied in work

Processor	Vector Lab Micro		Glatt GPCG-1		Niro MP4	
Experimental Case	Low Flow	High Flow	Low Flow	High Flow	Low Flow	High Flow
Flow rate of air (CFM)	1.2	3.2	10	33	53	175
Bed mass (kg)	0.15	0.15	2.0	2.0	45	45
Exhaust pressure (kPa)	101	101	101	101	92.5	92.5
Temperature (C)	35	35	35	35	35	35

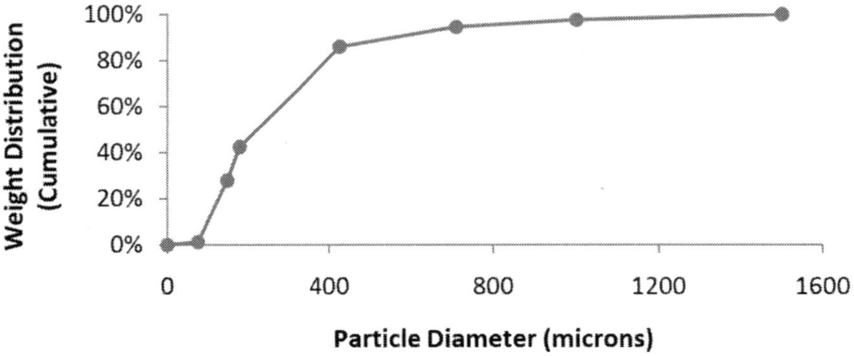

Figure 1 – Particle size distribution (PSD) of processor bed. The PSD was measured experimentally and used in the CFD model.

During operation, the dense phase bed height, elutriated particle trajectory, and patterns in the fluidization flow structure were observed and measured through sight glasses.

Computational Method

The numerical method employed in Barracuda is called Computational Particle Fluid Dynamics (CPFD®) and offers specific advantages that were essential to the success of this modeling effort. The CPFD method solves the fluid and particle momentum equations in three dimensions. The fluid is described by the Navier-Stokes equation with strong coupling with the discrete particles. The particle momentum follows the multi-phase particle-in-cell (MP-PIC) numerical description from O'Rourke (2009), Andrews (1996), Snider (1998), and Snider (2001) which is a Lagrangian description of particle motion described by ordinary differential equations with coupling with the fluid. The CPFD solution (Snider, 2001) as applied in Barracuda is aimed at solving commercial problems, which are generally physically large systems. In the CPFD method, a numerical particle is defined where particles are grouped with the same properties (species, size, density, temperature, etc.). The numerical particle is a numerical approximation which allows a large commercial system containing billions of particles to be analyzed using millions of numerical particles without losing the advantages of discretizing the solid phase in a Lagrangian frame of reference. As a result of CPFD's unique numerical approach, the following aspects of fluidized bed behavior are captured:

1. The ability to model full particle size distribution (PSD) for any number of solid species.
2. The capacity to model any solids loading from fully dilute up to close-packed (>60% solid by volume) regime in the same simulation and without prior knowledge of what the loading is likely to be.
3. Complete Lagrangian formulation for the solids, capturing mass, momentum, heat transfer, wear, etc.
4. The ability to model systems with physical particle counts over 1E16 particles.

Computational Model Setup

Three-dimensional CFD models were created for each fluidized bed processor model. As shown in Figure 2, the boundary conditions consisted of a uniform velocity boundary at the processor distributor and a pressure boundary at the processor filters. The velocity at the boundary is calculated from the distributor surface area and the flow rate listed in Table 1. The pressure boundary included a velocity dependent irrecoverable pressure drop consistent with the filter type and surface area.

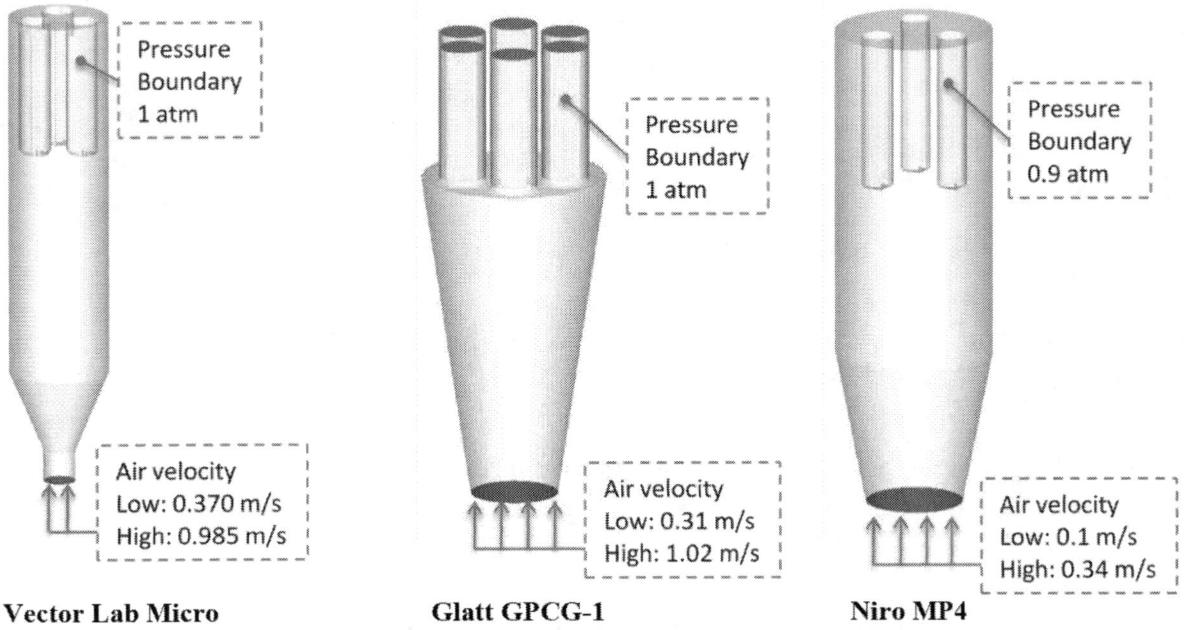

Figure 2 – Boundary conditions for CFD models of Vector Lab Micro, Glatt GPCG-1, and Niro MP4. For all models, a pressure drop existed at the pressure boundary (filters).

Each processor operating condition was simulated isothermally at 35 C. Particles were specified with the PSD shown in Figure 1 and were initially at rest in the processors. A close pack volume fraction of 0.55 was assumed, indicating a particle density of 818 kg/m^3.

Each processor was modeled on a grid consisting of 50,000 to 60,000 calculation cells and 300,000 to 500,000 Lagrangian computational particles. Sixty seconds of processor operation was simulated which included processor startup and steady-state operation. Calculations were completed in 1 to 2 days for each simulation on a 3.16 GHz processor.

Computational Results and Comparison to Experimental Data

Due to scale-dependent effects, the flow structure varied from a pattern approaching spouted flow to a homogeneously fluidized bed. This is illustrated in Figure 3 where typical simulated bed behavior at steady state is shown for each case.

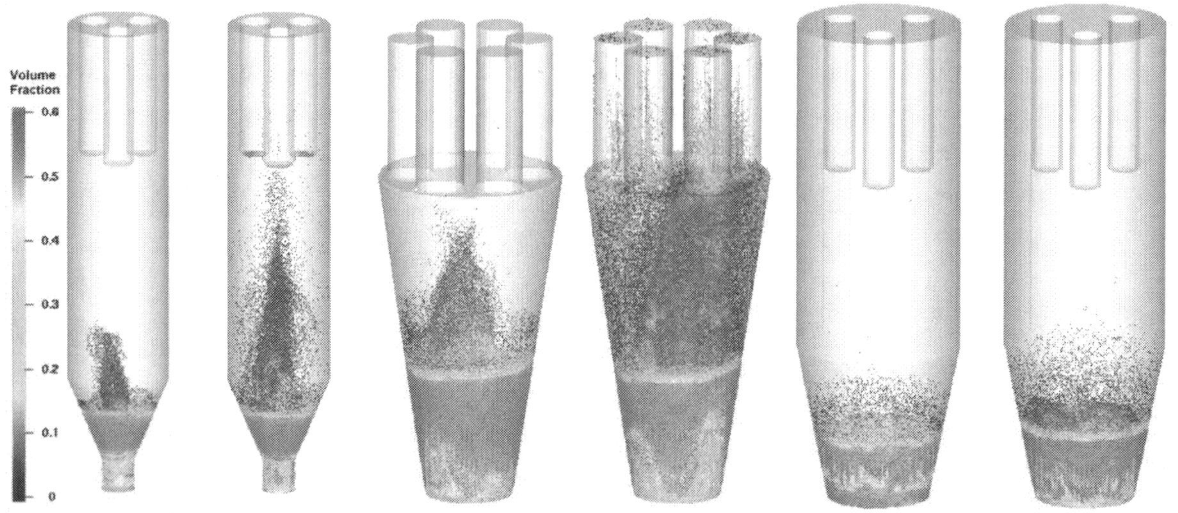

Figure 3 - Volume fraction of simulated cases: Vector Lab Micro (1.2 and 3.2 CFM, 71 cm height), Glatt GPCG-1 (10 and 33 CFM, 90 cm height), Niro MP4 (53 and 175 CFM, 280 cm height)

In Table 2, the flow structure observed within the fluidized bed processor is compared with the flow structure determined from simulation. As shown, there is good agreement between the simulated flow structure and experimental observations. The flow structure in the Vector Lab Micro was observed experimentally to be spouted with the height of the particle ejection into the freeboard being a function of gas flow rate. Identical flow structure was simulated in the Vector Lab Micro CFD model. In the Glatt GPCG-1, the simulation also closely matched the experimental flow structure. At low flow (10 CFM) in the GPCG-1, a slugging flow was observed in the center of the processor whereas at high flow (33 CFM), the flow was turbulent throughout the fluidized bed. Simulations with the CFD model also showed this dependence on flow rate in the GPCG-1. In the Niro MP4, a comparison was more difficult. The CFD simulation of the Niro MP4 showed a bubbling bed at low flow and significantly more turbulent conditions at high flow. The simulation also showed a gentle transport of particles into the freeboard. Experimental observation was difficult for the Niro MP4 but in the limited visibility some transport of particles into the freeboard was observed.

Table 2 - Flow observations for fluidized bed processors at three scales

	Experimental flow structure	Simulated flow structure
Vector Lab Micro Low Flow (1.2 CFM)	Spouted - material reaches the cylindrical section of the bed	Central spout reaching a height of 10 to 15 cm into freeboard. Little particle movement near cone edges
Vector Lab Micro High Flow (3.2 CFM)	Spouted - material reaches about halfway between the bed surface and the filters	Central spout reaching a height of 20 to 40 cm into freeboard
Glatt GPCG-1 Low Flow (10 CFM)	Flow of air is mostly in center of bed, material near walls is stationary	Slugging flow though bed. Particles typically ejected 12 to 20 cm into freeboard
Glatt GPCG-1 High Flow (33 CFM)	Good fluidization, Bubbling throughout bed	Turbulent flow in bed. Particles are hitting the top of the processor and moving into filters
Niro MP4 Low Flow (53 CFM)	Shadows could be seen periodically suggesting that material was being thrown about a meter above the bed. Material pressed against sight glass slowly flows downward	Bubbling flow exists in bed. No ejection of particles into freeboard.
Niro MP4 High Flow (175 CFM)		Turbulent flow in bed. Minimal spouting of particles into freeboard.

During the processor operation, the bed height data and fountain height were collected for the operation of the Vector Lab Micro and the Glatt GPCG-1. (No such data were available for the Niro MP4.) The fountain height is the distance which particles will be ejected into the processor freeboard. This is clearly shown in the simulation images in Figure 3 for both Vector Lab Micro cases and the Glatt GPCG-1 at 10 CFM. All measurements were made from the top of the processor distributor. Comparisons of the bed height and fountain height data with the simulated operation are shown in Figure 4 and Figure 5. During processor operation a typical fountain height was noted. This is compared to a distribution of fountain heights determined by simulation.

Examination of the bed height and fountain height shows that the simulation slightly over-predicted the bed expansion in the Glatt GPCG-1 and the fountain height in the Vector Lab Niro. While this discrepancy did not appear to affect the flow structure and prediction of scale up effects, it is something that would be further studied in a "non-blind test" situation. In this case, these discrepancies are likely due to assumptions made by the CFD analyst about the particle porosity. The information provided in the "blind test" required an assumption about the particle porosity and in turn, the close pack volume fraction of the solids. These parameters can have a significant effect on the bed expansion and flow data. In this case, the porosity was likely over-estimated, leading to the discrepancies shown in Figure 4 and Figure 5.

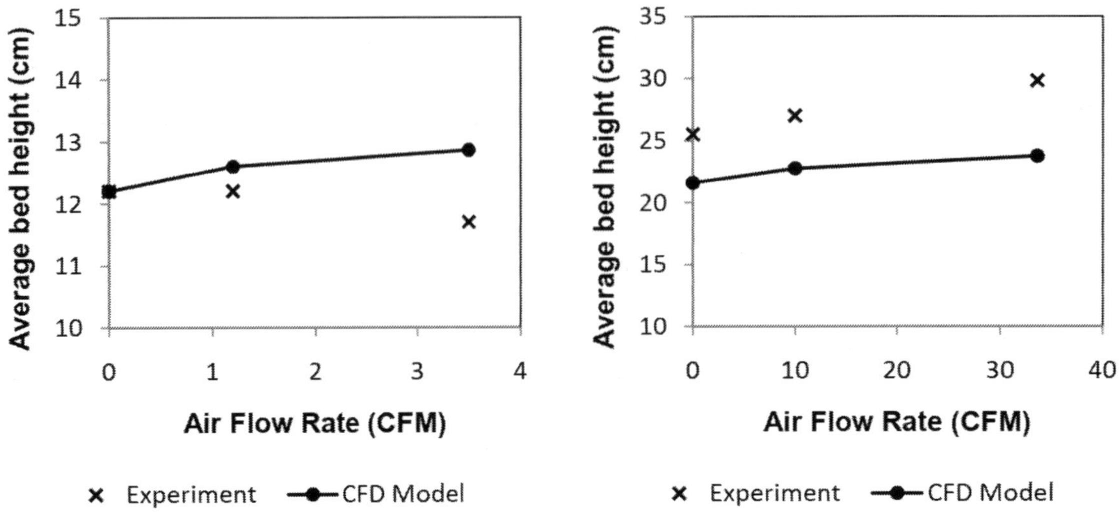

Figure 4 – Average bed height in Vector Lab Micro (left) and Glatt GPCG-1 (right) and comparison of experimental results with CFD model

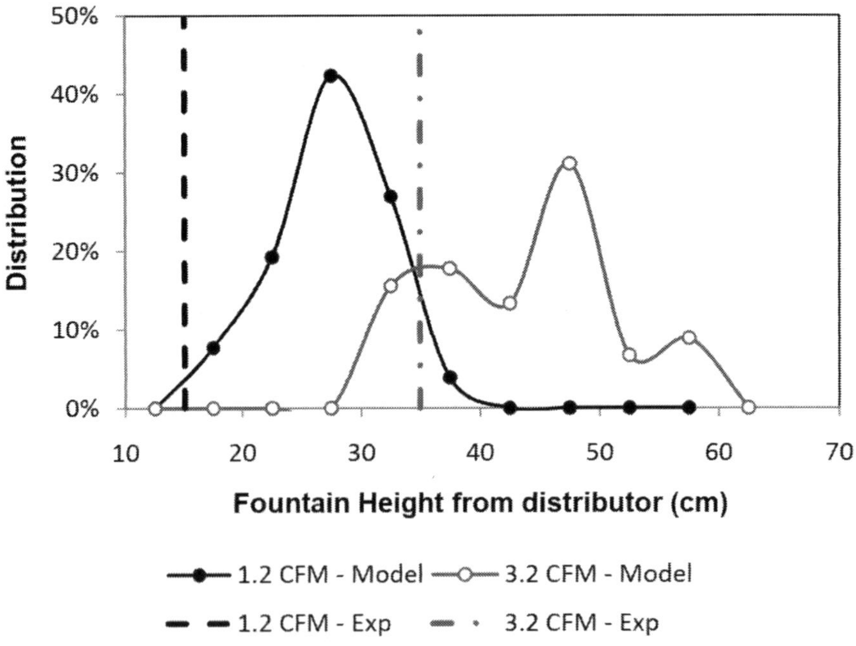

Figure 5 – Fountain height in Vector Lab Micro, comparison of typical observed height with distribution in simulation

Conclusions

The numerical method and simulation approach were found to successfully predict the flow behavior of several different sizes of fluidized beds, yielding qualitatively correlated information based solely on material properties, processor loading, inlet air flow rate and fluidized bed processor geometry. The simulations modeled both start-up and steady operation of the processors on a full 3D, time-transient basis within a reasonable calculation time (1-2 days). This implies that CFD of this kind may be used to assist in determining select operating parameters for fluidized bed processors commonly used in the pharmaceutical industry and in scaling processes to different sizes to meet production demands.

References

Andrews, M. J., and O'Rourke, P. J. (1996), "The Multiphase Particle-in-Cell Method (MP-PIC) Method for Dense Particle Flow", *Int. J. Multiphase Flow*, 22, 379-402.

O'Rourke, P. J., Zhao, P., Snider, D. (2009) A model for collisional exchange in gas/liquid/solids fluidized beds. *Chemical Engineering Science*. 64, 1784-1799.

Parker, J. (2011). Validation of CFD Model for Polysilicon Deposition and Production of Silicon Fines in a Silane Deposition FBR. *International Journal of Chemical Reactor Engineering*. 9, A40.

Snider, D. M., et al. (1998), "Sediment Flow in Inclined Vessels Calculated Using Multiphase Particle-in-Cell Model for Dense Particle Flows", *Int. J. Multiphase Flow*, 24, 1359-1382.

Snider, D.M. (2001). An Incompressible Three-dimensional Multiphase Particle-in-Cell Model for Dense Particle Flows. *Journal of Computational Physics*. 170, 523- 549

Snider, D.M. (2007). Three fundamental granular flow experiments and CPFD predictions. *Powder Technology*. 176, 36-46.

Snider, D., Guenther, C., Dalton, J., Williams, K. (2010). CPFD Eulerian-Lagrangian Numerical Scheme Applied to the NETL Bench-top Chemical Looping Experiment. 1st *International Conference on Chemical Looping*.

Zhao, P., Snider, D., Williams, K. (2006) Computational Particle-Fluid Dynamics Simulations of a Commercial-Scale Turbulent Fluidized Bed Reactor. *AIChE Annual Meeting*.

(131a) A Novel Approach to Crystallizing the Desired Form Through Solid-State Solvent Exchange

Srividya Ramakrishnan, M. Vamsi Krishna, Y. Sesha Reddy, M. Rajendar Reddy, Anil Patni, Surya Devarakonda, Sundaralakshmi Kanniah and Gautam Samanta, Dr. Reddy's Laboratories Ltd., Hyderabad, India

Discovering novel crystalline forms can be a very challenging task in the generic industry, with the search often resulting in solvates that are not pharmaceutically viable. In the case of Compound X, this task was particularly challenging as the innovator had claimed more than eight forms and other generic players had claimed several more including amorphous form. However, through determined pursuit, a novel crystalline form was discovered, but unfortunately this form contained about 5 % methanol. Not only was this form unstable, but removal of the methanol by conventional drying techniques resulted in conversion to amorphous. To achieve the ICH limit for methanol, a novel process was developed to reduce the methanol content in the solid state by solvent exchange with water without transformation of the crystal lattice arrangement, resulting in a pharmaceutically acceptable and stable form. It was found that the extent and kinetics of solvent exchange were dependent on both the crystallization and humidification conditions. The impact of various crystallization parameters such as temperature, mixing, anti-solvent addition rate etc. were studied through a systematic design of experiments (DOE), which revealed the interaction of parameters, their effect on the crystal morphology and its correlation to the solvent exchange dynamics. In addition, it was required to improve the bulk density and flowability of the API for formulation processing due to a very high drug loading. The process was tailored to achieve this by engineering the formation of spherical agglomerates. This talk will present a case study on how an unconventional crystallization process was developed to effectively transform the polymorph in the solid state with desired powder properties, and the challenges to scale-up and validate this process in manufacturing.

(Dr.Reddy's Laboratories Communication number: IPDO IPM - 00252)

See more of this Session: Advances and Case Studies In Crystallization and Post-Crystallization Processing
See more of this Group/Topical: Separations Division

(131b) Meeting Customer Needs: Tailoring the Physical Properties of An Active Pharmaceutical Ingredient Through Crystal Engineering

Christopher L. Burcham, Chemical Product Research and Development; Particle Design Laboratory, Eli Lilly and Company, Indianapolis, IN

A pharmaceutical compound recently in development exists as two enantiotropic anhydrate polymorphs and as a dihydrate, with the dihydrate being unstable in the formulated product. Patient expectations of a small tablet required a high concentration of drug substance in the product. A high shear wet granulation process cannot be used as part of tablet manufacture to avoid conversion to the dihydrate. As a result, a dry granulation process was required. The high drug load required the drug substance to have good flow properties as a dry powder. API produced by a typical temperature and anti-solvent driven crystallization process generated API with poor powder flow characteristics; tablets produced using this API had to be manually sorted to meet tablet weight requirements. The poor flow characteristics, resulting from a needle-like particle morphology and small particle size, led to the loss of tablet weight control.

Particle morphology was improved through the careful control of supersaturation, the incorporation of *in situ* milling during crystallization, and thermal cycling. The result of the crystal engineering process is particles having a rod-like morphology approaching cubes. The particles are three times wider on average than those produced from a conventional crystallization. The flow characteristics are greatly improved and have enabled tablets to be manufactured using a dry process at high (greater than 35%) drug loads with low tablet weight relative standard deviation. The control of polymorph selectivity will also be discussed.

See more of this Session: Advances and Case Studies In Crystallization and Post-Crystallization Processing
See more of this Group/Topical: Separations Division

(131c) Stochastic Optimal Control of Batch Crystallization Applying Ito Process

Kirti Maheshkumar Yenkie, Bioengineering, University of Illinois, Chicago, Chicago, IL and Urmila Diwekar, Vishwamitra Research Institute, Center for Uncertain Systems: Tools for Optimization and Management, Clarendon Hills, IL

Batch crystallization is one of the most important industrial separation and purification processes. Being a particulate process, crystallization is modeled in terms of population balance equations which include kinetics of nucleation, growth, aggregation and breakage. Analytical solutions to PBE's are available only for simplified kinetics and known initial size distribution and hence development of numerical methods for solving complex process kinetics was a major area of interest. It mainly consisted of estimation of kinetic parameters for efficient process operation using earlier experimental data. However, the uncertainties associated with these parameters were given little importance. Minimization of operation costs and the enhancement in product quality have been a major concern for all industrial processes. Predetermined operating conditions can help in achieving the goals for efficient production. These conditions can be determined using an optimal control analysis of the batch crystallization process. Optimal control problems are characterized by determination of time varying profiles for process parameters. Batch crystallization is associated with parameters like temperature, supersaturation and agitation. Some process parameters are functions of fundamental properties of the system into consideration like solubility, crystal lattice. Thus, the process parameters can have various physical and engineering sources of uncertainties associated which in turn would prevent the real process operation to be optimal. These uncertainties result in dynamic uncertainties due to the unsteady state nature of the process. This paper presents a novel approach to optimal control problems in batch crystallization. The uncertainties and variabilities are modeled by stochastic processes called Ito processes. The resulting stochastic optimal control problem is solved using Ito's Lemma, stochastic calculus, and stochastic maximum principle.

See more of this Session: Advances and Case Studies In Crystallization and Post-Crystallization Processing
See more of this Group/Topical: Separations Division

(131d) API Powder Property Control During Crystallization, Wet Milling, and Drying

Nathan Domagalski, **Brendan C. Mack**, Amanda Rogers, Jose Tabora, Chenchi Wang and Lindsay Hobson, Bristol-Myers Squibb Company, New Brunswick, NJ

Powder properties for an active pharmaceutical ingredient (API) are traditionally controlled during the final crystallization and drying unit operations. These unit operations, however, are often scale dependent and difficult to study in a laboratory setting. Consequently, the identification of a design space to control powder properties can prove challenging. This paper describes work at scales greater than 100 kg to study particle size and API saturation during cooling crystallization with simultaneous wet milling. Multiple parameters such as cooling rate, seed age, and turnover frequency were studied to determine their corresponding impact on final PSD and crystallization supersaturation profile. Further studies monitored form conversion, residual solvent, amorphous material production, and particle attrition during drying. The impact of dryer temperature, size, and agitation was assessed. Overall, these studies provided a wide breadth of knowledge that has been used to set process parameter ranges on both crystallization and drying operations to establish adequate ranges to control the physical attributes of the API.

See more of this Session: Advances and Case Studies In Crystallization and Post-Crystallization Processing
See more of this Group/Topical: Separations Division

(131e) Implementation of Supersaturation Control In Large Scale Crystallization

Brenda Remy[1], Jose E. Tabora[1], Amanda Rogers[1], Jacob Albrecht[1], Nathan Domagalski[1], Hamm Jason[2], Daniel Wasser[2] and George Armenante[2], (1) Chemical Development, Bristol-Myers Squibb Co., New Brunswick, NJ, (2) Chemical Development Operations, Bristol-Myers Squibb Co., New Brunswick, NJ

Development of robust and reliable crystallization processes is required to obtaine predictable and controlled physical properties of the crystallized material, particularly API's. Crystallization robustness requires control of mechanisms such as growth, secondary nucleation, agglomeration etc. These mechanisms are generally dictated, in large part, by the supersaturation profile of the crystallization. Crystallizations which control supersaturation to a uniformly low value are expected to provide maximum robustness of the final API quality and physical attributes. Previous work reported laboratory scale implementations of supersaturation control during crystallizations [1-3]. In this work we describe the development of a supersaturation control methodology in a 50 L scale by connecting the PAT (ATR-FFTIRS)directly to the pilot plant's process control system thus enabling constant supersaturation control of a temperature driven crystallization. The workflow incorporated automatic diagnostic of a standard crystallization, calibration of the IR spectra, and crystallization control at different levels of constant relative super saturation. This work also discusses the effect of different constant supersaturation crystallizations and seed protocols on the resulting crystal habit and size.

[1] Fujiwara, M.; Nagy, Z. K.; Chew, J. W.; Braatz, R. D. J. Process Contr.**2005**, 15, 493-504

[2] Cote el al., *Org. Process Res. Dev.*, **2009**, *13* (6), pp 1276–1283

[3] Yu, Z. Q.; Chow, P. S.; Tan, R. B. H. Ind. Eng. Chem. Res. **2006**, 45,438-444

See more of this Session: Advances and Case Studies In Crystallization and Post-Crystallization Processing

(131f) Inline Supersaturation Measurement and Control to Understand and Optimize Crystallization

Anjan P. Pandey, Des O'Grady, Simon Rea, Jennifer Andrews and Eric Dycus, METTLER TOLEDO AutoChem, Columbia, MD

In recent years process analytical technologies have developed as important tools to understand and optimize ideal process conditions and ensure consistent crystallization scale-up. Supersaturation, the driving force for crystallization, often varies with changes in seed load, agitation, and crystallizer operating conditions. Qualitative supersaturation control facilitates development of a process map where optimal cooling profiles are determined in the minimum number of experiments.

A new calibration free method of supersaturation control using *in situ* FTIR measurements and synchronized software provides rapid solubility and supersaturation determination without lengthy calibration. By tracking solute peak-height versus temperature, inline FTIR enables users to control the cooling rates to achieve an ideal supersaturation level. This ideal cooling rate can then be applied during scale up to ensure optimum batch repeatability.

See more of this Session: Advances and Case Studies In Crystallization and Post-Crystallization Processing
See more of this Group/Topical: Separations Division

(139a) Relationship Between Rheology and Morphology In Aqueous Dispersions of Hydrogenated Castor Oil Crystals

Dingzheng Yang, Chemical Engineering, McMaster University, Hamilton, ON, Canada and Andrew N. Hrymak, Chemical and Biochemical Engineering, University of Western Ontario, London, ON, Canada

Hydrogenated castor oil (HCO) dispersions are widely used as the rheological modifier in paint, cosmetics and personal care products. Three HCO crystal morphologies have been observed, which we refer to by their geometric shape as fiber, rosette and irregular crystal. Due to the difficulty of obtaining samples with a single morphology, rheology studies of suspensions containing mixtures of the three morphologies in a non-Newtonian surfactant solution have been undertaken. The viscometry of dilute suspensions has shown that the magnitude of intrinsic solution viscosity follows the sequence of samples having a majority of a particular crystal morphology, i.e. fiber > rosette > irregular crystal. Huggins coefficients have been found to be the function of crystal morphology, indicating that a solution with a fractional distribution of 0.43:0.40:0.17 of fibers:rosettes: irregular crystals, shows the highest particle/particle interaction. The shear dependent viscometry of concentrated dispersions shows that Cross model fits well for the data at low volume fraction while a power law model fits well for the data at high volume fraction. The volume fraction dependent zero shear viscosity of concentrated dispersions has been found to be a function of crystal morphology with the highest zero shear viscosity found in fiber-rich systems and least zero shear viscosity for systems rich in irregular crystals. The linear viscoelasticity has been investigated, showing fibers and rosettes contribute more to elastic properties than irregular crystals. There is also an aging phenomenon, where irregular crystals transform to short-branched rosettes over a period of two months. Both the steady shear viscosity and dynamic moduli have been found to increase, which may be significant in certain applications.

See more of this Session: Applications of Engineered Structured Particulates
See more of this Group/Topical: Particle Technology Forum

(139b) Drug Delivery Using Nanocarriers Self-Assembled by Triskelion Aromatic Molecules

Ying Chau, Chemical and Biomolecular Engineering, The Hong Kong University of Science and Technology, Hong Kong, Hong Kong and Weiping Wang, Chemical and Biomolecular Engineering, The Hong Kong University of Science and Technology, Hong Kong, CA, Hong Kong

Here we present a new drug delivery nanocarrier constructed by an efficient and facile self-assembling system. A simple triskelion Fmoc-conjugate is designed to rapidly self-assemble into nanoparticles in water. To stabilize the nanoparticles at physiological pH, an Fmoc-dipeptide is employed to bind to the nanoparticles. The fabricated two-component nanoparticles were found to be size-controllable within the range from 53 nm to 125 nm, well-tolerated by cells, and stable upon heating, long-time incubation and dilution. Moreover, these novel particles were readily decorated by short Fmoc-peptides as targeting ligands. We further demonstrated that the two-component nanoparticles could entrap poorly-soluble anticancer drug, paclitaxel (PTX), with a high loading efficiency. With Fmoc-FRGD modifying the nanocarriers to target integrin overexpressed on tumor cells, PTX-loaded nanoparticles showed better in vitro anti-cancer effects than free PTX, with IC_{50} value of about 3 nM against MDA-MB-435 breast carcinoma cells. The PTX-loaded nanoparticles were stable after 3 days of incubation at 37 ^{o}C with FBS-supplemented medium. Furthermore, Fmoc-FRGD stabilized nanoparticles were differentially taken up by MDA-MB-435 cells but Fmoc-FRDG stabilized nanoparticles not, which verifies the targeting property of Fmoc-FRGD stabilized nanoparticles. All these attractive characteristics make the novel nanoparticles a promising drug delivery carrier for anticancer therapy.

See more of this Session: Applications of Engineered Structured Particulates
See more of this Group/Topical: Particle Technology Forum

(139c) Reactive Microfluidic Production of Degradable Microgels for Drug Delivery

Leah Kesselman[1], Siawash Shinwary[2], Wen-I Wu[2], P. Ravi Selvaganapathy[2] and **Todd R. Hoare**[1], (1)Chemical Engineering, McMaster University, Hamilton, ON, Canada, (2)Mechanical Engineering, McMaster University, Hamilton, ON, Canada

INTRODUCTION: Microgels, best defined as hydrogel particles, have greatpotential to address many challenges in biomedicine. The facile injectability,high surface area to volume ratio, and ready tunability of both the surface andbulk of microgels present key advantages for a range of applications includingdrug delivery or drug scavenging vehicles[1], cell encapsulation matrices[2], or biosensors[3]. Microgels prepared using "smart"materials that can dynamically change their physical properties (e.g. size,charge, or colloidal stability) in response to external factors (e.g.temperature, pH, or the concentration of a particular chemical in the microgelenvironment) have particular utility in this regard, facilitating "on-demand"changes in microgel localization (i.e. aggregation versus circulation) or drugdiffusion in different environments[4]. Of particular note, microgel sizehas a significant impact on the ultimate biological performance of themicrogel; for example, smaller microgels release drugs faster and circulate foran extended period of time (ultimately localizing in the spleen or liver) whilelarger microgels release drug slower and typically remain at their injectionsite. However, conventional methods of fabricating environmentally-responsivemicrogels are significantly limited in terms of their capacity to controlmicrogel size. While monodisperse thermoresponsivemicrogels with sizes from 100-1000 nm are relatively easy to fabricate via aprecipitation-based technique[4], inverse emulsion strategies aretypically required to synthesize non-thermoresponsivemicrogels or any microgel with a diameter of >1 micron. Conventional methodsto generate inverse emulsions rely on sonication,homogenization, or other types of bulk shear forces to generate thewater-in-oil droplets, resulting in polydispersemicrogels that give unpredictable drug release kinetics and biologicalresponses. However, microgels with >1 micron diameters are of significanttechnological interest in that they are less likely to be endocytosedby macrophages or quickly sequestered within the body upon injection. Ofparticular utility for local drug delivery would be monodisperse populations oflarge (>1 micron diameter) microgels crosslinked via degradable linkages;such materials could be injected at a desired site of action, remain at thesite of injection to deliver drug locally over an extended period of time, andthen cleared from the body after use.

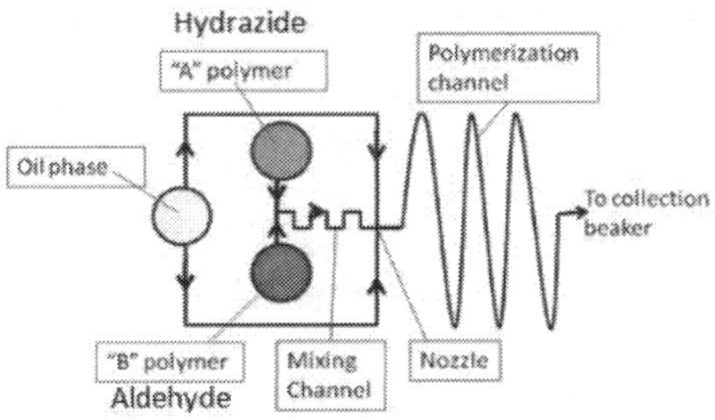

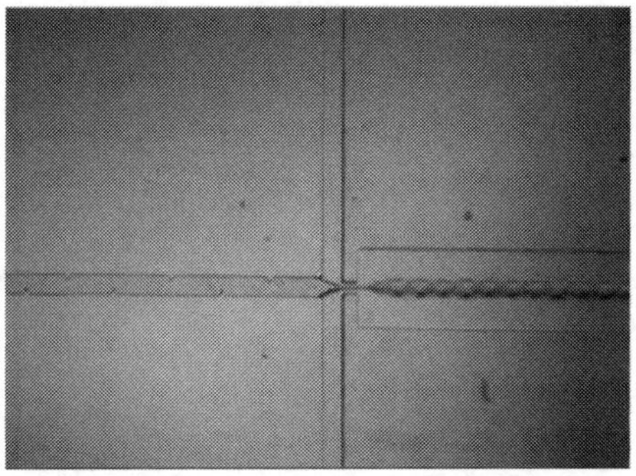

EXPERIMENTAL: Microfluidics chips were designed that facilitated on-chipcrosslinking between two reactive polymer precursors, pre-mixed prior to thedroplet formation step. Microgels were generated by reacting hydrazide-functionalizedpolymers (prepared via EDC-mediated grafting of adipicacid dihydrazide with a carboxylicacid-functionalized polymer) with aldehyde-functionalized polymers (preparedvia carbohydrate oxidation using sodium periodate) tospontaneously form a hydrazone-crosslinked microgel inside the microfluidicschip. The chip (pictured) contacts the two aqueous reactive polymer streams,passes the streams through a mixing channel, and then generates water-in-oildroplets using paraffin oil as the oil phase and Span 80 as the surfactant.Microgels were then extracted into an aqueous suspension via freeze-thawing andsubsequent solvent exchange. Particle size of the microgels was assessed vialight and fluorescence microscopy and surface charge was measured viaelectrophoretic mobility measurements. Drug loading/release experiments wereperformed by mixing purified microgels in a drug solution, centrifuging toisolate the microgels, resuspending the microgels inPBS inside a Float-a-Lyzer membrane bag (SpectrumLabs, 50,000 MWCO), and assaying drug release as a function of time. Microgelswith the same composition were prepared using sonicationto compare with microgels synthesized using microfluidics.

RESULTS: Reactive microfluidics facilitates successfully productionof microgels with significantly improved monodispersityand size control relative to traditional shear-based techniques. Microgels withparticle sizes ranging between 25-200 microns were generated with polydispersities <5% by varying both the relative flowrates of the oil and (aqueous) polymer streams or the total flow rate of theoil and polymer streams through the chip. Crosslinked microgels were generatedspontaneously upon droplet formation without the need for subsequent heating,UV photopolymerization, or the use of additionalinitiators, crosslinkers, or chain transfer agents, as required in previousmicrofluidic microgel generation systems[5]. Continuous particle production waspossible over at least 72 hours, with increased microgel yield possible throughthe use of parallel microfluidic reactors. Microgels were successfullysynthesized based on carbohydrates (carboxymethyl cellulose/dextran), mixturesof synthetic oligomers and carbohydrates (poly(N-isopropylacrylamide)/dextran) or synthetic oligomers only (poly(N-isopropylacrylamide)),illustrating the adaptability of the device to generate multiple types ofdegradable microgels with varying chemistries. Drug release for the model drugbupivacaine (a local anesthetic) was a weak functionof microgel size and followed first-order, diffusion-based kinetics, withsustained release over periods of up to two weeks.

CONCLUSIONS: Reactive microfluidics provides an effective route to thepreparation of monodisperse degradable microgels based on natural, synthetic,or mixtures of natural and synthetic polymers. Such materials are ideal forbiomedical applications in that the crosslink density, composition, size, andbiocompatibility can be tuned appropriately for the particular desired end-use.

ACKNOWLEDGEMENTS: This work was funded by the NaturalSciences and Engineering Research Council of Canada and 20/20: NSERC OphthalmicMaterials Research Network.

REFERENCES:(1) (a) Eichenbaum,G. M.; Kiser, P. F.; Dobrynin, A. V.; Simon, S. A.;Needham, D. *Macromolecules* **1999**, *32*, 4867(b) Hoare, T.;Pelton, R. *Langmuir* **2008**, *24*, 1005; (2) Bae, K. H.; Yoon, J. J.; Park, T. G. *BiotechnologyProgress* **2006**, *22*, 297; (3) Das, M.; Zhang, H.; Kumacheva, E. *Annual Review of Materials Research* **2006**,*36*, 117; (4) Hoare, T.; Pelton, R. *Macromolecules* **2004**, *37*,2544; (5) (a) Kim, J. W.; Utada, A. S.;Fernandez-Nieves, A.; Hu, Z. B.; Weitz, D. A. *Angewandte Chemie-InternationalEdition* **2007**, *46*, 1819(b) Zhang, H.; Tumarkin,E.; Sullan, R. M. A.; Walker, G. C.; Kumacheva, E. *Macromolecular Rapid Communications* **2007**,*28*, 527.

See more of this Session: Applications of Engineered Structured Particulates
See more of this Group/Topical: Particle Technology Forum

(139d) Stability and Release of H5N1 Hemagglutinin Antigens Encapsulated In Polyanhydride Nanoparticles

Kathleen A. Ross[1], Lucas Huntimer[2], Latrisha Petersen[1], Michael J. Wannemuehler[2] and Balaji Narasimhan[1], (1)Chemical and Biological Engineering, Iowa State University, Ames, IA, (2)Veterinary Microbiology and Preventive Medicine, Iowa State University, Ames, IA

Avian influenza A H5N1 is a respiratory viral pathogen that has the potential to be the next influenza pandemic threat. Human cases of H5N1 have proven to be approximately 60% fatal with a growing number of strains becoming resistant to medical treatments. At this point in time, the infection of H5N1 is mainly spread from avian species to humans and rarely from humans to humans. However, due to the H5N1 strain's ability to achieve rapid viral mutations, it is believed that human-to-human transmission of the virus may be coming in the near future, unleashing a possible global influenza pandemic. Therefore, there is a strong need to prepare for the prevention of H5N1 through vaccine technology.

Recombinant proteins are a new alternative in pandemic vaccines; however, they require adjuvants to enhance their immunogenicity. There are very few vaccines containing adjuvants available and licensed for human use, but the increasing ability to produce safe, purified, yet fragile and less immunogenic protein antigens calls for a need to design novel vaccine adjuvants. Biodegradable polymers have been explored as novel adjuvants with one of the most prominent class being polyesters. These polymers exhibit bulk erosion leading to exposure of the protein antigen to both water and acidic degradation products which can cause protein aggregation and degradation. Another class of biodegradable polymers, polyanhydrides, has shown potential for releasing stable protein antigens from nanoparticles. Polyanhydrides exhibit surface erosion and contain less acidic degradation products that preserve protein structure and function upon release. In this work, we have studied the release and stabilization of a recombinant H5N1 hemagglutinin (HA) from polyanhydride nanoparticles composed of sebacic acid (SA), 1,6-bis(*p*-carboxyphenoxy) hexane (CPH), and 1,8-bis(*p*-carboxyphenoxy)-3,6-dioxaoctane (CPTEG). Our results indicate that encapsulation of H5N1 HA antigen into amphiphilic nanoparticles based on CPTEG and CPH leads to stability thus retaining the antigenicity and biological function, in addition to providing sustained antigen release, which is vital to an induction of an immune response. These results are promising first steps in the development of an intranasal single dose vaccine against avian influenza.

See more of this Session: Applications of Engineered Structured Particulates
See more of this Group/Topical: Particle Technology Forum

(139e) Effect of Size and Surface Morphology of Pollen-Shape Carrier Particles In Dry Powder Inhalation Application

Meer Saiful Hassan and **Raymond Lau**, School of Chemical and Biomedical Engineering, Nanyang Technological University, Singapore, Singapore

For effective dry powder inhalation, drug particles need to be delivered to the alveoli region of the lung. The delivery of drug particles highly depends on their physical characteristics. In a previous study, pollen-shape drug carriers are compared with traditional carriers at different drug mixing ratios and flow rates. It is found that pollen-shape drug carriers can deliver large amount of drug particles and reduce drug losses especially at low flow rates and high drug mixing ratios. In this study, the effect of size and surface morphology of pollen-shape carriers on drug delivery performance is assessed. Pollen-shape carrier particles having various sizes and surface asperities are synthesized. Budesonide (Bd) is used as the model drug. The drug delivery performances of the pollen-shape carrier particles are investigated using an Andersen Cascade Impactor (ACI) equipped with a Rotahaler® at gas flow rates of 30 and 60 L/min. Three drug mixing ratios are considered. While an increase in the carrier particle size has a mild improvement on the ED, it significantly improves the FPF. A sparse surface asperity has negligible effect on the ED at low flow rates but it improves the FPF compared to a dense surface asperity under all experimental conditions.

See more of this Session: Applications of Engineered Structured Particulates
See more of this Group/Topical: Particle Technology Forum

(139f) Nanocomposites Preparation Through Aggregation and Controlled Breakage

Fabio Codari[1], Alexander Fischer[1], **Miroslav Soos**[1], Davide Moscatelli[2] and Massimo Morbidelli[1], (1)Institute of Chemical and Bioengineering, ETH Zurich, Zurich, Switzerland, (2)Chimica, Materiali ed Ingegneria Chimica G. Natta, Politecnico di Milano, Milano, Italy

Nanocomposites find application in different research and industrial fields ranging from food and cosmetic, electronics up to pharmaceutical products. Even if they present large interest, in the literature understanding of the physical processes involved in their production is still lacking. In this work we present an experimental study of production of nanocomposites prepared through aggregation and breakup process. Polymeric primary particles of different composition (different polymer) and inorganic nanoparticles (magnetite) represent building blocks used for nanoclusters (NCs) synthesis. Selected key parameters, such as NPs size and composition, type of surface charge groups and their density, inter-particles interactions and system fluid dynamics, and their effect on the final NC properties is systematically investigated. Developed methodology with proper tuning of inter-particle interaction during aggregation and breakage process and hydrodynamic conditions allow us to prepare NC of various sizes and compositions and therefore represent powerful tool for tailor made production of NCs for specific applications.

See more of this Session: Applications of Engineered Structured Particulates
See more of this Group/Topical: Particle Technology Forum

(139g) Alignment of Janus Particles In Dielectric Elastomer Films and Their Effect On Mechanical Actuation

Hsin-yu Chen[1], Ilona Kretzschmar[1] and Ashok Maliakal[2], (1)Department of Chemical Engineering, The City College of New York, City University of New York, New York, NY, (2)LGS Innovations, Florham Park, NJ

The ability of dielectric elastomers (DEs) to deform due to electrostatic interactions induced by external electrical fields has become the basis of newly designed actuators and sensors. The deformation, called compressive strain, is proportional to the overall dielectric constant of the DEs, which itself depends on the local dipole moments.

In the work presented here, Janus particles that carry a dipole moment due to their anisotropic surface are added into the DE system to enhance the local dipole moment of the DE. We observe that poly(ethylene glycol phenyl ether acrylate), p (EGPEA), films filled with gold-capped silica Janus particles are able to achieve a higher strain under a fixed, external electric field. In an attempt to enhance the overall value of dipole moment of the p(EGPEA) matrix the uncured composite is exposed to an external DC field during polymerization resulting in the alignment of the gold-capped Janus particles.

We will report on the fabrication, characterization and strain/electric field correlation of the p(EGPEA) matrix as a function of filler material and alignment. Preliminary data enable us to conclude that alignment of gold cap plays an important role in the electroactive response of our system. Further, the comparison of plain and Janus particle filled composite p(EGPEA) matrices shows a remarkable deformation (>10%) attributed to the polar nature of the injected filler material.

See more of this Session: Applications of Engineered Structured Particulates
See more of this Group/Topical: Particle Technology Forum

(139h) Study of PLA Particle Formation by Supercritical Fluid Technology for Oral Release of Active Agents

Priscila S. C. Sacchetin[1], Paulo T. V. Rosa[2] and **ngela M. Moraes**[1], (1)School of Chemical Engineering, State University of Campinas, Campinas, Brazil, (2) Institute of Chemistry, State University of Campinas, Campinas, Brazil

Increasing development of techniques to improve the bioavailability of various drugs and to reduce their dosage has been noticed recently, particularly approaches associating the drugs with different types of devices for controlled release. In this context, biodegradable polymers have received special attention due to their possible applications in the pharmaceutical, medical and biomedical engineering areas. Particulate systems based on PLA are already being used successfully for the release of active agents mostly intravenously, due to their high stability, which allows the release of the desired compound for extended periods. The successful use of such devices for intravenous drug delivery suggests that adopting similar strategies to administer drugs orally can be of interest, with the advantage of using a less invasive route. Among the techniques that may be employed to obtain PLA particles are the ones based on polymer precipitation through supercritical fluid technology, such as the method known as Supercritical Anti-Solvent (SAS). In the present work a statistically experimental design study was performed to analyze in detail the influence of pressure (8 to 16 MPa), concentration of PLA dissolved in dichloromethane (0.5 to 1.5 % w/v), and injection flow of polymer solution in the system (0.5 to 2.5 mL/min) on particles characteristics. It is noteworthy that the combination of flow reduction with increasing concentration of polymer solution resulted in the production of particles with smaller diameters (20 to 5um), since this condition favored increased solvent extraction combined with higher nucleation rates. The degree of crystallinity measured by X ray diffraction (XRD) and the thermal behavior evaluated by differential scanning calorimetry (DSC) of particles obtained in all tested conditions showed no change in comparison to the pure polymer. Regardless of the process operating conditions, strong tendency of particle aggregation was noticed, what was confirmed by zeta potential analysis, which showed low values (> ±5 mV) in the presence of aqueous solutions at different pH values (2.0, 4.5, 7.4 and 9.0). In general, particle morphology, size distribution, stability, thermal behavior and degree of crystallinity were not significantly affected by the distinct operational conditions tested. However, the residual amount of dichloromethane was lower (below 600 ppm) in particles prepared at high polymer concentration and low solution flow rate.

See more of this Session: Applications of Engineered Structured Particulates
See more of this Group/Topical: Particle Technology Forum

(151a) Applicability of the 3D Advection-Diffusion Model to a Bladed Granular Mixer

Avik Sarkar, Rahul Kumar and Carl R. Wassgren, Mechanical Engineering, Purdue University, West Lafayette, IN

The one-dimensional form of the advection-diffusion equation has been used to model axial mixing in rotating drums in a number of studies. This work attempts to examine whether granular mixing in more complex geometries can be modeled using the three-dimensional form of the advection-diffusion equation. A horizontal, cylindrical granular mixer with rotating impeller blades is the system under consideration. The inputs required for the advection-diffusion model (convective velocities and diffusion coefficients) are obtained from a discrete element method (DEM) simulation of the granular mixer. The variation in concentration of a tracer species with location is computed using a finite difference numerical solution of the 3D advection-diffusion equation. The results obtained from the advection-diffusion model are compared against measurements from the DEM simulation, which is considered as the "exact solution". This proposed methodology presents an efficient multi-scale modeling technique that uses measurements from short DEM simulations as inputs for predicting mixing performance over longer time scales.

See more of this Session: Dynamics and Modeling of Particulate Systems II
See more of this Group/Topical: Particle Technology Forum

(151b) DEM Simulations of Tablet Motion In a Vibratory Coating Pan

Rahul Kumar and Carl R. Wassgren, Mechanical Engineering, Purdue University, West Lafayette, IN

The motion of true biconvex tablet shapes in a vibratory continuous coater is studied using the Discrete Element Method (DEM). In contrast to rotating drums, motion is imparted to the tablets using vibrations about a pivot point offset from the center of the drum. A significant advantage of this approach is that the coater can be open from the top allowing for easy monitoring of the coating operation. A regime map identifying the range of amplitudes and frequency of vibrations required to create a tumbling motion of tablets inside the drum is obtained. It was found that for Froude numbers (Fr, based on the vibration frequency and amplitude at the drum center) greater than one, tumbling motion was achieved in the drum. For Fr slightly less than one, very slow tumbling motion occurs while for smaller Fr no motion was observed. The axial and transverse mixing of tablets was also studied for various values of Fr.

See more of this Session: Dynamics and Modeling of Particulate Systems II
See more of this Group/Topical: Particle Technology Forum

(151c) Parallelization of Three-Dimensional Discrete Particle Model for Dense Gas-Solid Flows On Unstructured Mesh

Chunliang Wu, Cain Department of Chemical Engineering, Louisiana State University, Baton Rouge, LA and K. Nandakumar, Chemical Engineering, Louisiana State University, Baton Rouge, LA

A parallel code for three-dimensional discreteparticle/element model (DPM or DEM) coupled with the computational fluiddynamics (CFD) model has been developed and used to simulate dense gas-solidflows in fluidized bed reactors. The gas-phase hydrodynamics is described inthe Eulerian framework while the dispersed phase is inthe Lagrangian one. The finite volume method isapplied to discretize the continuity and momentumequations of the gas phase on 3D unstructured mesh. The fractional time-marchingscheme is designed for the two-phase momentum coupling. The particle motion istracked by directly solving the dynamical equations of each individualparticle. Particle-particle and particle-wall interactions are taken intoaccount by a soft-sphere model. The numerical solution procedure in a time stepbegins by calculating the particle contact forces and integrating the particlemomentum equations over fractional DPM time steps, followed by solving thegas-phase hydrodynamics with SIMPLE algorithm. During integration of theparticle momentum equations, the hydrodynamic forces on particles arecalculated and the corresponding source terms of the gas momentum areintegrated.

The couplednumerical solution of the DPM-CFD problem, described above, poses newchallenges in parallelization. The total load on the DPM part of the code isdirectly related to the number of particles, which in our design is targeted tobe in the order of millions of particles. The computational load in the CFDpart can be distributed evenly on many processors using domain decompositionwithin the framework of MPI. However, if we use the same structure for DPM, theheterogeneous spatial distribution of particles will result in poor loadbalancing among the many processors.

Thehybrid parallel computing technique is used in handling the coupled CFD-DPMmodel to partly overcome the difficulties identified above and achieveefficient load balancing. Firstly at the coarse-grain level, the solutiondomain is decomposed into partitions using bisection algorithm to minimize thenumber of faces at the partition boundaries while keeping almost equal numberof cells in each partition. The solution of the gas-phase governing equationsis performed on these partitions. Particles and the solution of their dynamicsare associated with partitions according to their hosting cells. This makes nodata exchange between processors for calculating the hydrodynamic forces onparticles. By introducing proper data mapping between partitions, the cell voidfraction is calculated accurately even if a particle is shared by severaldecomposed partitions. Neighboring partitions are grouped by a gross evaluationbefore simulation, with each group having similar number of particles. Thecomputation task of a group of partitions is assigned to a compute node, whichhas multi-cores or multi-processors with a shared memory. Each core orprocessor in a node takes the computation of the gas governing equations in onepartition. Processors communicate and exchange data through Message PassingInterface (MPI) at the coarse-grain parallelism.

Secondly,the multithreading technique is used to parallelize the computation of thedynamics of the particles in each partition. The number of compute threads isdetermined according to the number of particles in partitions and the number ofcores or processors in compute node. Thread pooling isemployed to assign computing threads to cores or processors. In such a way thereis almost no waiting of the processors in a compute node. Since the particlenumbers in all compute nodes are almost the same, the above strategy yields anefficient load balancing among compute nodes. Test numerical experiments on HPCcluster show that the developed model is scalable up to 128 CPUs and can beused to simulate dense gas-solid flows in fluidized bed with more than 10 millionsof particles.

See more of this Session: Dynamics and Modeling of Particulate Systems II
See more of this Group/Topical: Particle Technology Forum

(151d) Numerical Study On the Pneumatic Conveying of Gas-Solid Two-Phase In Vertical-to-Horizontal Bends

Subhashini Vashisth and John R. Grace, Department of Chemical and Biological Engineering, University of British Columbia, Vancouver, BC, Canada

Pneumatic conveying is extensively used for the transportation of different types of dry and free-flowing powdered and granular materials. Owing to its wide application in chemical, cement, food, pharmaceutical, mining and power industries, the proper design of pneumatic conveying systems is of great importance. Bends and elbows often used in the pneumatic conveying systems offer flexible routing and distribution but at the cost of various problems such as high pressure loss, particle segregation and degrading, particle roping, and pipe wall erosion. In the past, several efforts have been made to understand the internal flow in such geometries and minimize the above stated problems by employing different bend designs such as T-bend, ankle-bend, ribbed bends and bends with flow mixers.

The important factors which influence the design of such systems are properties of gas and particles, conveying gas velocity and solid mass flow rate. It is of economical interest to convey solids at lowest possible gas velocity so as to reduce the power consumption, pipe erosion and particle attrition. However, below certain gas velocity the particles cannot remain in suspension leading to the particle deposition and pipe blockages thereby affecting the operating efficiency and safety of the conveying system. Hence, objective of the present work is to numerically investigate the gas-solid flow in a 90° vertical-to-horizontal bend for low conveying gas velocities. The knowledge about the effects of gas flow velocity, gas to solids mass flow rate ratio and bend radius to pipe diameter ratio on particle concentration, velocity profiles and particle roping characteristics would further aid in design purposes.

The numerical investigation was conducted using the commercial Computational Fluid Dynamic (CFD) software Ansys-Fluent 12.1, employing the Euler-Lagrangian approach. The computational model was consisted of a circular elbow geometry with 5D long vertical straight duct, a circular 90° elbow with $R/D = 1.5$,

3.0 and 5.0 and a 20D long horizontal straight duct, where R is the bend radius and D is the diameter of the pipe. The predictions were taken at y = 3D, 5D and x = 1D, 5D, 10D, 15D, 20D upstream of the bend exit in the horizontal duct. The gas and particle data were collected at six different angular locations (0°, 30°, 45°, 60°, 75° and 90°), where x and y are the transverse distances in pipe cross-section. All calculations were performed on structured hexagonal grids created in Ansys-Workench 12.1. The grid was sufficiently refined at the wall and was found to produce a grid-independent solution. Pulverized coal particles, having an average particle diameter of 50 mm and particle density of 1680 kg/m^3, was used as the solid material. The solid particles were released into the duct at the same velocity as the gas phase, i.e., zero slip, and from random locations on the inlet plane.

Euler-Discrete phase methodology (DPM) along with renormalization group (RNG) k-ε model was employed to perform numerical calculations of the particle-laden gas flow in 90° elbow with different bend ratios. This approach computes the Navier-Stokes equation for the gas phase and the motion of individual particles by the Newtonian equations of motion. Computations were performed for the gas phase followed by the solution of the solid phase. Hence, various hydrodynamics forces acting on the particles were determined based on the prior knowledge of predicted gas flow. The renormalization group (RNG) k-ε model was used to simulate the gas phase turbulence (Yin et al., 1996). DPM treats the interaction between the gas and solid phases using the particle-source-in-cell method (Crowe et al., 1977). The conservation equations include appropriate source terms resulting from the dispersed phase (i.e. two-way coupling). Stochastic method was used to account for the influence of gas turbulence on the solid phase (Gosman and Ioannides, 1981). The interaction between particles and the wall was modeled using a coefficient of restitution. The effect of particulate phase on the gas flow, turbulent dispersion, lift forces and particle-wall collisions were incorporated in the model. Fully developed turbulent flow was assumed at the inlet. A 'no-slip' boundary condition was employed for the gas velocity at the wall.

The pressure based segregated solver was used, with the SIMPLE scheme for pressure-velocity coupling. All the diffusion terms used second-order discretization. The second-order upwind interpolation was used for the convection terms. Under-relaxation factors were usually left at the Fluent default settings. To handle the nonlinearity of the equations, the treatment of all transported variables involved two nested levels of iterations referred to as inner and outer iterations. Outer iterations were repeated until the problem satisfied a convergence criterion. A converged solution of the coupled two-phase flow system is obtained by successive solution of the Eulerian and discrete phase, respectively. For the simulations performed in this study, the convergence of outer iteration was judged by how accurately the continuity equation was satisfied by the current values of the dependent variables. The solution procedure was considered converged when the ratio of the summation of absolute mass source residuals to the total rate of mass inflow fell below a prescribed tolerance (1x 10^{-6}).Thereafter, a large number of particles were tracked through the flow field (typically 100000 or more). Each case was simulated for 2 to 5 s.

For the purpose of the model validation, the parameters used in the simulation are mainly taken from the experiments of Akilli et al. (2005). The solid particles and the gas segregate in the bend due to centrifugal forces. It was observed that most of the particles aggregate towards the outer wall of the duct, referred to as ropes. This can as well observed from the particle concentration profiles. As the particle ropes flows along the horizontal duct, they falls from the top-outer wall towards the bottom-inner wall of the duct due to gravity. Similar observations were also made by McCluskey et al. (1989) and Akilli et al. (2001, 2005). They reported that particle deposition was a consequence of rope formation in bends. Further, parametric investigations were performed to study the effect of operating variables on particle roping and the results henceforth can offer improved insight and initial design modelling capability for gas-solid pneumatic conveying systems. The comparison of the mean streamwise velocity profiles for both air and pulverized coal particles and pressure drop for with the experimental data set of Akilli et al. (2005). The fundamental features like particle roping phenomena, particle segregation, secondary flow and recirculation, gas-particle interaction were reasonably captured and compared for different circular bends. The results show secondary flows disperse the particle rope by carrying particles around the pipe circumference while turbulence disperses the rope by localized mixing of particles. The results presented in this work are directly applied to engineering applications such as development of coal gasification plants, fluidized pneumatic conveying.

REFERENCES

Akilli, H., Sahin, B. and Levy, E.K., (2001), "Gas–solid flow behavior in a pipe after a 90◦ vertical-to-horizontal elbow", *Powder Technol.,* 116, 43–52.

Akilli, H., Sahin, B. and Levy, E.K., (2005), "Investigation of gas–solid flow structure after a 90◦ vertical-to-horizontal elbow for low conveying gas velocities, *Advanced Powder Technol.,* 16(3), 261-274.

Crowe, G. T., Sharma, M. P. and Stock, D. E.: 1977, 'The particle-source-in-cell model for gas droplet flows, J. Fluid Eng. 99, 325–332.

Gosman,A., E. Ionnides, Aspects of computer simulation of liquid fuelled combustors, AIAA paper. No. 81-0323, 1981.

McCluskey, D. R., Easson, W. J., Greated, G.A. and Glass, D. H., (1989), "The use of particle image velocimetry to study roping in pneumatic conveyance", Particle Particle Syst. Character. J. 6, 129–132.

Yin,M., F. Shi, Z. Xu, Renormalization group based C turbulence model for flows in a duct with strong curvature, Int. J. Eng. Sci. 34, 1996. 243–248.

See more of this Session: Dynamics and Modeling of Particulate Systems II
See more of this Group/Topical: Particle Technology Forum

(151e) Effects of Concentration, Interaction and Shape Anisotropy On Dynamics of Colloidal Mixtures

Tianying Jiang, Chemical and Biomolecular Engineering, University of Illinois Urbana-Champaign, Urbana, IL and C.F. Zukoski, Department of Chemical & Biomomlecular Engineering, University of Illinois at Urbana-Champaign, Urbana, IL

Rules for equilibrium and non-equilibrium phase behavior of colloidal mixtures are not well understood. Theory and simulation have been used to predict phase diagrams and dynamics of colloidal mixtures formed by spherical particles with varied size ratios and interaction energies. Limited experimental work has been reported that can be used to compare with these theoretical and modelling predictions. In addition, neither theory nor experiment have investigated the behavior of mixtures with shape anisotropy incorporated. Here a seeded emulsion polymerization method is used to produce large quantities of uniform particles with similar surface chemistry but varied shape and size. By coating the particles with a stabilizer, these particles experience weak attractions ionic strength ionic strength is increased. In our studies the mechanical properties of mixted dispersions are investigates for different number density ratios at increasing volume fraction for varied strengths of attraction and shape anisotropies. Our particular interest lies in characterizing these materials in the glassy and weak gel region. Mixtures are found to alter gel volume fractions and volume fraction dependencies of moduli, their frequency depends and yielding of glassy systems. This behavior will be discussed in terms of anisotropic interactions and changes of free volume in suspension introduced by mixing.

See more of this Session: Dynamics and Modeling of Particulate Systems II
See more of this Group/Topical: Particle Technology Forum

(151f) Modeling of a Spray Coating Process In a Forberg Style Mixer/ Coater: Particle Size Effect On the Coating Preference

Jianfeng Li[1], Carl R. Wassgren[2] and James D. Litster[1], (1)Chemical Engineering, Purdue University, West Lafayette, IN, (2)Mechanical Engineering, Purdue University, West Lafayette, IN

The discrete element method (DEM) is applied to investigate the heterogeneous particle flow of a spray coating process with polydisperse particles in a Forberg style mixer/coater. Simulation results of particle movement demonstrate radial convective flow with the rotating paddle and axial dispersion flow along the shaft. A spray zone area is specified at the top bed surface, in which particles are being coated proportional to the surface area exposed under the spray. Different segregation phenomena are observed when different ranges of particle sizes are used, resulting opposite coating preference over polydisperse particles. Further DEM simulations are performed altering particle density, in order to understand that whether the segregation is induced by shape or by gravity. Finally, a lumped growth rate expression is derived to be applied in the population balance (PB) modeling of particle coating distribution.

See more of this Session: Dynamics and Modeling of Particulate Systems II
See more of this Group/Topical: Particle Technology Forum

(151g) Numerical Simulation of Gas-Solid Two-Phase Flow and Separation In An Spray Granulating Tower

Wang Juan, Yu Mao and Jiangyun Wang, College of Chemical Engineering, China University of Petroleum,Beijing, Beijing, China

A full-size numerical simulation of the gas-solid flow in a spray granulating tower was carried out by applying Model and Eulerian multiphase Model. All details of gas and solid in the flow domain were obtained. And the effects of structure parameter and operating condition on the separating efficiency of the spray granulating tower were studied. It was showed that the flow of gas-solid in the spray granulating tower was rotary, which could be separated into two sections, inner upper flow and outer downer flow. Additionally, the parameters such as operating gas load, grain physical property, and distributor structure have significant relation with separation efficiency in the spray granulating tower.

See more of this Session: Dynamics and Modeling of Particulate Systems II
See more of this Group/Topical: Particle Technology Forum

(154a) Eulerian-Eulerian Simulations of Dual Fluidized Bed Reactor for CO2 Capture

Kringan K. Saha, Computational Sciences Division, National Energy Technology Laboratory, Morgantown, WV and **E. David Huckaby**, Computational Science Division, National Energy technology Laboratory, Morgantown, WV

Eulerian-Eulerian simulations are performed for a recently constructed low-temperature sorbent based CO_2 capture system (Shadle *et al.*). The system, which is designed to process about 115 *L*/min of simulated flue gas, consists of a riser with mixing zone for absorption and a bubbling bed for regeneration. The results of both full loop and partial loop simulations are compared to non-reacting operation of the system with 200 μ m polystyrene particles. The sensitivity of the results to sub-models for drag, turbulence and frictional flow are also reported.

See more of this Session: Fluidization and Fluid-Particle Systems for Energy and Environmental Applications
See more of this Group/Topical: Particle Technology Forum

(154b) In-Situ, Simultaneous Milling, Coating and Curing of Brittle Particulates Pre-Coated with a LED UV-Curable Formulation in a Fluid Energy Mill

Huiju Liu[1], Chunmeng Lu[2], Linjie Zhu[2], Subhash Patel[2], Ming-Wan Young[2] and **Costas G. Gogos**[1], (1)New Jersey Institute of Technology, Newark, NJ, (2) Polymer Processing Institute, Newark, NJ

In conjunction with UV technology, a fluid energy mill (FEM) was demonstrated to *simultaneously* and in-situ achieve several functions, namely: size reduction of pre-coated coarse micron-sized particles with UV-curable chemicals into smaller (ca. 1~10 μ m) particles, coating of UV chemicals onto the milled particles, and curing of the UV chemicals. Potassium chloride (KCl) was chosen as the matrix material, and acrylate was chosen as the UV-curable formulation. The hold-up mass method was utilized to estimate the average residence time of particles in the FEM, the results of which showed that the average residence time decreases with increasing grinding pressure and solid feed rate, whereas it was not significantly affected by feeding pressure. Fluorescent microscopy showed that the UV-curable chemicals were evenly transferred to almost every single particle during FEM milling without UV irradiation, whereas they were extensively, but transferred to a lesser extent during FEM milling with UV irradiation. Thermal analysis (TGA and DSC) and IR analyses were employed to characterize the conversion of the free radical polymerization. IR analysis showed that the double bond conversion was up to 71%, denoting extensive curing.

Corresponding authour: gogos@polymers-ppi.org

See more of this Session: Fluidization and Fluid-Particle Systems for Energy and Environmental Applications
See more of this Group/Topical: Particle Technology Forum

(154c) Experimental Study of Coal Topping Process Coupled with a Coal-Fired CFB Boiler

Songgeng Li, Wenli Song, Weigang Lin, Lin Du and Ze Wang, Institute of Process Engineering, Chinese Academy of Sciences, Beijing, China

Coal is the primary energy in China, which provides ~75% of the total energy consumption. However, reserves of petrolium oil and natural gas in China is limited. The amount of imported oil exceeds over 50% of the domestic need in 2010 due to the rapid economic development. Production of liquid fuels from coal is considered as a solution to reduce the dependency on imported crude oil. The coal topping process patented by Institute of Process Engineering(IPE), Chinese Academy of Sciences (CAS) is an integrated multi-product process aiming to obtain a high yield of light liquid fraction and gasesou products by flash pyrolysis. Heat and electricity can be co-generated simutaneously in the process when it is coupled with a CFB boiler.

Extensive work on the process has been conducted in a laborotary bench scale facilty by Energy Conversion Group in IPE, CAS. Coals from different locations of China have been tested. Combustion properties of char derived from the coal topping process and effects of pyrolysis temperature and particle size on the yields of gas, char and liquid products were investigated. Based on the bench-scale experimental results, the process capacity has been scaled up to 2 t/h, which is integrated into a 75t/h CFB boiler in a thermal power plant. A comparison of the experimental results on the bench scale and the pilot-scale facilites has been given.

See more of this Session: Fluidization and Fluid-Particle Systems for Energy and Environmental Applications
See more of this Group/Topical: Particle Technology Forum

(154d) Numerical Investigation of Flow/ Chemistry Interactions During Biomass Gasification In a Fluidized Bed Reactor

Perrine Pepiot, National Bioenergy Center, National Renewable Energy Laboratory, Golden, CO and Mark R. Nimlos, National Renewable Energy Laboratory, Golden, CO

A current limitation to the efficiency of biomass gasification for the production of biofuels is the high level of tars usually found in the product gas, causing fouling of downstream equipment. Tar formation is a relatively slow process involving bi-molecular reaction pathways, and is therefore likely to be affected by local flow conditions. In this context, developing primary tar mitigation strategies requires a better understanding of the coupled chemical and physical processes occurring inside the reactor. In this work, two- and three-dimensional simulations of biomass gasification in a lab-scale fluidized bed reactor are conducted using a Lagrangian approach to handle the solid phase. The biomass gasification chemistry is described using a semi-detailed kinetic model assembled from literature data and recent experimental measurements performed at the National Renewable Energy Laboratory. Statistical analysis of the particles and velocity fields is conducted to investigate the effect of biomass volatilization on the dynamics of the bed. A posteriori analysis of products distribution, gas mixing and residence time inside the reactor is used to explore tar formation pathways during gasification, and how those are impacted by the dynamics of the bed, especially the bubbling process.

See more of this Session: Fluidization and Fluid-Particle Systems for Energy and Environmental Applications
See more of this Group/Topical: Particle Technology Forum

(154e) Mixing and Segregation of Biomass Particles In a Bubbling Bed

J. S. Halow[1], Benjamin Crawshaw[1], Stuart C. Daw[2] and Charles E. A. Finney[2], (1)Chemistry, Waynesburg University, Waynesburg, PA, (2)Feerc, Oak Ridge National Laboratory, Knoxville, TN

Introduction

Fluidized beds are being widely considered for convertingsolid biomass into liquid fuels, syngas, and chemicalproducts. While fluidized beds havebeen extensively studied for many applications, the detailed hydrodynamics andmixing patterns in systems with disparate types of bed particles is still notfully understood. A number of biomassgasification and pyrolysis processes now being considered employ bubbling bedsof fine sand into which larger biomass particles are fed. At steady state, the biomass particlesusually constitute only a small fraction of the bed so that the dynamicsinvolve a few coarser, lighter particles circulating in a finer, densermedium. This disparity in particle sizesand densities creates the potential for segregation, especially at lowfluidizing velocities. The extent ofsegregation can be extremely important, because it determines the intimacy andduration of contacting between released gases and bed material. In turn, the degree of gas-particlecontacting can greatly affect product yield and composition, especially whenthe bed material has catalytic properties. We report results from visual observation of bubbling laboratory bedscombined with magnetic particle tracking to measure detailed mixing andsegregation patterns for simulated biomass particles. Our ultimate goal is to develop correlationsand statistical models which can be used to predict biomass mixing andsegregation patterns in bubbling bed reactors.

ExperimentalMaterials and Methods

We use our previously reported (1) magnetic particletracking method in a 55 mm three-dimensional bed fluidized with air at ambientconditions. Safe, inexpensive magnetictracers and detectors are used, thus avoiding the issues associated with pastparticle-tracking methods. In thisstudy, we used neodymium magnets embedded in wooden spherical and cylindricalparticles and externally positioned magnetic field detectors to continuouslylocate the position of a tracer particle. The method takes advantage of the tendency of the tracer particles toalign their magnetic axes with the earth's magnetic field like a compassneedle. Proper positioning of the probescan take advantage of this tendency, which simplifies the data analysis. We also collected high-definition slow-motionvideos of the bed surface to observe top-segregated tracer motion.

Both beds of approximately 200 micron glass beads and 100 to200 micron sand were studied. Singlebiomass tracer particles were dropped into the beds and their positionsdetermined over a 5-minute run time. Tracers had diameters of 3 to 5 mm and densities from 0.55 to 1.2g/cc. Tracer positions were determinedusing the algorithm we previously reported on. Statistical analysis on the time series position data was used tocharacterize the tracer behavior. Fluidization velocity varied from 1.1 to 6 timesminimum fluidization. Slumped bed depthto diameter ratio was held constant at 1.

Results andDiscussion

Frequency distributions of the vertical position for variousconditions were determined. In somecases these show segregation at the top or bottom and under other conditions auniform distribution across the bed. Wefound that these distributions could be reasonably well fitted with thetwo-parameter Weibull distribution function and thatthe two parameters correlated with tracer density and superficialvelocity. Vertical velocitydistributions were also determined. Inaddition, typical three-dimensional trajectories showed typical bubble-inductedsolids circulation at low velocities, top and bottom segregation behaviors, andmixing at higher velocities. It wasfound that a tracer with a density of 0.89 g/cccontinued to circulate (appeared to be neutrally buoyant) at even very lowfluidization velocities (1.1 times minimum fluidization velocity) while heaviertracers segregated to the bottom and lighter ones to the top at lowvelocities. At high velocities, alltracers in the 0.55 to 1.2 g/cc density range mixed and circulated. Higher-density tracers, however, did not alwayscirculate, even at the highest velocities studied. The implications for biomass processing isthat appropriate processing conditions will likely depend on the type ofbiomass, degree of dryness, and whether it is processed or raw.

1. Patterson *et al.*, Ind. Eng. Chem. Res. 2010, **49**, 5037-5043.

Abstract submitted forpresentation at the 2011 Annual A.I.Ch.E. meeting in Minneapolis, MN

See more of this Session: Fluidization and Fluid-Particle Systems for Energy and Environmental Applications

(154f) Simulating Biomass Fast Pyrolysis In Fluidized-Bed Reactors for Bio-Oil Production

Qingluan Xue[1], Theodore J. Heindel[2] and R. O. Fox[1], (1)Department of Chemical & Biological Engineering, Iowa State University, Ames, IA, (2) Department of Mechanical Engineering, Iowa State University, Ames, IA

A numerical study is conducted to investigate optimal operating conditions of biomass pyrolysis in a fluidized-bed reactor for condensable bio-oil (tar) yield. A MFIX CFD model is presented for biomass fast pyrolysis in a fluidized-bed reactor via coupling the multifluid model with a global biomass particle pyrolysis model. The multifluid model is derived from the kinetic theory of granular flows. The kinetic model is based on superimposed cellulose, hemicellulose, and lignin reactions for micro-particles (< 1000 μm). The pyrolysis products are categorized into three groups: tar vapor, producer gas, and char. The CFD model uses 8 solid species to differentiate general biomass feedstocks by species mass fractions with 2 additional gaseous species released into the gas phase. The interaction between the biomass pyrolysis and reactor environment is represented through mass, heat, and momentum exchange terms. The model is used to investigate the effect of various operating parameters on the tar yield in a fluidized-bed reactor. The model simulates sufficient physical time (200 secs) to reach a steady-state condition. Results indicate that at a fixed temperature and superficial gas velocity, the biomass particle size determines the particle residence time inside the bed, the heating and reacting processes, and ultimately influences the tar yield. At a fixed particle size, the temperature is the prominent parameter that influences tar yield. For the range of studied fluidized gas temperatures, the predicted optimum tar yield at steady state is achieved near 450°C. The product yields for different biomass feedstocks show a material with more cellulose mass fraction favors tar yield and more lignin mass fraction favors char production. It is also found that shallow fluidized beds and higher fluidized gas velocities have a positive effect on tar yield due to short tar vapor residence time in the reactor.

See more of this Session: Fluidization and Fluid-Particle Systems for Energy and Environmental Applications
See more of this Group/Topical: Particle Technology Forum

(154g) Bimetallic Chemical Looping Gasification with Oxygen Uncoupling - Feasibility Study and Process Analysis

Fanxing Li, Chemical and Biomolecular Engineering, North Carolina State University, Raleigh, NC

In atypical chemical looping gasification (CLG) process, an iron oxide based oxygencarrier is used to indirectly convert carbonaceous fuels into separate streamsof sequestrable CO_2, hydrogen,and heat/electricity. Although the novel process concept has the potential tobe both clean and efficient, the reactions between the oxygen carrier particlesand fuels are often kinetically limited, especially when solid fuels such ascoal and/or biomass are used. In order to enhance the rates of the aforementionedreactions, it is proposed that a secondary metal oxide be incorporated to the ironoxide based oxygen carrier. For instance, when copper (II) or manganese(III/IV) oxide is used as the secondary metal oxide, molecular oxygen can bereleased from the bimetallic oxygen carrier through "oxygen uncoupling", i.e. decompositionof the secondary metal oxide at high temperature. As a result, both the in-situsolid fuel gasification and iron oxide reduction reactions can be kineticallypromoted. The present study explores the technical feasibility of this novelbimetallic CLG process concept. A potential configuration for the bimetallicCLG process is first proposed. This is followed by thermodynamic analysis ofkey reactors to arrive at the maximum fuel and oxygen carrier conversionsconsidering thermodynamic equilibrium limitations. Finally, ASPEN Plussimulation is carried out to evaluate the energy conversion efficiency of thebimetallic CLG process. Preliminary kinetic data for coal char conversion withbimetallic oxygen carrier are also presented. The present study indicates thatthe incorporation of the oxygen uncoupling metal oxide has the potential toenhance the fuel and oxygen carrier conversions without adversely affecting theoverall efficiency of the CLG process.

See more of this Session: Fluidization and Fluid-Particle Systems for Energy and Environmental Applications
See more of this Group/Topical: Particle Technology Forum

Encapsulation of catalytically active solids in calcium alginate microparticles and their reaction – diffusion properties

Jiří Dohnal, František Štěpánek

Institute of Chemical Technology Prague
Department of Chemical Engineering
Laboratory of Chemical Robotics
Technicka 5, 16628 Prague, Czech Republic
Jiri.Dohnal@vscht.cz, Frantisek.Stepanek@vscht.cz

Introduction

Calcium alginate gel is a biomaterial most often used for the encapsulation of enzymes, living cells and proteins. The aim of this work is to use this material for encapsulation of inorganic solids to produce beads or microparticles with different usage as a photocatalysts or controlled release application. The first class of these capsules are microparticles which were fabricated by InkJet Drop on Demand technique. These particles with diameter from 65 to 90 μm (dependent on the solids concentration) can be used for artificially started controled release of active substances. The internal solids are reservoirs containing drugs or other active substance. The second class of beads are particles with diameter around 2 mm containing TiO_2 nanostructures. These beads can be used for photodecomposition of pollutants in water.

Fabrication and properties of microparticles with solids inclusions

We have used the InkJet Drop on Demand technique (Microfab Tech., Plano, Texas) for generation of small microdrops (Figure 1). A control unit drive causes contraction of piezoelectric crystal in printhead which pump the solution from capillary orifice where the droplet is created and tears off. For filling of the piezoelectric capillary and stability of printing process the system is connected to a pressure control unit.

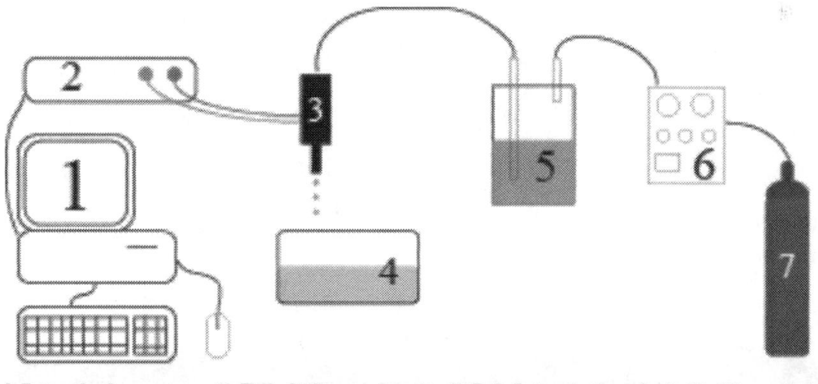

Figure 1: InkJet printing setup: 1) PC; 2)Control unit; 3) InkJet printhead; 4) $CaCl_2$ pool; 5) Reservoir with TiO_2 dispersion in sodium alginate solution; 6) Pressure control unit; 7) Nitrogen

Dispersion of solids in sodium alginate solution (1% w/w) was dropped to a calcium chloride pool. Alginate droplets are gelling in a Ca^{2+} solution to produce gel beads. As inner inclusions were used TiO_2 particles with a diameter around 0.5 μm which were stabilised by NaOH (pH 9) in 1% (w/w) of sodium alginate solution. For mapping of the dropping process was used concentration of TiO_2 particles in the range from 0.5 to 2% (w/w) which corresponds to a volume ranging from 1.95 to 7.8% (v/w). Settings of the InkJet printhead for smooth droplets printing, dependence of droplet size on system settings and viscosity of the solution and dependence of microparticle size on solids concentration were investigated and are showed in Figures 2-4.

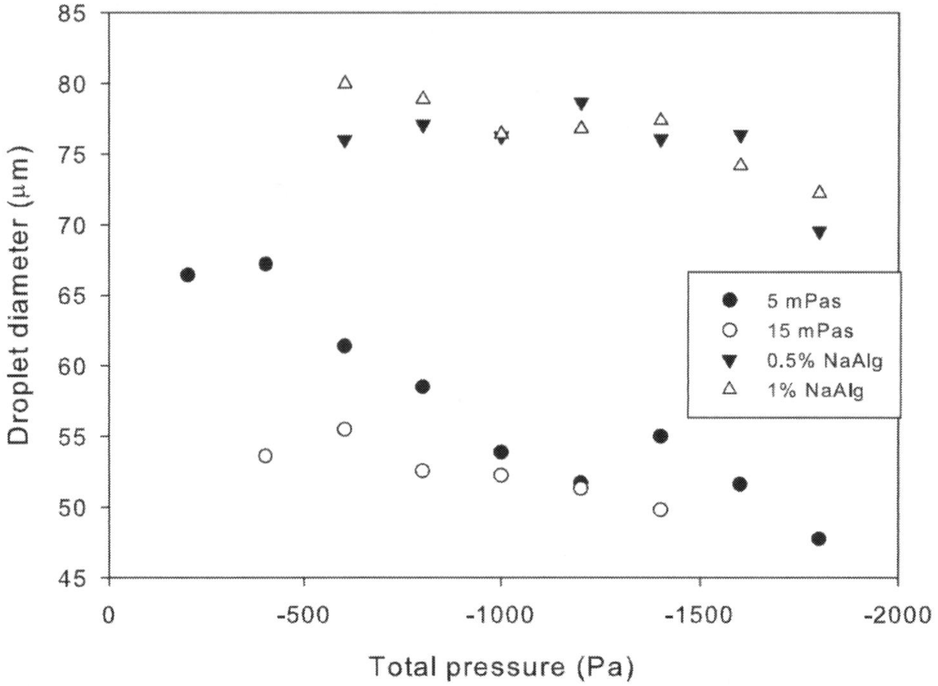

Figure 2: Dependence of droplet dimeter on viscosity of liquid for mixture of water and glycerol compared with 0.5 (η=6.5mPas) and 1 % (w/w) (η=13 mPas) of sodium alginate solution. Printhead setup: Voltage ±130V, frequence 100 Hz

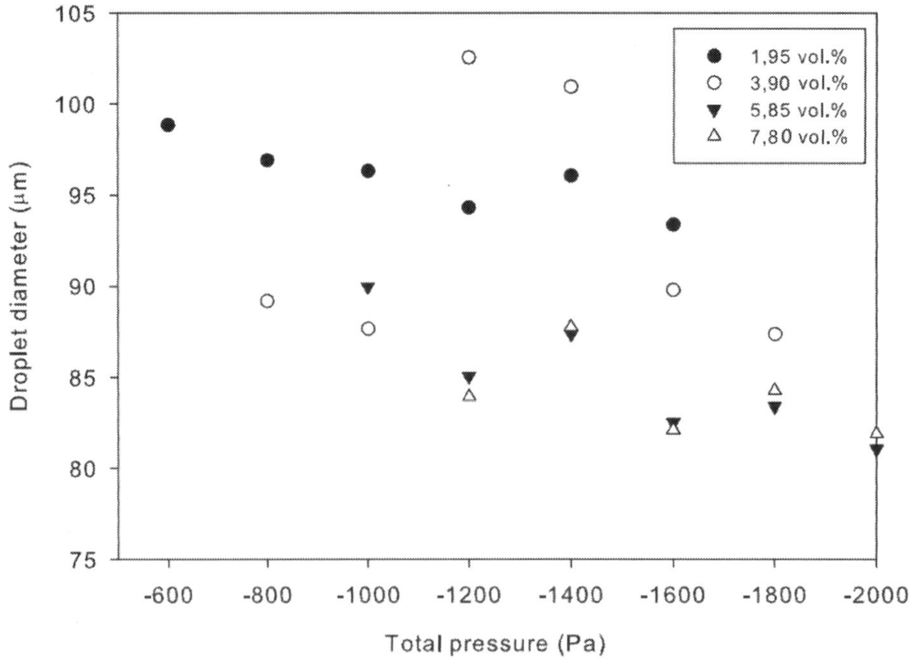

Figure 3: Dependence of droplet diameter on pressure in the system and volume concentration of solids in the 1% (w/w) sodium alginate solution

These graphs shows that droplet diameter is dependent on pressure settings, viscosity of solution and composition of solution. Alginate solution with similar viscosity as a mixture of water and glycerol have different droplet size because droplet diameter is influenced by viscosity and surface tension. Droplets from dispersion of solid in alginate solution is influenced by concentration of solids. Higher concentration of TiO_2 particles caused smaller diameter of droplet.

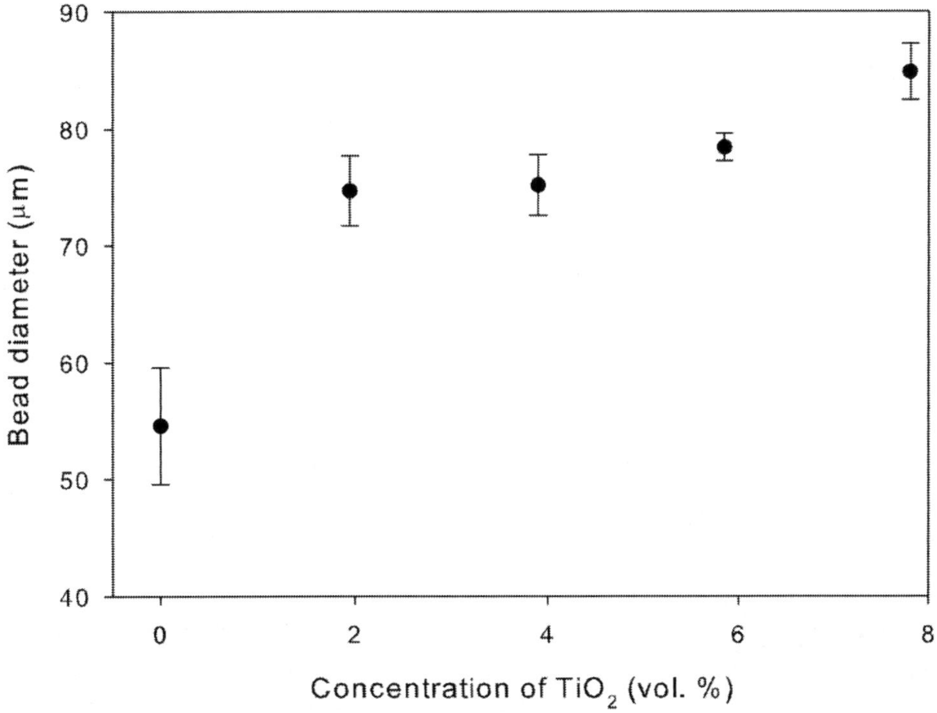

Figure 4: Dependence of bead diameter on concentration of TiO_2. Printhead setting: Voltage ±130V, frequence 100Hz, pressure -1200 Pa

Figure 4 show diameter of particles from alginate solution and dispersion of solids in this solution. Diameter of these droplets is dependent on concentration of inner inclusions. Higher concentration of solids caused less shrink ratio.

Calcium alginate/TiO_2 composite particles for photodecomposition of pollutants in water

1% sodium alginate solution with 1% TiO_2 particles was dropped by syringe needle to calcium chloride solution to produce calcium alginate composite beads with diameter around to 1.5 mm (particles type 1). 10g of these particles were freezed by liquid nitrogen and lyophilized (particles type 2). These particles were of the same size but different density as particles type 1. 10 g of particles type 1 were freeze in freezer (-18°C) where the shrunk then lyophilized. These particles have different size and shape.

All types of particles were put into 15ml of methylene blue solution (1.5_x10^{-5} % w/w) and exposed to UV light (λ=325 nm). Photodecomposition kinetics of methylene blue with the particles described above was measured by UV-VIS spectrometer (Specord 205). All kinetics were compared with results obtained from experiment where TiO_2 particles weren't encapsulated in calcium alginate gel (Figures 5, 6).

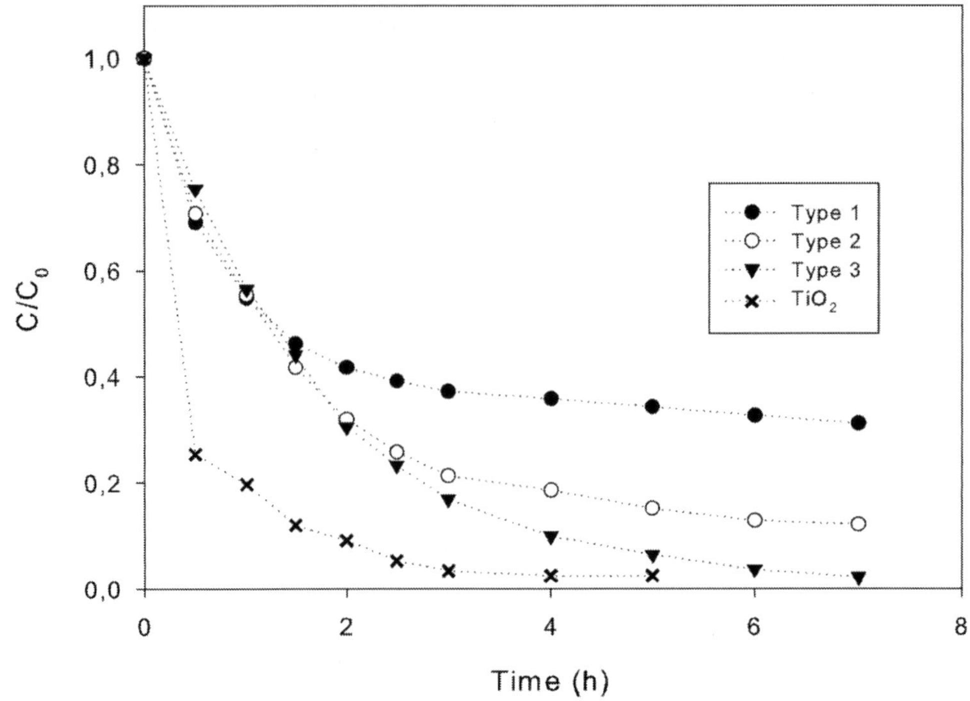

Figure 5: Photodecomposition kinetics of methylene blue

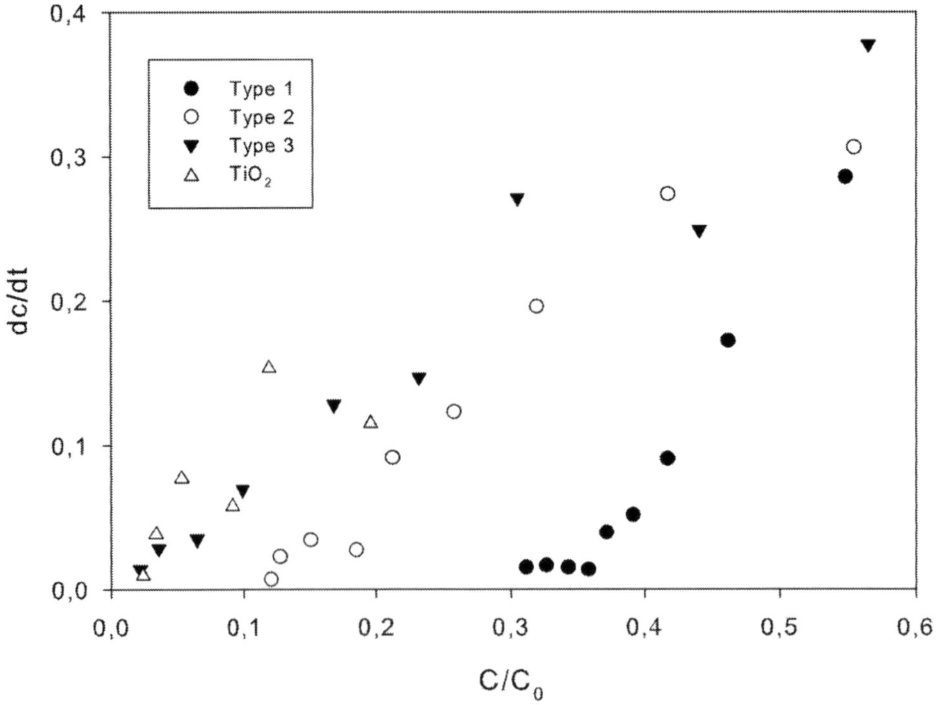

Figure 6: Dependence of decrease concentration of methylene blue on relativ concentration of dye in solution

Photocatalytic decomposition affected not only on methylene blue but also on calcium alginate gel. Therefore particles stability of each type of particles was measured as a weight decrease during light exposure for 10 days. Particles 2 and 3 were each 24 hours washed by distilled water dryed and weighted. Particles 1 were washed and weighted only after 10 days. While particles 2 and 3 were slightly decomposed particles 1 were almost fragmented. Results of these experiments are showed in Figure 7.

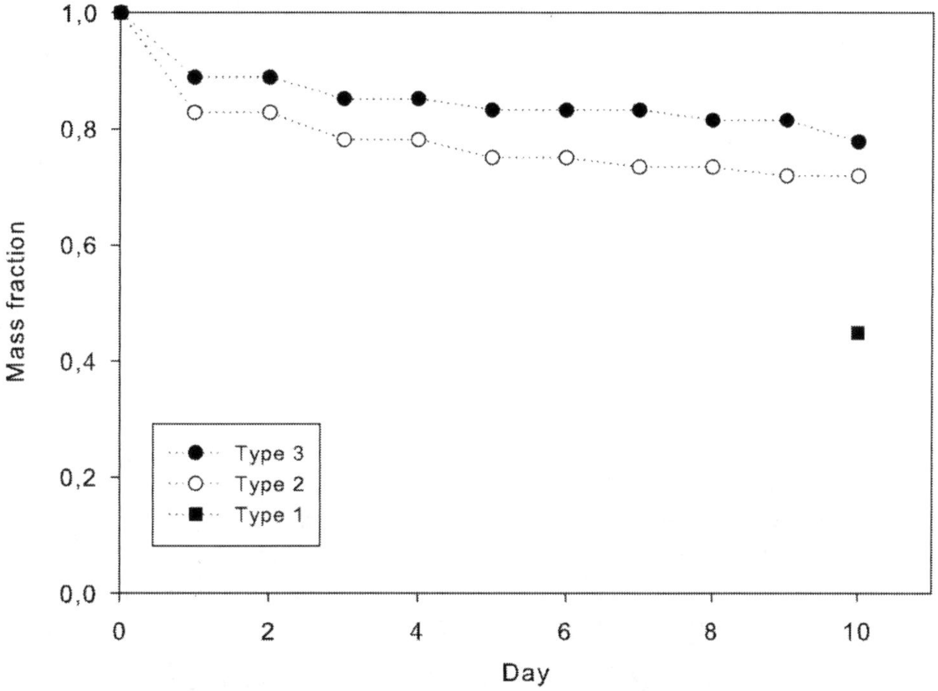

Figure 7: Particles stability during 10 days of light exposure

Type 1 **Type 2** **Type 3**

Figure 8: Different types of CaAlg /TiO2 beads: Scale = 500 μm

Literature:

1) Dohnal J., Štěpánek F.; Fabrication of composite microcapsules by drop-on-demand inkjet: Effect of precursor composition on the process limits; Chemical Engineering Science 66 (2011) 3829–3835

NANO-STRUCTURED CALCIUM SILICATE HYDRATE PRODUCTION FROM RICE HULL ASH

Paper Number: 222796

*O. Taspinar**, *T. Nalbantoglu**, *I. Ozturk**, *O. Ozcan***

*Istanbul Technical University, Chemical Eng. Dept., Maslak, Istanbul 34469, Turkey

** Istanbul Technical University, Environmental Eng. Dept., Maslak, Istanbul 34469,Turkey

Introduction

Rice hull ash (RHA) has a high silica content over 60 % SiO_2 (w/w) and is used in the production of several silicon compounds (1, 2). Kamath and Proctor (3) have extracted the silica in RHA with NaOH to obtain a sodium silicate solution as an alternative process to the conventional high temperature metallurgical process as in the following reaction.

$n\ SiO_2(s) + 2\ NaOH(aq) \rightarrow Na_2O.n\ SiO_2(aq) + H_2O$ ....(1)

Pure silica xerogels, synthetic magnesium silicates with high surface area of over 650 m^2/g, calcium, barium, zinc and aluminum silicates with also high surface areas were produced through further studies from sodium silicate solutions extracted by utilizing the Kamath and Proctor's method (4-9). Various micro structural forms of silicas such as, individual particles (fume silicas), precipitated silicas and silicate networks, and random closed packed structures of silica gels are well known and used in industrial applications. Their production, characterization, properties, structures and uses are presented in text books (10) and numerous research publications and patents (11).

Nano-structured calcium silicate with an open framework structure has been formed from sodium silicate solution by Johnston et al. (12) with a large surface area of about 350–600 m^2/g depending upon the synthesis method. The preparation, functionalisation and use of a novel calcium silicate from the sodium silicate solution were also patented in another study of Johnston et al. in 2008 (13). The use of a spacer compound during the filtration/washing of the precipitate to maintain the framework structure in the novel calcium silicate product is one of the important parts of that invention. The same novel structured calcium silicate was produced in our laboratories recently from rice hull ash (14, 15). In the present study, diethylene glycol was also used as spacer compound to obtain the framework structure of the nano-size calcium silicate which was produced from the reaction of rice hull ash origin Na_2SiO_3 and $CaCl_2$. The possible reaction between the Na_2SiO_3 and $CaCl_2$ is as follows:

$Na_2SiO_3\ (aq) + CaCl_2\ (aq) + xH_2O \rightarrow CaSiO_3.xH_2O\ (ppt.) + 2NaCl\ (aq)$.......... (2)

Experimental

Materials

Rice hull ash used in the experiments was acquired from an industrial rice company in Thrace region in Turkey. This industrially burnt ash was re-burnt in a muffle furnace at 600^{0}C for 6 h to increase the SiO_2 content. Then, 100 g of the ash was extracted with 600 ml of 1 N NaOH solution by boiling for 1 h in a beaker on a hot plate. The suspension was filtered; the filtrate and washing solutions were put in a plastic container and used as stock SiO_2 solution, and the waste ash was discarded. The SiO_2 content of this stock solution was determined as 11.78 g $SiO_2.H_2O$/100 ml in the previous study which was done in the same laboratory using the Kamath and Proctor's method (3, 8, and 9). Calcium chloride, $CaCl_2.2H_2O$ used for precipitation of CSH, and NaOH and HCl used for pH adjustments were all analytical grade. Distilled water was used for dissolution and washing purposes. The spacer compound, diethylene glycol (2-ethoxyethanol) was a technical product.

Methods

10 ml of the above mentioned Na_2SiO_3 stock solution (which has a pH value around 12) was treated with 8 % molar excess of $CaCl_2$ solution. This particularly necessary molar excess amount of $CaCl_2$ was taken from Johnston et al.'s study (13) which showed 5-10 % molar excess for formation of the nano-sized CSH particles in the process. The calculated amount of $CaCl_2$ (1.81 g) was dissolved in 14 ml distilled water, and the pH of this solution was also adjusted to 12 with NaOH. Some additional water, i.e. 14 ml was also added to sodium silicate solution before reaction to avoid co-precipitation and guarantee the dissolution of impurities such as NaCl etc. The sum of these two amounts of water was calculated as the twice of the necessary amount (i.e. 28 ml total) for the dissolution of the NaCl impurity in stoichiometric ratios. Then, the $CaCl_2$ solution was added onto Na_2SiO_3 solution very quickly by rigorously mixing on a magnetic stirrer. After 5 min rigorously stirring, the mixing rate was decreased to a medium level and the mixture was kept by stirring for 6 h ageing. The reaction is very fast and the milky colored colloidal suspension of CSH was formed suddenly. And by stirring in a medium level for 6 h in ageing, the framework structured nano-sized porous clusters are formed. The particle size determination was done with a part of this suspension using Mastersizer 2000 (Malvern Instruments, UK).

The suspension then was divided in two parts. One half parts was filtered under vacuum and washed with spacer compound, diethylene glycol and the other half part was washed with water after filtration. The two solid paste samples were then put a side in the laboratory for drying in open air. After drying, the samples were analyzed with a high resolution TEM (JEOL-2100 HRTEM, Japan) apparatus. For getting the high resolution, the particles were dispersed in ethyl alcohol and then one-two drops of this dispersion were put on the copper coated plates of the TEM apparatus. The whole process steps, reaction/precipitation/ageing, stirring, filtering, washing and drying were done at room temperature. The schematic representation of the process is given in Fig. 1.

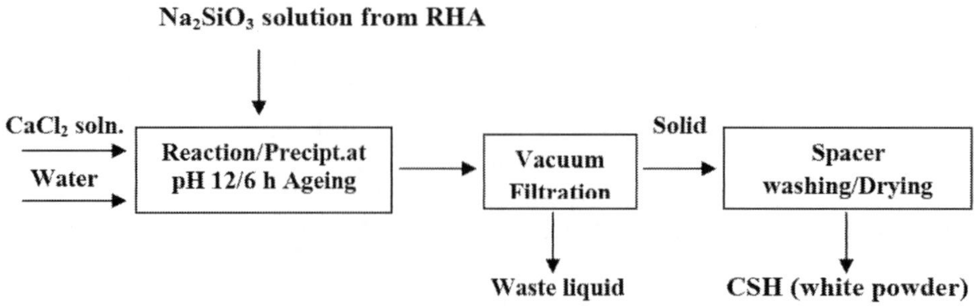

Fig. 1. Schematic representation of the production of nano-sized calcium silicate hydrate, CSH.

Results and Discussions

CSH samples produced with the process shown in Fig. 1 were analyzed using a high resolution TEM apparatus to see their framework and nano-sized structures. TEM pictures of the samples washed with water and diethylene glycol obtained with 100 nm scale are given in the Figs. 2 and 3. The general production conditions of these samples are outlined in Table 1. Particle size distribution of these samples is also given in Fig. 4.

Table 1. The production conditions of the produced CSH samples.

Product Nr.	Na_2SiO_3 amount, ml	$CaCl_2$ molar excess, %	$CaCl_2$ amount, g	Water for $CaCl_2$ dissolution, ml	Additional water for dissolution of impurities, ml	Washing reagent
Sample 1	10	8	1.81	14	14	Water
Sample 2	10	8	1.81	14	14	Diethylene glycol

As it is very clearly seen in Figs. 2 and 3 , the framework structure was formed when the sample was washed with the spacer compound, diethylene glycol. When the same sample was washed with water however, the framework collapsed. Similar differences are shown in the Fig. 5 which was obtained at 200 nm (0.2 μ) scale for diethylene glycol. As soon as water enters into the washing, the formed framework structure collapses and appearance of the particles becomes agglomerate with less porosity. The two samples were the same particle size about 5 μ in their suspension before washing operation as seen in Fig.4.

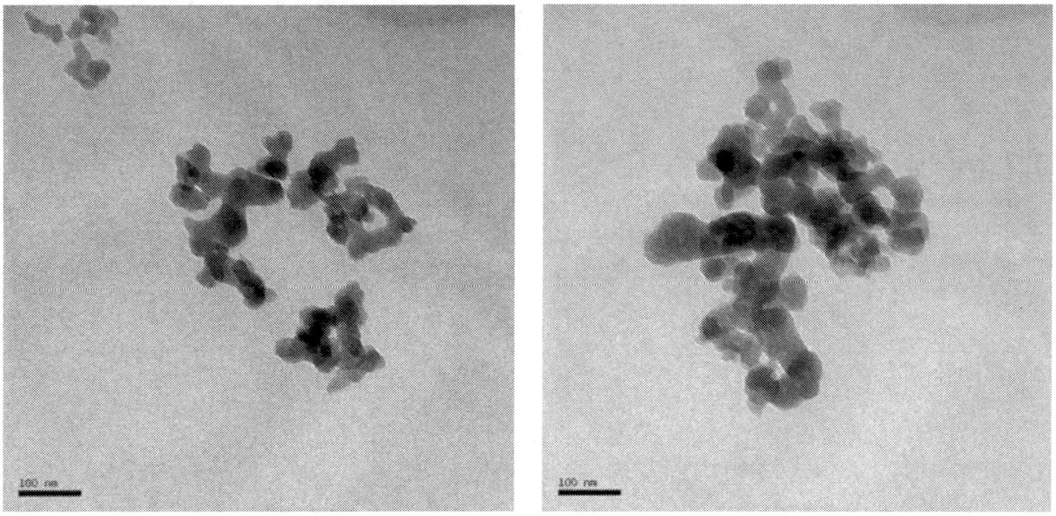

Fig. 2. TEM micrographs of CSH samples washed with water, Sample 1 in Table 1 (100 nm).

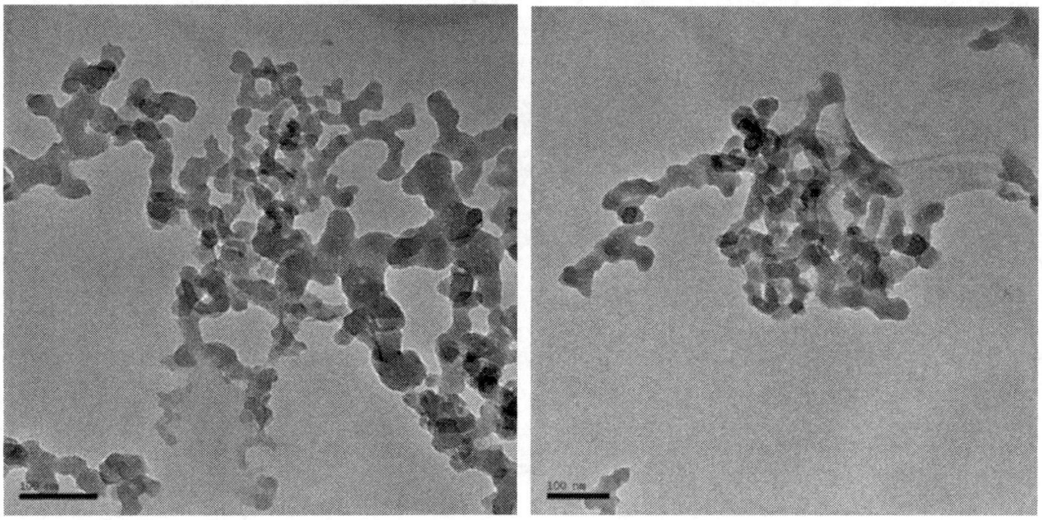

Fig. 3. TEM micrographs of CSH samples washed with diethylene glycol,
Sample 2 in Table 1, (100 nm).

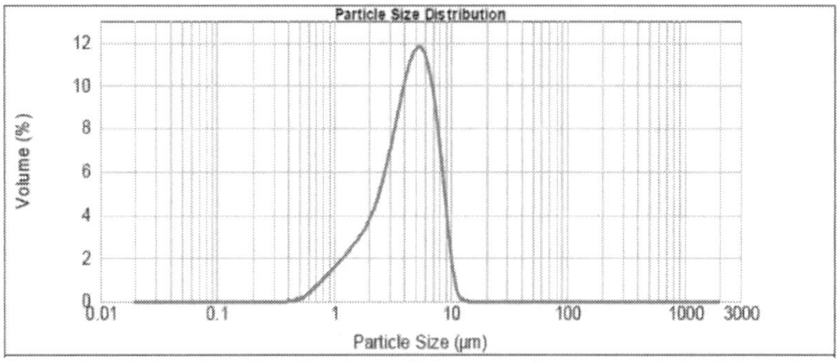

Fig. 4. Particle size distribution of the samples 1 and 2 in Table 1.

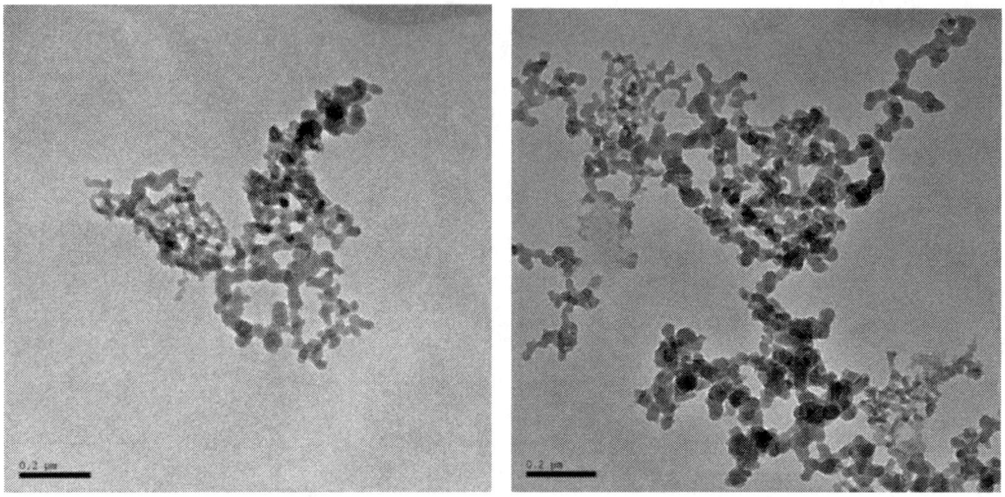

Fig. 5. TEM micrographs of CSH samples washed with diethylene glycol,
Sample 2 in Table 1, (200 nm).

Conclusion

It is possible to produce a highly porous framework structured nano-sized calcium silicate from rice hull ash. The most important parameter of the formation of nano-sized framework structure is the controlled ageing of the suspension after treatment of sodium silicate with calcium chloride at pH 12. To avoid the collapsing of this framework structure, a spacer compound, diethylene glycol must be used during filtration/washing operations.

References

1. Sun, L. and Gong, K. (2001) Review, silicon based materials from rice husks and their applications. *Ind. Eng. Chem. Res.* 40:5861-5877.
2. Proctor, A. (2005) Rice hull & rice hull silica utilization. Socrates Intensive Programme "Renewable Biomaterials", Toulouse, France, April 8-25, (2005).

3. Kamath, S.R. and Proctor, A. (1998) Silica gel from rice hull ash: Preparation and characterization. *Cereal Chem*. 75:484–487.
4. Kalapathy, U., Proctor, A. and Shultz, J. (2000) A simple method for production of pure silica from rice hull ash. *Bioresource Technology* 73:257-262.
5. Kalapathy, U., Proctor, A. and Shultz, J. (2002) An improved method for production of silica from rice hull ash. *Bioresource Technology* 85:285-289.
6. Taspinar, O.O. and Ozgul-Yucel, S. (2008) Lipid adsorption capacities of magnesium silicate and activated carbon prepared from the same rice hull ash. *Eur. J. Lipid Sci. Technol*. 110:742–746.
7. Bilgin, N. (2008) Determination of the Parameters Effecting the Surface Area of Magnesium Silicate Produced from Rice Hull Ash. *M. Sc. Thesis*, Istanbul Technical University, Turkey.
8. Sadic, E. (2008) The Production of Various Silicates from Rice Hull Ash. *M. Sc. Thesis*, Istanbul Technical University, Turkey.
9. Sadic, E., Taspinar, O., Ozcan, O. (2010) Comparison of surface areas and oleic acid adsorption characteristics of various silicates produced from rice hull ash. *Fresenius Environmental Bulletin* 19 (12): 2823-2831.
10. Iler, R.K., (1979) *The Chemistry of Silica*. Wiley-Interscience, New York.
11. Harper, R.T., Johnston, J.H. and Wiseman, N., (1997) Controlled precipitation of amorphous silica from geothermal fluids or other aqueous media containing silicic acid. *US Pat.,* No. 5,595,717.
12. Johnston, J.H., McFarlane, A.J., Borrmann, T. and Moraes, J., (2004) Nano-structured silicas and silicates-new material and their applications in paper. *Current Applied Physics,* 4, 411-414.
13. Johnston, J.H., McFarlane, A.J. and Borrmann, T., (2008) Nano-structured silicate, functionalised forms thereof, preparation and uses. The Patent Description & Claims data from USPTO Patent Application 20080305027, Class: 4233309 (USPTO).
14. Taspinar, O., Ozturk, I., Nalbantoglu, T., (2009) Nano-size calcium silicate production. *Proceedings of Nanomats,* International Conference on Nanomaterials and Nanosystems, İTÜ Ayazağa Kampüsü, İstanbul, 10-13 August, 2009.
15. Nalbantoglu, T., Ozturk, I., (2009) Nano Structured Calcium Silicate Production from Rice Hull Ash. *Graduation Project* (in Turkish), Istanbul Technical University, Chemical Engineering Department, Maslak/Istanbul, Turkey.

(208c) Investigation of Voids In Nanostructured RDX-Based Compositions Using Ultra-Small-Angle X-Ray Scattering

Victor Stepanov, Munitions Engineering and Technology Center, US Army, RDECOM-ARDEC, Picatinny, NJ

Investigationof Voids in Nanostructured RDX-based Compositions using Ultra-Small-Angle X-rayScattering

Victor Stepanov[1],Trevor M. Willey[2] and Jan Ilavsky[3]

[1]US Army, RDECOM-ARDEC, MunitionsEngineering Technology Center

Picatinny Arsenal, NJ 07806, USA

[2]LawrenceLivermore National Laboratory, 7000 East Avenue, Livermore, CA 94550, USA

[3]AdvancedPhoton Source, Argonne National Laboratory, Argonne, IL 60439, USA

Introduction

The structure of voids within novelcyclotrimethylene trinitramine (RDX)-based nanocomposites was investigated. Voidsare an important structural element due to their strong influence on propertiessuch as sensitivity and performance. In order to probe the "sealed" voids,synchrotron-based Ultra-Small-Angle X-ray Scattering (USAXS) with Bonse-Hart configurationwas employed. This powerful technique enabled probing voids with sizes rangingfrom *ca.* 1 nm to 3 mm with statistically meaningfulsample sizes. Scattering data on three compositions with varying RDX crystalsize, including 200 nm, 500 nm, and 2 mm, was measured.Modeling of the scattering intensity profiles revealed bimodal, log-normal voidsize distributions for all samples. The mean void size was found to increasewith the RDX crystal size and was found to be of similar dimensions to RDXcrystals. Inthis work, Ultra-Small-Angle X-ray Scattering (USAXS) was employed tocharacterize the voids within cyclotrimethylene trinitramine (RDX)-based nanocomposites.Nanostructured explosive compositions are gaining interest in regard tolowering the initiation sensitivity and improving performance. Recent work withnano-

RDX-based compositions revealed that the shock and impact sensitivitieswere significantly lower compared to compositions with micron- scale HEcrystals [Stepanov, Qiu]. Moreover, an important attribute of such nanocompositeswas low shock sensitivity even at high porosities, ca. 10 %. It is expectedthat such behavior is a consequence of a very small mean voids size and theabsence of large voids. Characterization of voids in these compositions iscritical in explaining the observed behavior. Compositionswith 88 wt. % RDX and 12 wt. % binder were prepared by slurry coating of 200nm, 500 nm, and fluid energy milled (FEM) RDX specimens. Formulated RDXspecimens were pressed into nominally 5.08 mm x 0.8 mm pellets for USAXSanalysis. The materials were pressed at 110 MPa and room temperature. **Results** Experimentaldata reduction and analysis were performed using the Indra[website] and Irena[Ilavsky2009] codes respectively, which are subroutines created within IGOR Prosoftware specifically for treating scattering data. Representative desmearedscattering intensity profiles for the three compositions are shown in Figure 1.

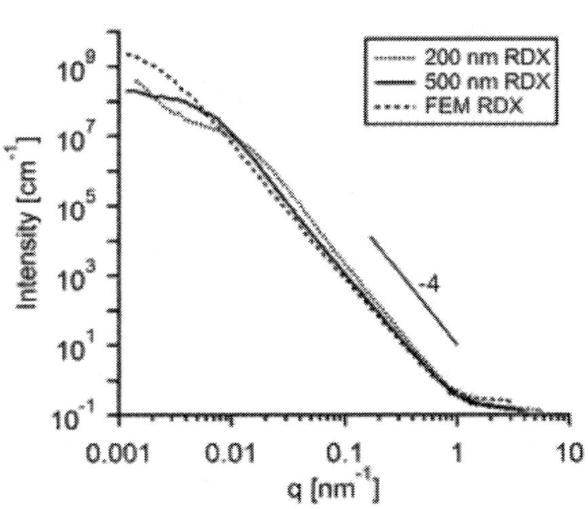

Figure 1. USAXS intensityas a function of

q

"for 200 nm, 500 nm, and FEM RDX-basedcompositions. A reference line with a slope of -4 is included." TheMaximum Entropy algorithm within the Irena package was used to evaluate thevoid size distributions. This inversion method converges to a unique size distributionfor a given form factor by maximizing the configurational entropy of thecalculated size. Among the required input parameters were shape and aspectratio, range of diameters, and the scattering length density contrast. Thecalculated void size distributions for the three specimens are shown in Figure 2.All materials exhibit bimodal size distributions with log-normal peaks.

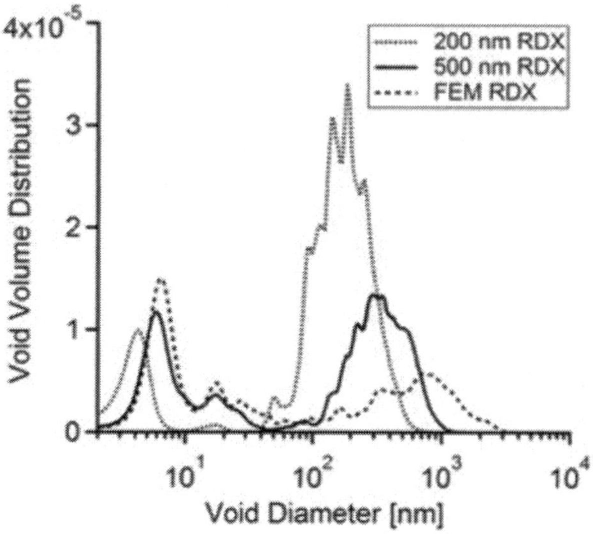

Figure 2. Void size distributions of 200 nm, 500 nm, and FEM RDX-based compositionscalculated from USAXS data. Based on thecalculated size distributions it can be concluded that the mean void size isproportional to the crystal size.

See more of this Session: Characterization of Engineered Particles and Nanostructured Particulate Systems

See more of this Group/Topical: Particle Technology Forum

(208d) Structure Determination of MFI and MWW Zeolite Nanosheets

Xueyi Zhang[1], Kumar Varoon[2], K. Andre Mkhoyan[3] and Michael Tsapatsis[2], (1) Chemical Engineering & Materials Science, University of Minnesota, Minneapolis, MN, (2)Chemical Engineering and Materials Science, University of Minnesota, Minneapolis, MN, (3)Materials Science and Engineering, University of Minnesota, Minneapolis, MN

MFI and MWW zeolite nanosheets are building units for state-of-the-art zeolite thin films for gas separation. In this study, the structures of exfoliated MFI and MWW zeolite nanosheets were determined using a combination of experimental and simulation methods. Based on characterization results from atomic force microscopy and transmission electron microscopy, the structures and thicknesses of the exfoliated zeolite nanosheets were proposed. After optimization with Car-Parrinello molecular dynamics, X-ray diffraction patterns and electron diffraction patterns were simulated from these structures. The agreement between experimental and simulated characterization data suggested that the proposed structures represented the actual structures of the exfoliated zeolite nanosheets. The methods used in this study can be employed to determining structures of other zeolite nanostructures.

See more of this Session: Characterization of Engineered Particles and Nanostructured Particulate Systems
See more of this Group/Topical: Particle Technology Forum

(208e) Characterization of the Durability of Super-Hydrophobic Surfaces Produced by Nanoparticle Vapor Deposition

Md. A Mahmud[1], Jeff Chinn[2], Christopher B. Roberts[3], William R. Ashurst[4] and Wanda Jones[2], (1)Chemical Engineering, Auburn University, Auburn University, AL, (2)Integrated Surface Technologies (IST), Menlo Park, CA, (3)Department of Chemical Engineering, Auburn University, Auburn, AL, (4)Chemical Engineering, Auburn University, Auburn, AL

Characterization of durability of super-hydrophobic surfaces produced by nanoparticles vapor deposition.

One of the main reasons for early failure of personal electronic gadgets is stiction of water droplets to circuit boards within the equipment. To mitigate this problem, a number of super-hydrophobic films were deposited on silicon wafers produced by different vapor deposition techniques and were tested for their durability. These coated wafers were subjected to a gravity driven water jet as a means of examining the durability of the nanoparticles coating, and the contact angle of the eroded films were measured by contact angle goniometry. Static contact angles above 160^0 are classified as super-hydrophobic. As such the time to failure for the films from the water erosion tests was determined as the time at which the static contact angle fell below 160^0. The properties which make the surfaces super-hydrophobic and the degree of durability of the coatings were characterized by surface imaging techniques such as Atomic Force Microscopy (AFM), Scanning Electron Microscopy (SEM). Based on these findings, the effects of surface roughness and surface coverage on the durability of the super-hydrophobic films will be discussed.

See more of this Session: Characterization of Engineered Particles and Nanostructured Particulate Systems
See more of this Group/Topical: Particle Technology Forum

(208f) Raman Spectroscopy On Size Segregated Nanoparticles From a Flame Generator

Mikko Aromaa[1], Heikki Suhonen[1], Turkka Salminen[2], Tapio Niemi[2], Juha-Pekka Nikkanen[3], Erkki Levnen[3] and Jyrki M. Mkel[4], (1)Aerosol Physics Laboratory, Tampere University of Technology, Tampere, Finland, (2) Optoelectronics Research Centre, Tampere University of Technology, Tampere, Finland, (3)Department of Materials Science, Tampere University of Technology, Tampere, Finland, (4)Aerosol Physics laboratory, Tampere University of Technology, Tampere, Finland

Flame methods are widely used among nanoparticle synthesis via aerosol routes. Many flame methods include spraying of liquid precursor followed by evaporation and nucleation into product species. Liquid Flame Spray (LFS) works in this way. The precursor solution is sprayed into a turbulent, high temperature, high velocity H_2-O_2 flame. Final particle size is adjusted with the precursor concentration and feed rate. A characteristic of these spray methods is that the sprayed droplets are not completely evaporated and so called residual particles are formed. Mainly the synthesised nanoparticles form via gas-to-particle conversion. However, the slightly larger residual particles are formed via liquid-to-solid conversion. This might lead to different crystalline forms. Conventional way to study the crystalline forms is X-Ray diffraction from powder samples. However, it is only an average of the whole sample. In this study, we combine aerosol sampling in impactor with Raman spectroscopy in a novel way. We synthesise nanoparticles with the LFS and collect the particles in a cascade impactor. Different size fractions are segregated in the impactor. Afterwards, the impactor plate samples are studied using Raman spectroscopy to distinguish the crystalline form of different parts in the aerosol distribution. Iron oxide and titanium dioxide particles are studied separately. Iron oxide is found to be mainly magnetite but the residual particles are hematite. Titanium dioxide is anatase as produced and there can be found residual particles of rutile.

See more of this Session: Characterization of Engineered Particles and Nanostructured Particulate Systems
See more of this Group/Topical: Particle Technology Forum

(208g) Particle Shape and Roughness Characterization by Focused Ion Beam Tomography for Adhesion Simulation

Dave K. Balachandran and Stephen P. Beaudoin, Chemical Engineering, Purdue University, West Lafayette, IN

Particle cohesion and adhesion interactions have far-ranging impacts within the pharmaceutical industry with regard to both the manufacturing of particulate solids and the delivery of active pharmaceutical ingredients. The ability to measure and model these interactions can aid material selection, improve the design of process equipment, and increase the accuracy of various process simulations and models.

The accuracy of particle adhesion models and simulations is highly dependent on particle geometry and roughness which are often either oversimplified estimates or hybrids of estimates and measured data. A novel method for the simultaneous characterization of particle geometry and roughness was developed in which the particle of interest is serially sectioned using a focused ion beam (FIB) and the cross sections are imaged with a field emission electron microscope (FESEM). Subsequently, the cross-sectional data are used to synthesize a 3-dimensional particle surface mesh which contains micron- and nanometer-scale information.

Colloidal probe microscopy (CPM) using an atomic force microscope (AFM) was used to examine particle-substrate adhesion interactions between individual 10 μm diameter silicon dioxide particles and a polished stainless steel substrate. Individual particles were mounted as probes on tipless AFM cantilevers. The force of adhesion was measured in various locations across the polished stainless steel substrate in an electrostatically neutral and humidity controlled environment. Finally, the particles were serially sectioned and imaged in microscope housing both a FIB column and FESEM column. The cross sectional data were used to synthesize 3D mesh surface models for each particle, and a simulation assisted particle adhesion model was used to estimate the value of the Hamaker constant.

See more of this Session: Characterization of Engineered Particles and Nanostructured Particulate Systems
See more of this Group/Topical: Particle Technology Forum

(223a) Experimental and Numerical Investigation of Flame-Spray-Assisted TiO2 Synthesis From Titanium Tetraisopropoxide

Sebastian Hardt[1], H. Wiggers[2], I. Wlokas[1] and C. Schulz[2], (1)University of Duisburg-Essen, Duisburg, Germany, (2)CeNIDE, Center for NanoIntegration Duisburg-Essen, University of Duisburg-Essen, Duisburg, Germany

sebastian.hardt@uni-due.de

Titania nanoparticles (TiO_2) are of major interest for catalytical and electronic applications, especially as a major material for photocatalysis and dye-sensitized solar cells. It is accessible via processing of either $TiCl_4$ or titanium tetraisopropoxide (TTIP). Up to now, industrial processes producing TiO_2 are based on the combustion of $TiCl_4$ leading to particles in a size regime between a few ten and a few hundred nanometers. With respect to high activity based on a high surface area, more dilute reaction systems must be applied for the formation of very small titania nanoparticles. A promising route also enabling for large-scale synthesis is flame-assisted spray pyrolysis of a flammable organic solvent containing the titania precursor.

In our approach, a mixture of isopropanol and TTIP is used as the precursor material. An oxygen jet around a hollow injector needle atomizes the liquid and forms a spray. Isopropanol as a solvent allows for the formation of dilute mixtures within a wide range and also suppresses the premature decomposition of TTIP inside the supply system. To enable operation points between 200 and 2000 mbar the burner is assembled in a pressure-controlled stainless steel housing.

Our work focuses on the investigation of the synthesis route within a wide pressure regime, the characterization of the synthesized nanoparticles and the numerical simulation of the spray flame reactor. Therefore, ex-situ particle-diagnostic techniques such as nitrogen adsorption (BET), transmission electron microscopy (TEM), x-ray diffraction (XRD) and UV-VIS spectroscopy are applied. Spray formation and droplet evaporation is modeled in an Euler-Lagrange formulation. The mechanisms for TTIP combustion and TiO_2 formation described by Tsantilis et al. are used to calculate the particulate species [1]. The highly turbulent spray flame is simulated in 3D using simplified reaction kinetics. The population balance equations, that describe the particle dynamics, are solved using a monodisperse approach. Our experimental setup enables the formation of crystalline titania nanoparticles at sizes of 10 nm and below depending on pressure and precursor concentration. It is found that the materials exhibit a narrow particle-size distribution and that they mostly consist of the anatase phase. The simulation of the spray flame reactor shows the temperature distribution, the particle number concentration as well as particle diameters.

[1] S.Tsantilis, H.K.. Kammler, S.E.Pratsinis, Population Balance modeling of flame synthesis of titania nanoparticles, Chem. Eng. Sci. 57 (2002)

See more of this Session: Gas Phase Synthesis of Nanoparticles
See more of this Group/Topical: Particle Technology Forum

(223b) Model-Assisted Scale-up of Nanoparticle Synthesis by Flame Spray Pyrolysis

Arto J. Grhn, Sotiris E. Pratsinis and **Karsten Wegner**, Particle Technology Laboratory, Swiss Federal Institute of Technology (ETH Zurich), Zurich, Switzerland

Flame Spray Pyrolysis (FSP) aerosol reactors have become an increasingly promising method for nanoparticle production and a broad range of new nanoparticles has been developed in the past decade (Teoh et al., 2010). Most of this development took place at the lab-scale and now requires transfer to an industrial manufacturing environment. In order to facilitate scale-up of the FSP technology to a cost-efficient production process a better understanding of the fluid and particle dynamics underlying particle growth in spray flames is still required.

Here, the flame spray process is investigated computationally and experimentally by the example of ZrO_2 nanoparticle synthesis. For a lab-scale FSP reactor (Madler et al., 2002), spray, fluid, combustion and aerosol dynamics are solved to predict product nanoparticle characteristics by using a commercial CFD solver without adjustable parameters or need of experimental input data. Fourier-transform infrared spectroscopy, Phase-Doppler Anemomentry and nanoparticle sampling in-situ the flame are employed for characterizing the temperature, velocity and species profiles of the flame reactor and are used for validation of the simple and robust computational code. Based on the model, design correlations for process scale-up are developed and it is shown how the model can assist in process optimization and reactor development.

References:

Mädler, L., Kammler, H.K., Mueller, R. and Pratsinis, S.E. (2002), *J. Aerosol Sci.* **33**, 369-389.

Teoh, W.Y., Amal, R. and Mädler, L. (2010), *Nanoscale* **2**, 1324-1347.

See more of this Session: Gas Phase Synthesis of Nanoparticles

(223c) Single-Step Aerosol Synthesis and Deposition of Au Nanoparticles with Controlled Size and Separation Distributions

Elijah Thimsen, Material Science Division, Argonne National Laboratory, Argonne, IL

Immobilized noble metal nanoparticles are being explored for a variety of applications where control over the particle size and separation distance on the substrate is important for performance. A proof-of-concept is presented that Au nanoparticles can be deposited in a single step with control over the size and separation distributions using an aerosol process. Samples were deposited with mean particle diameters in the range from 15 to 43 nm, and mean separation distances from 11 to 39 nm. Depending on the separation distance, particles exhibited localized surface plasmon resonance (LSPR) dominated by either intra- or inter-particle resonances, as determined by *uv*-visible extinction spectroscopy. Ultrathin TiO_2 shells of different thicknesses, in the range from 0 to 24 nm, were deposited on the Au nanoparticles by atomic layer deposition (ALD) to determine the sensing distance into the surrounding dielectric medium, which was estimated to be 10 nm.

See more of this Session: Gas Phase Synthesis of Nanoparticles
See more of this Group/Topical: Particle Technology Forum

(223d) Development of a New Aerosol Reactor for the Synthesis of Metal Nanoparticles

William J. Scharmach[1], Vasilis Papavassiliou[2], Perry Pacouloute[2], Raymond D. Buchner[1] and Mark T. Swihart[1], (1)Chemical and Biological Engineering, University at Buffalo (SUNY), Buffalo, NY, (2)Praxair Technology Center, Tonawanda, NY

We present a new flame-based aerosol reactor configuration that combines thermal decomposition and hydrogen reduction to produce metal nanoparticles. This approach uses a fuel-rich hydrogen flame as a source of low-cost energy to initiate particle synthesis, but separates the flame chemistry from the particle formation chemistry. Hot combustion products pass through a nozzle to produce a high-temperature reducing jet. A liquid precursor solution is rapidly atomized, evaporated, and decomposed by the expanding jet, initiating particle formation. Particles are then quenched and collected downstream via filter. We have produced several different metal nanoparticles including pure metals such as copper, silver, and gadolinium; alloys and multi-crystal particles such as copper/silver hybrids. By modifying the liquid precursor injected into the reactor we can also produce controlled coatings on the particles. An example of such a coating includes carbon on copper or silver. Nanoparticles are characterized by aerosol mobility distribution measurements, electron microscopy, and x-ray diffraction. Silver and copper serves here as a prototype for non-oxide material that is generally difficult to produce in flame-based reactors. This work demonstrates that such materials can be produced in substantial quantities with particle diameters below 20 nm using this new reactor technology.

See more of this Session: Gas Phase Synthesis of Nanoparticles
See more of this Group/Topical: Particle Technology Forum

(223e) Hybrid DNS-LES of Nanoparticle Nucleation: The Effects of Small-Scale Fluctuations On Metal Nucleation

Sean C. Garrick and Andrew Fager, Mechanical Engineering, University of Minnesota, Minneapolis, MN

Hybrid DNS-LES of Nanoparticle Nucleation: The effects of small-scale fluctuations on metal nucleation

Vapor-phase particle formation occurs in a wide variety of processes of engineering and scientific interest. In a number of these processes, gas-to-particle conversion via homogeneous nucleation is a dominant mechanism. Additionally, a number of these processes typically occur under turbulent flow conditions. In industrial processes, rapid mixing is often used to increase the saturation which drives the nucleation process. It is well known that homogeneous nucleation has a highly non-linear dependence on both temperature and concentration. As a result, turbulent fluctuations in temperature or concentration can result in orders of magnitude changes in the local nucleation rate.

Methods used for simulation of fluids containing nanoparticles fall under three general categories: Direct numerical simulation (DNS), Reynolds-averaged Navier-Stokes simulations (RANS) and large-eddy simulation (LES). DNS solves explicitly for all scales of motion but is computationally prohibitive for all but the simplest flows. RANS simulations are the most computationally affordable and as a result have been used for analysis of many engineering flows. LES has the advantage of DNS in that it captures the unsteady evolution of large-scale flow features while using models similar to those found in RANS to simulate the small, more uniform, scales of motion. Both RANS and LES requires models for all non-linear interactions. These "turbulence models" have been widely developed and utilized for turbulent reacting flows. However, the effects of the small-scale, unresolved, fluctuations on homogeneous nucleation aren't entirely clear.

In this work we perform simulations of zinc nanoparticle nucleation in turbulent jets. The flows consist of a high-speed jet of zinc metal vapor issuing into a cooler stream. As the two streams mix, the metal vapor becomes super-saturated and nanoparticles form (and grow via condensation and coagulation). We utilize a Navier-Stokes + nodal general dynamic equation solver to perform a variety of

DNS and hybrid DNS-LES studies to isolate and characterize the effects of mixing on homogeneous nucleation. In the DNS all length and time-scales are resolved. This serves as a benchmark simulation. In the hybrid DNS-LES approach, a combination of fully-resolved variables, and large-scale variables are used to obtain the particle field. More precisely, when computing the nucleation rate, and critical diameter, the temperature, pressure, and vapor concentrations are filtered to remove the small-scale features. This approach allows us identify the effects of the small, unresolved scales on nucleation, and illustrates the effects of insufficient resolution in the modeling of particle formation processes.

See more of this Session: Gas Phase Synthesis of Nanoparticles
See more of this Group/Topical: Particle Technology Forum

(223f) Design of Aerosol Coating Reactors by CFD

Beat Buesser and Sotiris E. Pratsinis, Department of Mechanical and Process Engineering, ETH Zurich, Zurich, Switzerland

Core-shell particles facilitate incorporation of functional particles intohost (e.g. liquid or polymer) matrices like silica-coated TiO_2pigments[1] , carbon coated Cu for sensors[2] or superparamagnetic[3] Fe_2O_3. Typicallycore-shell particles are made in the liquid phase[4] but there is keen interest todevelop gas-phase or aerosol coating processes that do not generate liquidby-products, offer fewer process steps, easier particle collection and hermetic[3] shells. Coating of particles in thegas phase, however, is challenging, as particle motion and growth are muchfaster than in liquids. As a result, it is difficult to control and develop ascalable gas phase coating process. So, even commercially produced particlesmade by aerosol routes (e.g. pigmentary TiO_2 made by the °Chlorideprocess) are coated by wet processes[4] .

Here[5] , gas-phase (aerosol) coating iselucidated in considerable detail, for the first time to our knowledge, bycomputational fluid and particle dynamics for core particles (TiO_2)and coating shells (SiO_2). Emphasis is placed on understanding theinfluence of process variables (coating weight fraction and mixing intensity(Figure 1) and geometry of core aerosol & shell precursor vapor) oncore-shell product characteristics by a trimodal aerosol particle dynamicsmodel[6] accounting for SiO_2 monomergeneration, coagulation and sintering. The predicted extent of complete (orhermetic) coating shells is compared to the measured photocatalytic oxidationof isopropanol by such particles[7,8] and release of acetone. As hermetic SiO_2shells prevent the photocatalytic activity of TiO_2, the performanceof coated particles is explained by the spatial distribution of shell thicknesson core particles with detailed reactor flow field analysis.

Financial support from the Swiss National Science Foundation (SNF) grant #200021-119946/1 and European Research Council is gratefully acknowledged.

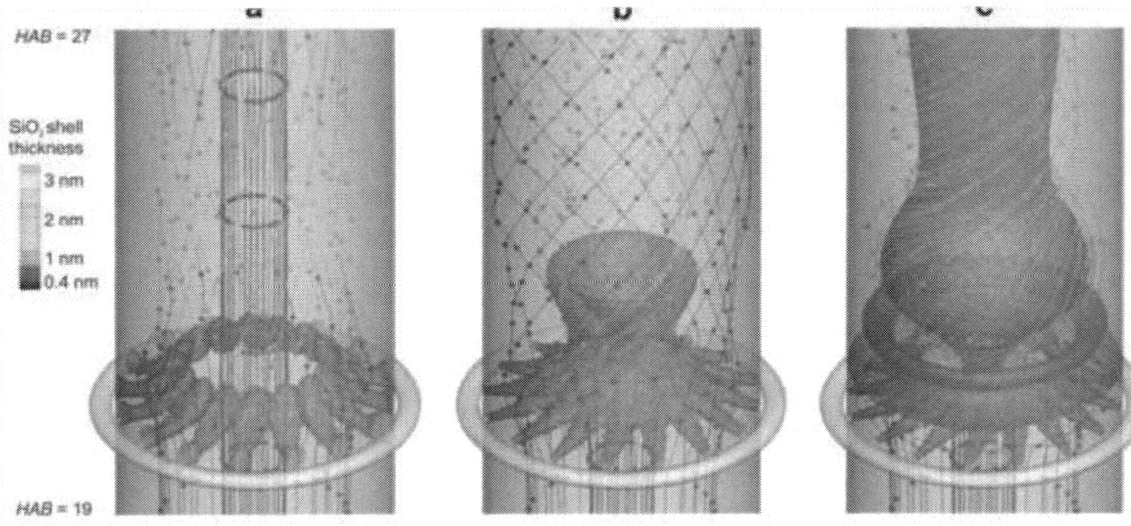

Figure 1 Influence of nitrogenflow rate (mixing intensity) a) 5.8 l/min, b) 15.8 l/min and c) 30.8 l/min onthe coating precursor and coating shell thickness distribution inside theaerosol coating reactor.

1. Subramanian NS, Diemer RB, Gai PL; E. I. du Pont de Nemours and Company(Wilmington, DE, US); Process for making durable rutile titanium dioxidepigment by vapor phase deposition of surface treatment. US patent200627303(A1). 2006.

2. Athanassiou EK, Grass RN, Stark WJ. Large-scale production of carbon-coatedcopper nanoparticles for sensor applications. *Nanotechnology*. 2006; 17,(6), 1668-1673.

3. Teleki A, Suter M, Kidambi PR, Ergeneman O, Krumeich F, Nelson BJ, PratsinisSE. Hermetically coated superparamagnetic Fe_2O_3 particleswith SiO_2 nanofilms. *Chem. Mater.* 2009; 21, (10), 2094-2100.

4. Egerton TA. The modification of fine powders by inorganic coatings. *KONA*. 1998;16, 46-59.

5. Buesser B, Pratsinis SE. Design of gas-phase synthesis of core-shell particlesby computational fluid aerosol cynamics. *AIChE J.* 2011; DOI: 10.1002/aic.12512,

6. Buesser B, Pratsinis SE. Design of Aerosol Particle Coating: Thickness, Textureand Efficiency. *Chem. Eng. Sci.* 2010; in Press, doi:10.1016/j.ces.2010.07.011,

7. Teleki A, Heine MC, Krumeich F, Akhtar MK, Pratsinis SE. *In-situ* coatingof flame-made TiO_2 particles by nanothin SiO_2 films. *Langmuir*.2008; 24, (21), 12553-12558.

8. Teleki A, Buesser B, Heine MC, Krumeich F, Akhtar MK, Pratsinis SE. Role ofgas-aerosol mixing during in situ coating of flame-made titania particles. *Ind. Eng. Chem. Res.* 2009; 48, (1), 85-92.

See more of this Session: Gas Phase Synthesis of Nanoparticles
See more of this Group/Topical: Particle Technology Forum

(230a) Living Polymerization/Reaction Control: Innovative Strategies for Enhancing Drug Delivery From Imprinted Thin Films

Vishal D. Salian and Mark E. Byrne, Biomimetic & Biohybrid Materials, Biomedical Devices, & Drug Delivery Laboratories, Department of Chemical Engineering, Auburn University, Auburn, AL

Molecular imprinting allows for greater control of drug transport from hydrogels by the creation macromolecular memory for the drug/template molecule. Using living polymerization to create imprinted polymer networks is a more recent technique to enhance the effect of molecular imprinting. Here we will demonstrate that living polymerization leads to significantly enhanced binding parameters and delayed transport properties for drug molecules in weakly crosslinked aqueous polymers. Poly(HEMA-DEAEM-PEG200DMA) polymer gels were used as the drug delivery system with diclofenac, a non-steroidal anti-inflammatory drug, as the template molecule. Multiple gel compositions were created using both living and conventional free radical polymerization by varying the drug concentration present during polymerization, concentration of crosslinking monomer and solvent content in the pre-polymerization mixture. Template binding affinity and capacity were calculated using the Langmuir isotherm while the analysis of the polymer network was carried out using digital scanning calorimetry and gel permeation chromatography.

We demonstrate that using living polymerization results in a 38% increase in binding affinity and a 67% increase in capacity over the conventionally created hydrogel. By controlling the conversion further increases in capacity or affinity can be achieved adding greater tailorability to the hydrogel properties. Lastly, the drug release profile was extended two fold over conventionally created imprinted hydrogels and four fold over the non imprinted hydrogels. The observed enhancement in binding and transport properties may be explained by an extension of the chemically controlled chain propagation mechanism by the living polymerization technique. Thus, the results confirm that improved structural homogeneity of template binding sites and a global energy minimum of the spatial arrangement of polymer chains lead to enhanced binding parameters and delayed template release. This work will lead to a new generation of tailorable drug delivery biomaterials which can be designed from the molecular scale to yield exquisite control over controlled release of therapeutic drug molecules.

See more of this Session: Innovations In Drug Delivery Technology I
See more of this Group/Topical: Food, Pharmaceutical & Bioengineering Division

(230b) Stripfilms As Oral Dosage Platform for the Delivery of Nano-Particles of Active Pharmaceutical Ingredients (APIs)

Rajesh N. Dave, Chemical Engineering, New Jersey Institute of Technology, Newark, NJ, Lucas Sievens-Figueroa, New Jersey Center for Engineered Particulates, New Jersey Institute of Technology, Newark, NJ and Ecevit Bilgili, Chemical, Biological and Pharmaceutical Engineering, New Jersey Institute of Technology, Newark, NJ

We describe formation of stripfilms loaded with nano and low micron sized API particles to achieve desired delivery properties through maintaining of the high surface area of particles via particle engineering. As a platform to incorporate particle engineering, this approach will allow for a fundamental shift in the design of pharmaceutical products containing nano-particles without the need for first drying or recovering nano-particles in dry form. Our approach consists of forming stable nano-suspensions of BCS Class II drugs via three top-down and bottom-up approaches to achieve size range of 100–2000 nm, with API loading of 0.1 to 20 wt% in the suspensions. The API suspensions are stabilized via judicious use of polymer-surfactant combinations to achieve size stability ranging from 2 hours to 2 days. These suspensions are further processed to increase their polymer content by mixing with higher polymer containing aqueous gels or solutions. These film-precursors containing nano-APIs are formed into films, whose thickness is in the range of 50 to 200 microns, via the solvent-casting approach utilizing convective drying.

We have analyzed the retention of large API surface area in these films via re-dispersion in water, showing that this integrated particle engineering approach allows for maintaining high surface area of the API particles, which has been a non-trivial problem faced by the pharmaceutical industry. Films characterization confirms good API uniformity with API content linearly dependent on film thickness. Dissolution of films also proves to be much faster than other dosage forms mainly due to the high surface area associated with thin films in combination with fast, full-recovery of small API particles during dissolution. This approach may be suitable for BCS Class II drugs requiring enhancements in bioavailability for producing clinical supplies. Stripfilms can also serve as a quick-to-market platform for flexible, uniform dose manufacturing of BCS Class II drugs.

See more of this Session: Innovations In Drug Delivery Technology I
See more of this Group/Topical: Food, Pharmaceutical & Bioengineering Division

PARTICLE SCALE STUDY OF THE DISPERSION OF DRY POWDER IN PHARMACEUTICAL AEROSOL INHALERS

Z.B. Tong, University of NSW, Sydney, AU
R.Y. Yang, University of NSW, Sydney, AU
A.B. Yu, University of NSW, Sydney, AU
S. Adi, University of Sydney, Sydney, AU
H.-K. Chan, University of Sydney, Sydney, AU

Introduction

Inhalation aerosols offer the distinct advantage of rapid and convenient delivery of bioactive substances to the site of action. Significant research effort has been directed towards the development of innovative technologies for the generation and delivery of dry powder aerosols for inhalation. Dry powder aerosols require no propellant (detrimental to the environment), have superior chemical stability and bacterial growth resistance compared with solution, and are easy to use (hence good patient compliance).

Existing DPIs have low efficiency and high variability of dosing, and generally less than 30 % of the dose loaded to the device can deposit in the lungs [1]. This is due to the strong cohesion between pharmaceutical particles which are usually less than 5 μm and hence make them difficult for dispersion. In drug delivery, the agglomerates have to be dispersed to form a fine respirable aerosol cloud in order to minimise deposition in the mouth and upper airways, and to maximise delivery to the more distal parts of the respiratory tract. To improve the aerosol performance, much research effort in the past two decades has been directed at the powder formulation [2]. However, powder dispersion also depends strongly on the prevailing flow conditions in an inhaler device and their effects on powder deagglomeration remain unclear because it is difficult to obtain experimental details and quantitative information about the dispersion process due to the small agglomerate size (~1 mm) and the short duration (5s <) of dispersion.

We have developed a coupled computation fluid dynamics (CFD) and discrete element method (DEM) model to simulate the dispersion process in DPIs. The model incorporated powder properties obtained from experiments and was calibrated carefully. Here we briefly review these studies, aiming to provide a base for future development in this area.

Model development

CFD-DEM model

The model treats the dispersion process as a multiphase flow where the discrete powders and the continuous air flow are modelled by DEM and CFD, respectively. In DEM model, the translational and rotational motions of a particle in a system are governed by its interactions with neighbouring particles, walls and surrounding fluid described by Newton's second law of motion. The governing equations for the continuous fluid in CFD model are the conservations of mass and momentum in terms of the local mean variables over a computational cell. The CFD-DEM coupling is achieved by combining an in-house built DEM with the commercial software ANSYS Fluent®. At each time step, DEM calculates particle-related information based on fluid flow field, such as the positions and velocities of individual particles, to determine porosity and volumetric particle-fluid interactions in the individual computational cells. CFD then uses these data to determine the air flow field which then yields the particle-fluid forces acting on individual particles. Incorporation of the resulting forces into DEM produces information about the motion of individual particles for the next time step. Such procedure continues until the whole process is finished [3-5].

Model validation

The numerical model was validated by comparing the simulated results, such as the trajectory of the agglomerates and the amount of fine particles less than 5μm (FPF) generated, with those from the experiments using the setup as shown in Fig. 1. In the setup, air of different flow rates entered the impaction throat from the right side air inlet and an agglomerate was introduced via the feed apparatus (as indicated by "B" in Fig. 1). The agglomerate moved along with the air flow and impacted with throats containing two angles (Figs. 1a and 1b) or a single angle (Figs. 1c, 1d and 1e).

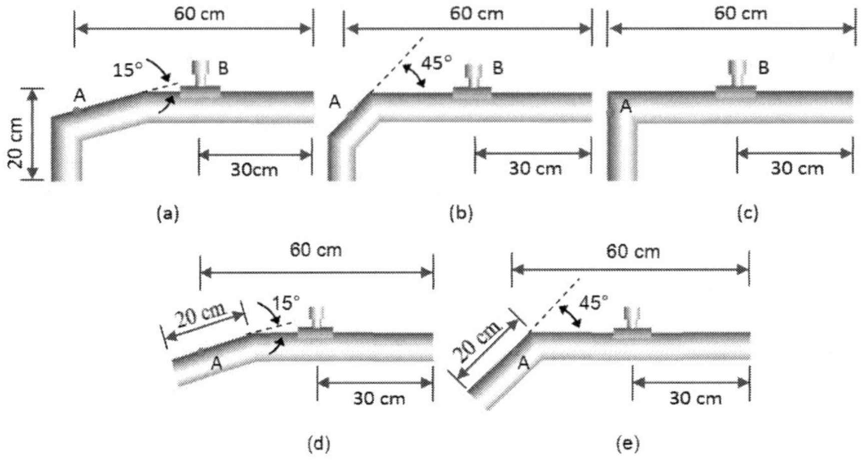

Fig. 1. Schematics of impaction throats A-E.

Fig. 2 shows the simulated and experimentally recorded trajectories of the agglomerate in the 90° impaction throat at the flow rate of 120 l min^{-1}. The trajectories are similar, both indicate that mannitol agglomerate initially drops to the bottom of the pipe after being fed into the impaction throat, then is accelerated and lifted by the airflow and finally impacts on the throat.

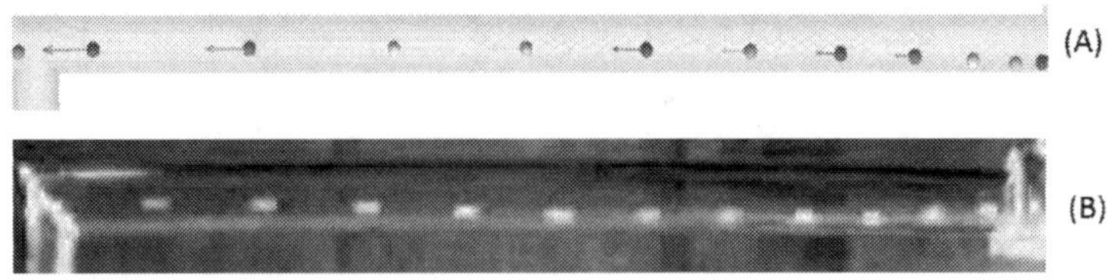

Fig. 2. Comparison of the trajectory of the agglomerate from simulation (A) and experiment (B) in the one angle impaction (90° and 120 l min^{-1}).

Fig. 3 shows the simulated and measured FPF and powder depositions under different impaction conditions. Again the simulations are quite comparable with experimental results, further confirming the validity of the model. For the two-angle throats (A and B) and the single angle throat C with 90-degree impaction angle, FPF increases with impact angles at low flow rates (60 and 120 l min^{-1}) but decreases at the high flow rates of 150 l min^{-1}. For the single angle throats D and E, FPF increases monotonically with flow rate. As shown later on, the decrease in FPF at high flow rate for throats A, B and C is caused by the increased powder deposition. In general, the throats with two angles (A and B) result in better dispersion than those with a single angle (C, D and E), showing that the multiple impactions can improve powder de-agglomeration.

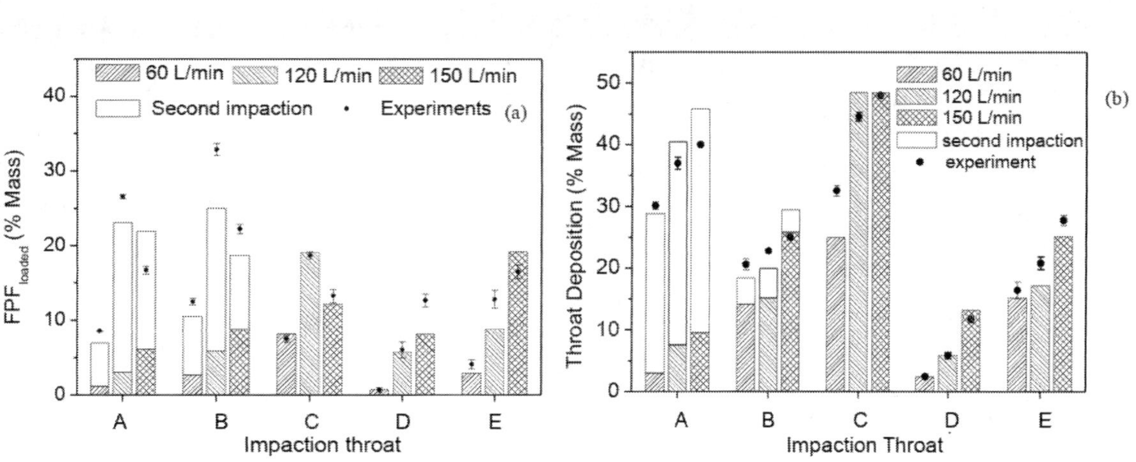

Fig. 3. (a) FPF generated at different impaction conditions and flow rates; (b) Mass fraction of powder deposition at different flow rates. The experimental data [6] are also plotted for comparison.

Dispersion mechanisms

Particle-wall interaction

The powder dispersion due to the mechanical impaction was simulated by impacting an agglomerate with a target wall with different velocities and angles [7]. The agglomerate was formed with 5 μm mannitol powders and had final diameter of about 100 μm with a packing fraction of 0.552 and a relatively weak strength. When impacted with the wall, the agglomerate experiences a large plastic deformation on impact before disintegrating into many small pieces (Fig. 4a). By analyzing the structural and force evolutions (Fig. 4b) of the agglomerates during the impacts and the final breakage patterns, the work showed that the impact can be divided into three periods in which the microscopic structures of the agglomerates (i.e. porosity, velocity profile and force) evolve differently.

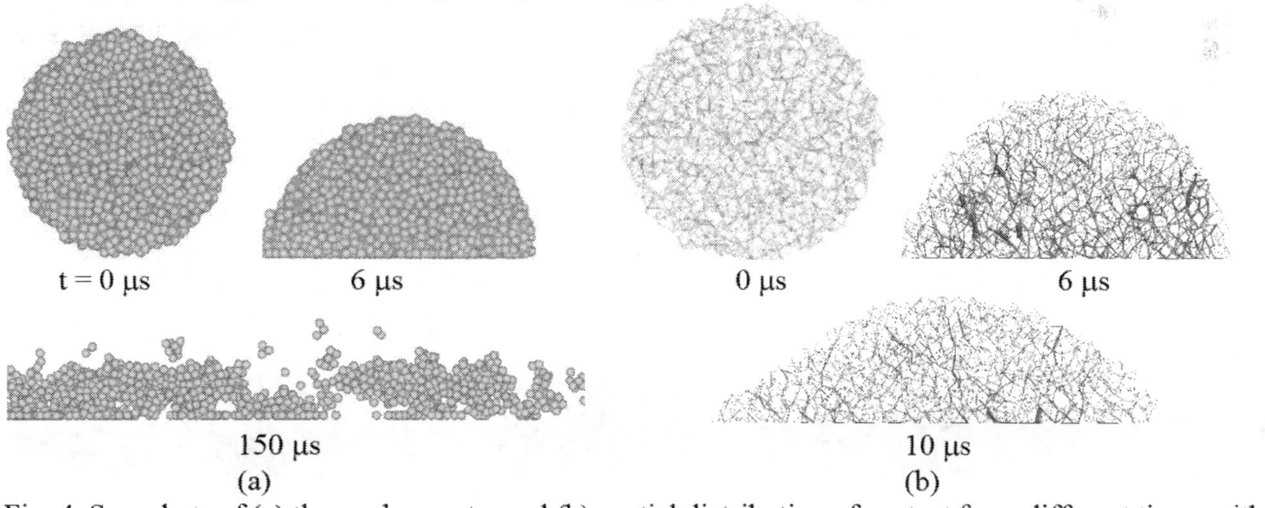

Fig. 4. Snapshots of (a) the agglomerate; and (b) spatial distribution of contact force different times with $\theta = 45^{\circ}$ and v = 10 m/s (only part of the fragments are shown at t = 150 μs).

Unlike strong agglomerates, a shear band can be observed in the velocity fields of the agglomerates. Larger forces are developed within the agglomerate near the contact region and propagate upwards. The size distribution of fragments shows that while increasing impact velocity generally has better breakage performance, impact angle at 45-degree can give the maximum breakage performance for a given impact velocity. The particle-wall impact energy can be used to interpret the breakage pattern of agglomerates. For breakage of loose agglomerates, the impact energies in normal and tangential directions are comparable and both should be considered in order to describe breakage performance properly. As shown in Fig. 5, the breakage of loose agglomerate is closely related to the total impact energy from the wall, and the breakage efficiency is governed by the combined effect of normal and tangential interactions between the particles and the wall.

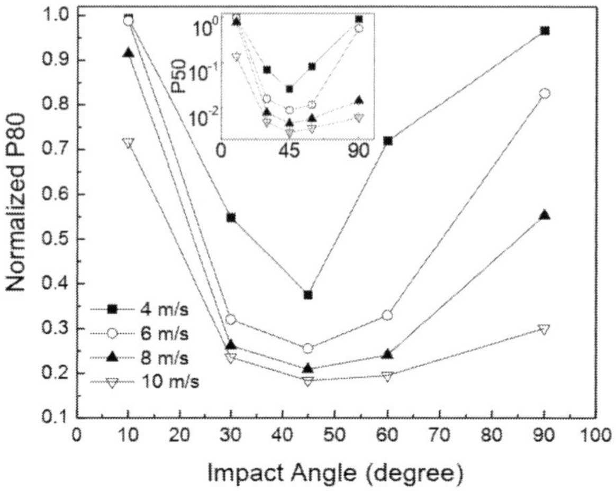

Fig. 5. Normalised P_{80} as a function of impact angle for different impact velocities. The inset shows the variation of the normalised P_{50} values.

Particle-flow interaction

This work [8] investigated the dispersion process of single agglomerate in a cyclonic model which has two air inlets and one outlet (Fig. 6a). In the simulation, an agglomerate is introduced via the left inlet with zero initial velocity while air flow enters from both inlets with a pre-set velocity. The agglomerate starts to accelerate due to the flow drag force and, due to flow shearing and turbulence and impact with the device, is dispersed into small fragments which are then discharged from the outlet. Fig. 6b shows size distributions and morphology of formed agglomerates. Agglomerates A, B and E have similar spans (0.412) with decreasing D_{50}, (4.14 – 2.6μm) while agglomerates B, C and D have similar D_{50} (3.27 μm) with increasing span (0.412 – 1.168) [9].

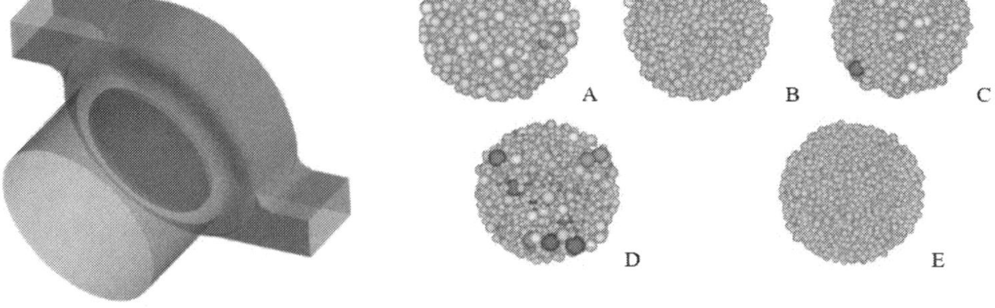

(a) (b)

Fig. 6. (a) 3-dimensional view of the cyclonic model; and (b) Morphology of the formed agglomerates (the colours represent particle diameter).

Fig. 7a indicates a typical cyclonic flow pattern with the velocity having the maximum of 34 m/s near the two inlets, gradually slowing down along the flow direction and reaching the lowest velocity of 2 m/s at the centre. Fig. 7b shows the velocity and breakage pattern of agglomerate B at the flow velocity v = 30 m/s. It shows that increasing flow velocity decreases the agglomerate residence time t_e which is defined as the duration of the agglomerate in the device before being discharged from the outlet.

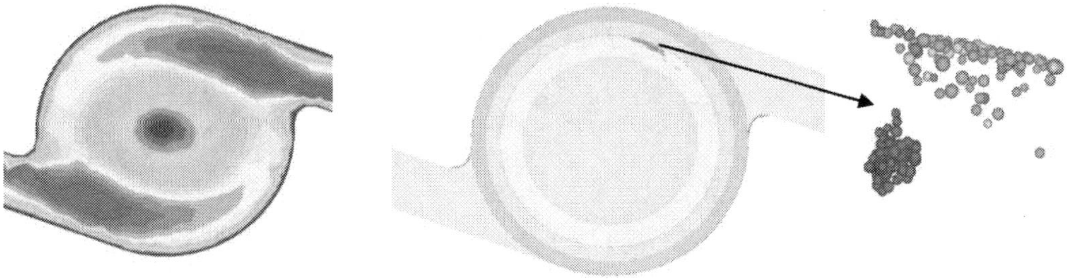

(a) (b)

Fig. 7 Top views of (a) fluid velocity; and (b) particles velocity and breakage pattern of an agglomerate at t = 0.0015 s. Air flow velocity v = 30 m/s

Fig. 8 shows the FPF of the agglomerates at various flow velocities. In general, larger flow velocities generate larger FPFs for an agglomerate. However, different agglomerates behave differently with flow rates. While agglomerate A has the best dispersion efficiency at the low flow velocity of 20 m/s, agglomerates B and E, which have similar spans but large D_{50}, generate more fine particles at the median and high flow rates, respectively (Fig. 8a). Fig. 8b shows the agglomerates with smaller span (narrower size distribution) have larger FPF. However, the differences in FPF for the three agglomerates are very small, indicating that the effect of powder polydispersity on dispersion efficiency is less significant than particle size.

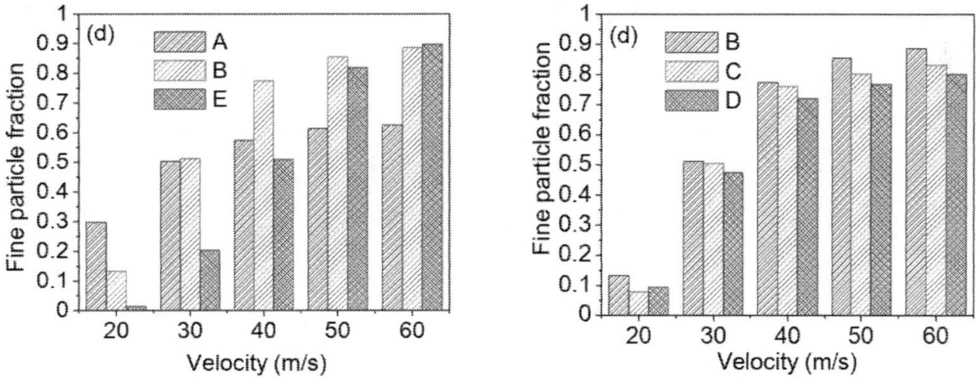

(a) (b)

Fig. 8 Effects of (a) particle size; and (b) span of distribution on the dispersion process at different flow velocities.

Particle-particle interaction

This work simulates powder dispersion in a T-shape pipe with two air inlets and one outlet [10]. In the simulations, the air enters from both inlets and two agglomerates are introduced separately via the left and right inlets. The simulations are considered to study the effects of airflow rate, agglomerate strength and agglomerate-agglomerate impact angle. Fig. 9 shows the dispersion of the agglomerates in the pipe at different times with the air flow velocity 15 m/s. The two agglomerates start to accelerate due to the flow drag force, then due to agglomerate-agglomerate interaction, are broken into small fragments, and finally are discharged from the outlet. The fluid flow and agglomerates are coloured by the magnitudes of the airflow velocity and the radius of primary particle, respectively. The influence of agglomerates on the surrounding fluid flow can be observed from the change of fluid velocity at different times.

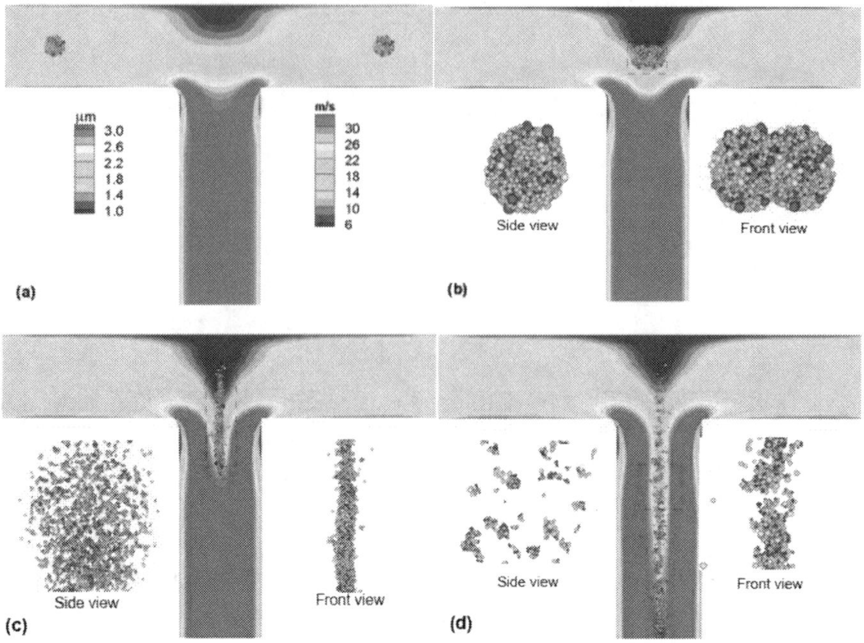

Fig. 9. Contour of airflow velocity (only central slices are shown here) and breakage pattern of agglomerate at airflow velocity of 15 m/s: (a) $\tau = 0$; (b) $\tau = 0.15$; (c) $\tau = 0.26$; and (d) $\tau = 0.33$.

Fig. 10 shows the FPF of the agglomerates at different flow rates (Fig. 10a), agglomerate strength (Fig. 10b) and impaction angles (Fig. 10c). In general, FPF increases with increasing flow rates, decreasing agglomerate strengths and increasing impaction angles.

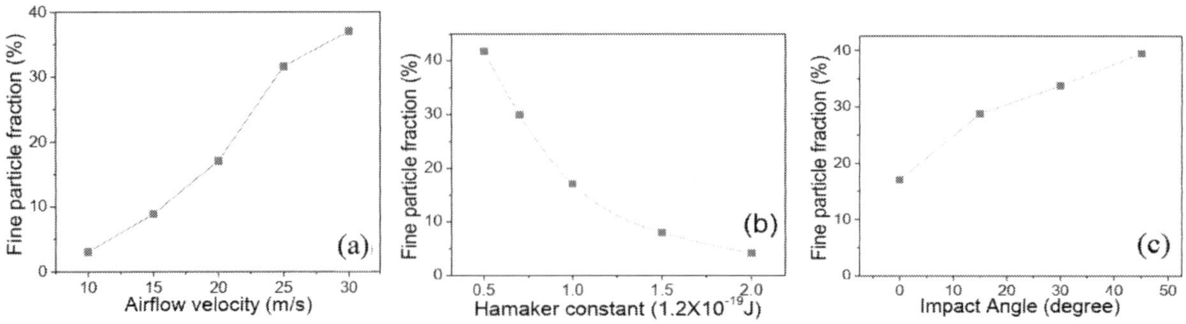

Fig. 10. Fine particle fraction after dispersion with different (a) airflow velocity; (b) agglomerate strength; and (c) impact angle.

Dispersion in dry powder inhaler

This work focuses on the dispersion process of multiple agglomerates in a dry powder inhaler (Fig. 11a) which includes all types of interaction mentioned above. In the simulations, multiple agglomerates were sprayed out from the capsule. Then agglomerates were accelerated due to the fluid drag force, and due to flow shearing and turbulence and impact with the device were dispersed into small fragments which were then discharged from the outlet. Fig. 11b shows the flow field at the flow velocity of 40 m/s. This indicates a typical cyclonic flow pattern with the velocity field inside the chamber having significant change with rotating capsule.

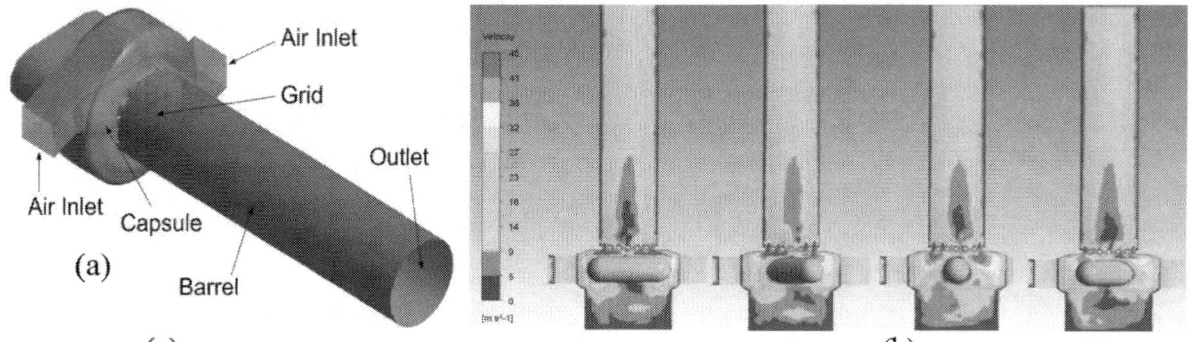

(a) (b)

Fig. 11. (a) 3-dimensional view of the device; and (b) Side view of the velocity pattern at different times schematic and numerical grid representation.

Agglomerate dispersion normally is more efficient with increased flow rate as the total impact energy increases (Fig. 12a). However, FPF_{laoded} sharply decreases with further increasing to 50 m/s as the throat deposition increases sharply. As shown on Fig. 12b, FPF_{laoded} also increases with the number of agglomerates from 2 to 8 agglomerates, then has little changes with further increasing the number of agglomerate. So FPF_{laoded} can be explained by a balance between de-agglomeration rate and deposition with increasing number of agglomerates and airflow rate.

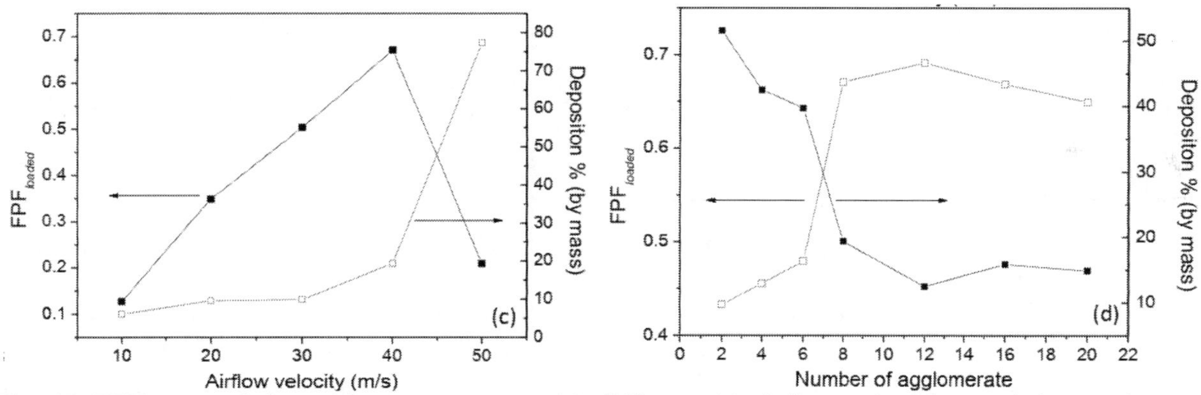

Fig. 12. FPF_{loaded} and deposition percentage with different (a) airflow velocities; and (b) numbers of aggloemrates.

Dispersion quantification

As demonstrated above, agglomerate breakage is mainly determined by particle-particle cohesion and particle-wall impact. Since the particle-wall impact is governed by the particle-fluid interaction related to flow velocity, the effect of flow rate can also be reflected from the impact interactions. Therefore, establishing a quantitative link between the interaction forces and dispersion performance can provide a way to characterise and optimise dispersion processes based on simulation results. Fig. 13 shows FPF_{loaded} obtained under different conditions as a function of the ratio of impact and tensile energies. The previous results under the swirl flow [8], T-pipe and throats [11] are also plotted for comparison. It is observed the results collapse on a single master curve. The results further indicate that agglomerate breakage is governed by the same mechanisms even the breakages occur at different conditions.

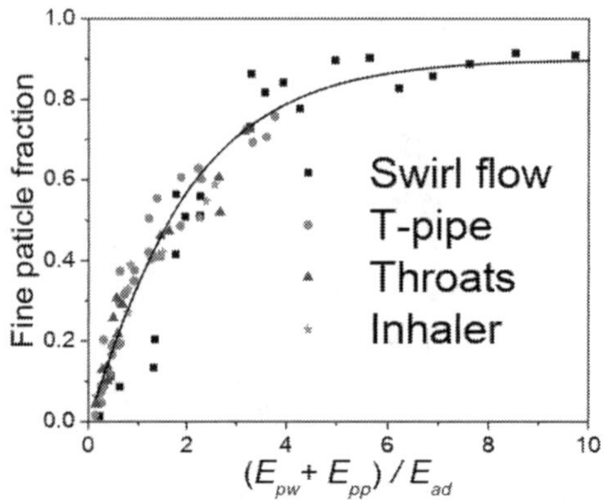

Fig. 13. Correlation between relative fine particle fraction (FPFLoaded) and the ratio between the impact energy and agglomerate cohesion energy (ϕ) in the inhaler and previous models [7, 8, 11].

Conclusions

CFD-DEM model has been developed to simulate the dispersion process of pharmaceutical powders in dry powder inhalers (DPIs). The model can generate the flow and powder related information and reliably describe the dispersion of drug powders. Analysis of microdynamic results such as particle-particle, particle-fluid and particle-inhaler interactions along with trajectory of and transient forces acting on individual particles are useful to understanding mechanisms governing particle agglomeration and de-agglomeration in relation to flow conditions and inhaler geometry. It also has the wide applicability to a variety of inhalers and operational conditions. The results can be used to improve the performance of existing DPI or designing new inhalers with desirable dispersion.

Acknowledgements

Authors are grateful to the Australia Research Council (ARC) for the financial support for this work.

References

1. I. Gonda. Physicochemical principles in aerosol delivery. Topics in pharmaceutical sciences 1991. Stuttgart, Germany: Medpharm Sci Publ 1992:95-115.
2. H.-K. Chan. Dry powder aerosol drug delivery-opportunities for colloid and surface scientists. Colloids and Surfaces, A: Physicochemical and Engineering Aspects 284-285 (2006) 50-5.
3. R.Y. Yang, R.P. Zou, A.B. Yu. Computer simulation of the packing of fine particles. Physical Review E 62 (2000 Sep) 3900-8.
4. K.W. Chu, B. Wang, A.B. Yu, A. Vince. CFD-DEM modelling of multiphase flow in dense medium cyclones. Powder Technol 193 (2009) 235-47.
5. Z.Y. Zhou, S.B. Kuang, K.W. Chu, A.B. Yu. Discrete particle simulation of particle-fluid flow: model formulations and their applicability. Journal of Fluid Mechanics 661 (2010 Oct) 482-510.
6. S. Adi, Z.B. Tong, H.K. Chan, R.Y. Yang, A.B. Yu. Impact angles as an alternative way to improve aerosolisation of powders for inhalation? Eur J Pharm Sci 41 (2010 Oct) 320-7.
7. Z.B. Tong, R.Y. Yang, A.B. Yu, S. Adi, H.K. Chan. Numerical modelling of the breakage of loose agglomerates of fine particles. Powder Technol 196 (2009) 213-21.
8. Z.B. Tong, R.Y. Yang, K.W. Chu, A.B. Yu, S. Adi, H.K. Chan. Numerical study of the effects of particle size and polydispersity on the agglomerate dispersion in a cyclonic flow. Chem Eng J 164 (2010) 432-41.

9. S. Adi, H. Adi, H.-K. Chan, W.H. Finlay, Z. Tong, R. Yang, A. Yu. Agglomerate strength and dispersion of pharmaceutical powders. Journal of Aerosol Science 42 (2011) 285-94.
10. Z.B. Tong, R.Y. Yang, A.B. Yu, S. Adi, H.K. Chan. Numerical Study of Effects of Powder Size and Polydispersity on the Dispersion of Fine Powders in a Cyclonic Flow. In: Nakagawa M., Luding S., eds. Powders and Grains 2009. Melville: Amer Inst Physics 2009:807-10.
11. Z.B. Tong, S. Adi, R.Y. Yang, H.K. Chan, A.B. Yu. Numerical investigation of the de-agglomeration mechanisms of fine powders on mechanical impaction. Journal of Aerosol Science In Press, Accepted Manuscript (2011).

(230d) A Microfluidic Platform for Aerosol and Nanoparticle Drug and Gene Delivery

Aisha Qi, Anushi Rajapaksa, Peggy Chan, James Friend and **Leslie Y. Yeo**, Micro/Nanophysics Research Laboratory, Monash University, Clayton, VIC, Australia

Pulmonary drug administration requires the direct delivery of drug formulations into the deep lung region in the form of inhaled particles or droplets, providing a distinct advantage over other methods for the treatment of respiratory diseases: the drug can be delivered directly to the site of inflammation, thus reducing the need for systemic exposure and the possibility of adverse effects. Nevertheless, it is difficult to produce aerosol droplets of a drug solution within a narrow polydisperse size range between 1 and 5 μm to achieve maximum deposition in the lower pulmonary tract and alveoli. We first demonstrate the use of surface acoustic wave (SAW) microfluidic atomization as an efficient means to generate aerosols containing the short-acting β_2-agonist salbutamol for the treatment of asthma. The 2.84 μm mean diameter of the aerosol droplets produced lies well within the optimum size range. This is subsequently confirmed using a twin-stage impinger *in vitro* lung model, in which we observe that approximately 70% to 80% of the drug is deposited within the lung; such lung dose efficiencies are considerably higher than those possible through currently available metered dose inhalers and nebulizers. Thus, even with a modest power range of around 1 W, which is a tenth of that required with ultrasonic nebulizers, the SAW atomization technology provides a viable and efficient generic nebulization platform for drug delivery via the pulmonary route. In addition, the potential for the platform to be used for non-viral gene delivery is demonstrated through high levels of gene expression observed in Western blot analysis of the COS-7 cells transfected with post-atomised plasmid DNA encoded with a merozoite surface protein 4/5, which is a potential malaria vaccine candidate. This is subsequently verified through the expression of yellow fluorescent protein in lung sections of Swiss mice following intratracheal instillation of the plasmid DNA containing the protein. Further, we show that the technology is a very rapid, efficient and straightforward means for the synthesis of biodegradable polymeric particles with dimensions down to 100 nm within which therapeutic molecules such as nucleic acids, proteins and peptides can be encapsulated. Finally, the ability to synthesize multiple coatings of polyelectrolyte layers encapsulating these biomolecules is demonstrated using sequential atomization-suspension steps as a fast and efficient alternative to conventional layer-by-layer polyelectrolyte assembly. These multilayer nanocapsules offer the exciting possibility for tuning the drug release profile for controlled delivery over a prolonged period or for targeting the delivery to a specific location within the body. In all of the above, the control offered over the aerosol/particle size, the possibility for flexible and multilayer encapsulation, the low power requirement, the high delivery efficiency, and the miniaturization of the system altogether suggest that the proposed SAW atomization platform represents an attractive alternative to current nebulizers and inhalers, which we envisage could constitute next-generation devices that revolutionize pulmonary drug and gene delivery for needle-free vaccination or for the treatment of various diseases in the near future.

See more of this Session: Innovations In Drug Delivery Technology I
See more of this Group/Topical: Food, Pharmaceutical & Bioengineering Division

(230e) Development of Anti-Tubercular Nanoscale Drug Cocktails for Targeted Therapy

Suzanne M. D'Addio[1], Venkata Reddy[2], Philip Kwok[3], Leo Einck[2], Hak-Kim Chan[3] and Robert Prud'homme[1], (1)Chemical Engineering, Princeton University, Princeton, NJ, (2)Sequella, Inc, Rockville, MD, (3)Faculty of Pharmacy, The University of Sydney, Sydney, Australia

While tuberculosis (TB) is a curable intracellular disease, it remains a global health issue with over 1 million deaths reported in 2007. Causes of failed treatment include patient non-compliance and the emergence of drug resistance over the 6 month therapeutic course. Advanced technologies are needed to improve the efficiency of drug delivery to reduce dosing frequency and limit drug resistance, while ensuring patient compliance. Block copolymer directed rapid precipitations by Flash NanoPrecipitation enable formation of nanoparticles which have two unique advantages: (1) multiple drugs can be simultaneously assembled in the same nanoparticle and (2) targeting ligands can be introduced efficiently in drug formulations. Using nanoparticles as a drug delivery platform, we have created nanoscale drug cocktails to colocalize a novel, potent anti-tubercular drug with an efflux pump blocker to ensure therapeutic doses are achieved intracellularly. Targeted polyethylene glycol protected nanoparticles with variable mannose density on the surface have been studied *in vitro* to achieve optimized active targeting vehicles for anti-TB drug delivery specifically to macrophages. To promote patient compliance, aerosol carrier particles have been developed for inhalable targeted delivery of nanoparticles to the lung, with predictable fine particle fractions. The combination of these functionalities is a promising route to overcome some of the current treatment limitations in TB therapy.

See more of this Session: Innovations In Drug Delivery Technology I
See more of this Group/Topical: Food, Pharmaceutical & Bioengineering Division

(230f) Engineering Multifunctional Nanoparticles for Overcoming Drug Resistance In Multiple Myeloma

Tanyel Kiziltepe[1], Jonathan D. Ashley[2], Jared Stefanick[2] and **Basar Bilgicer**[2], (1) Advanced Diagnostics and Therapeutics / Chemical and Biomolecular Engineering, University of Notre Dame, Notre Dame, IN, (2)Chemical and Biomolecular Engineering, University of Notre Dame, Notre Dame, IN

Multiple myeloma (MM), a B-cell malignancy characterized by proliferation of monoclonal plasma cells in the bone marrow (BM), is the second most common type of blood cancer in the U.S. Despite the recent advances in treatment strategies and the emergence of novel therapies, it still remains incurable due to the development of drug resistance with a median survival of 4-5 years. A major factor that contributes to development of drug resistance in MM is the interaction of MM cells with the BM microenvironment. It has been demonstrated that the adhesion of MM cells to the BM stroma via $\alpha 4\beta 1$ integrins leads to cell adhesion mediated drug resistance (CAM-DR), which enables MM cells to gain resistance to drugs such as doxorubicin (Dox)–a 1st line chemotherapeutic in the treatment of MM. The overall objective of this proposed project is to engineer "smart" nanoparticles that will deliver and exert the cytotoxic effects of the chemotherapeutic agents on MM cells, and at the same time do it in such a manner to overcome CAM-DR for improved patient outcome. To enable this, we engineered micellar nanoparticles that are (i) functionalized with $\alpha 4\beta 1$-antagonist peptides as well as Dox conjugates, and (ii) designed to show the adhesion inhibitory and the cytotoxic effects in a temporal sequence. When the nanoparticles are delivered to the MM cells, as a first step they interact with the cell surface $\alpha 4\beta 1$ integrins and inhibit MM cell adhesion to the stroma, thereby preventing development of CAM-DR. In the second step, Dox exerts its cytotoxic effects after cellular uptake, as the nanoparticles will be designed to require a low pH environment such as the endocytic vesicles, to release active Dox. This way, the "smart" nanoparticles act on the MM cells in a temporal fashion and prevent development of CAM-DR for improved patient outcome.

See more of this Session: Innovations In Drug Delivery Technology I
See more of this Group/Topical: Food, Pharmaceutical & Bioengineering Division

(230g) Melt Mixing Active Pharmaceutical Ingredient and Polymer to Achieve Different Release Profiles

Adeyinka Adegoke, Ashish Sarode and **Peng Wang**, Department of Chemical Engineering, University of Rhode Island, Kingston, RI

Traditional blending processes in the pharmaceutical industry are operated at room temperature. In recent years, melt mixing has received a great deal of interest, mainly due to the challenge associated with the ever increasing number of poorly soluble drugs. In addition to the dissolution enhancement, melt mixing also possesses several other advantages. It is a solvent free process; it can be realized by using a continuous extruder, which is commonly used in the plastic and processed food industries. Our experimental data show that API and polymeric excipients can form one phase at the elevated temperature. The drug may remain in amorphous form or recrystallize after the mixture is cooled to room temperature, depending mostly on the formulation. In addition, our results suggest that melt mixing can be applied to significantly increase or slow down drug's release rate, which cannot be achieved if the same components are blended at room temperature.

See more of this Session: Innovations In Drug Delivery Technology I
See more of this Group/Topical: Food, Pharmaceutical & Bioengineering Division

(238a) Influence of Particle Size On Mixing In a Simple Bladed System

Kevin J. Brennan and Carl R. Wassgren, Mechanical Engineering, Purdue University, West Lafayette, IN

The single-bladed mixing experiments of Bridgwater *et al.* (1968/69) are modeled using the discrete element method (DEM). The DEM simulations are performed over a wider range of parameters than those studied by Bridgwater *et al.*, and the measurements of the particle movement are much more precise. The goal of the studies is to determine how particle size influences their mean and diffusive displacement in a simple mixing system.

The principle results of Bridgwater *et al.* are confirmed: mean particle displacement is independent of particle size, but the standard deviation is dependent on size. Variations in Froude number have little effect on particle displacement, especially after the critical value of Fr = 1.0; however, variations in the bed depth-to-blade height ratio are strongly correlated with the mean and standard deviation of particle displacement. These results indicate that it may be possible to use large-particle DEM simulations to predict the macroscopic flow of smaller particles in some systems.

See more of this Session: Mixing and Segregation of Particulates I
See more of this Group/Topical: Particle Technology Forum

(238b) Dynamics of Monoshaped Non-Spherical Particles In a Cylindrical Tumbler

Olivier Dub[1], Jamal Chaouki[2] and Francois Bertrand[2], (1)Chemical Engineering, Ecole Polytechnique de Montreal, Montreal, QC, Canada, (2)Department of Chemical Engineering, Ecole Polytechnique de Montreal, Montreal, QC, Canada

The dynamics of granular materials in cylindrical tumblers have been studied extensively. However, even though it has been known for a while that the shape of particles may affect significantly their dynamical properties, most reported studies on this topic have involved so far spherical or quasi-spherical particles. This is no doubt in part due to the difficulty in producing monoshaped non-spherical particles with given properties, and singling out the effect of particle shape on particle dynamics.

The objective of this talk is twofold: first, to present an optimized approach for the production of identical non-spherical particles suitable for radioactive particle tracking (RPT) experiments, and then to use RPT as a characterization method to investigate the dynamics of these particles in a cylindrical tumbler. The particles are in fact tablets that were produced using a high-performance Korsh XL100 press, and irradiated in the Slowpoke reactor of Ecole Polytechnique de Montral. These tablets were coated with an aqueous solution of Opadry II (Colorcon) to alleviate friability problems in the tumbler. Results obtained in the case of seven different tablet shapes with significantly different sphericity indices in a cylindrical tumbler rotating at various speeds will be presented. In particular, velocity, trajectory and occurrence fields, as well as radial and axial mixing efficiencies, will be discussed in detail to highlight the effect of particle shape on flow and mixing dynamics.

See more of this Session: Mixing and Segregation of Particulates I
See more of this Group/Topical: Particle Technology Forum

(238c) Effect of Powder Cohesion and Mixer Operation Parameters On Blend Uniformity Using a New Low Shear Continuous Tumble Mixer

Miguel Florian, Chemical Engineering, Engineering Research Center for Organic Particulate System, University of Puerto Rico Mayaguez, Mayaguez, PR, Luis Obregon, Chemical Engineering, University of Puerto Rico Mayaguez, Mayaguez, PR, Carlos Velazquez, Chemical Engineering, ERC for Structured Organic Particulate Systems, University of Puerto Rico, Mayaguez, PR and Rafael Mendez, Chemical Engineering, University of Puerto Rico, Mayaguez, PR

The study of flow and mixing of granular materials has increased exponentially in computational and experimental studies in the last years especially in batch systems. In most of pharmaceuticals industries powder mixing is an important operation used daily, and due to the Process Analytical Technology (PAT) initiative these industries need to move on to continuous systems and to be in the capacity to control instabilities on line to make processes more efficient.

In this work, a new continuous low shear mixing device consisting of two, short, acrylic, hollow concentric cylinders has been developed to study granular mixing. The inside cylinder has multiple orifices of 4 mm of diameter and the complete mixer moves counter clockwise making part of the powders to form an avalanche while at the same time powders escape from the inside cylinder to the outside cylinder where the sample is collected to obtain the degree of mixing achieved. The complete system consists of two feeders that deliver the active pharmaceutical ingredient and the excipients. The experimental design was set to obtain three different formulations (2.5, 10.5 and 20% API) each one at three flow rates and four mixer different velocities. Results show an effective mix with values of RSD lower than 6% demonstrating that the operating conditions and powder properties such as particle size distribution and powder cohesion affect considerably the homogeneity, hold up and the average residence time of the mixer.

The experimental part will be completed with simulations of the continuous mixing process using DEM (discrete element method) to study the powder phenomena including the pattern flows, avalanche deformation and flow through tumble orifices to relate these with the properties of the materials characterized previously.

See more of this Session: Mixing and Segregation of Particulates I
See more of this Group/Topical: Particle Technology Forum

(238d) Discrete Element Simulation of Particle Mixing and Segregation In a Saddle Blender

Ebrahim Alizadeh[1], Francois Bertrand[2] and Jamal Chaouki[2], (1)Chemical Engineering, Ecole Polytechnique de Montreal, Montreal, QC, Canada, (2) Department of Chemical Engineering, Ecole Polytechnique de Montreal, Montreal, QC, Canada

Tumbling blenders are widely used for granular mixing operations in many industries such as pharmaceutical, cosmetics and food. These blenders exist in different sizes, are easy to operate and can handle materials with various properties. In particular, the V-blender has been one of the most common tumbling blenders used to mix granular materials in the pharmaceutical industry. One aspect that must be addressed when designing such mixing systems is the axial mixing efficiency, which can lead to nonhomogeneous mixtures, especially when the physical and flow properties of the particles that are brought into play vary.

We recently got interested in the so-called tetrapodal mixing device, patented in 1964 (U.S.A. Patent Office, 3134578), which consists of a container with one axis of rotation going through its center of gravity, one leg extending from it in one direction and three other legs organized as in a tripod and extending in the opposite direction. In other words, it can be thought of as a blender with two V-shaped pairs of legs. Compared to the V-blender, this geometry then has an increased ability to pour, split and recombine powder materials in one cycle, and eventually lead to better (axial and radial) mixing efficiency. The objective of this work is to investigate using the discrete element method (DEM) particle mixing and segregation in both the V- blender and the tetrapodal mixing device, which we have dubbed "saddle" blender. The results obtained will be compared on the basis of mixing time and mixing uniformity, both axial and radial, for different loading patterns, fill levels and rotational speeds.

See more of this Session: Mixing and Segregation of Particulates I
See more of this Group/Topical: Particle Technology Forum

(238e) Rate-Based Separation In Collisional Particle Flows

Tathagata Bhattacharya and J. J. McCarthy, Department of Chemical and Petroleum Engineering, University of Pittsburgh, Pittsburgh, PA

Separation of particles is extremely important in a number of industries involved in the handling of solids. Rate-based separation processes hold promise as both more environmentally benign as well as less energy intensive as convention particle separations technologies. This approach is based on differences in the kinetic properties of the components of a mixture, such as the velocity of migration in a medium or diffusivity. In this work, two examples of novel rate-based separation devices will be demonstrated. The first example will analyze a ratchet mechanism where a directed current of particles can be produced in a direction perpendicular to the direction of energy input. In this setup, a vibrating saw-toothed base is employed to induce different mobility for different types of particles. The effect of different operating conditions and design parameters on the separation efficiency will be discussed. The second rate-based separation example will involve the study of the dynamics of gravity-driven particles through an inclined chute fitted with an ordered array of obstacles normal to the chute surface. Dissipative collisions (both with the particles themselves and with the obstacles) give rise to a diffusive motion of particles perpendicular to the flow direction. The extent of separation (i.e., how far one type of particle is removed from another) depends on the different distances traversed by the two types of particles and equals $\langle v_2 \rangle$ t - $\langle v_1 \rangle$ t, where t is the time allowed for migration.

Thus the extent of separation is directly proportional to the time of migration in the gravity field (i.e., time-dependent), a feature that distinguishes such separations from other separation techniques.

See more of this Session: Mixing and Segregation of Particulates I
See more of this Group/Topical: Particle Technology Forum

(238f) Quantifying Mixing and Segregation In a Fluidized Bed with a Particle Segregation Number

Norman K. Keller[1], Wei Bai[2], Rodney O. Fox[2] and Theodore J. Heindel[1], (1) Department of Mechanical Engineering, Iowa State University, Ames, IA, (2) Department of Chemical & Biological Engineering, Iowa State University, Ames, IA

Mixing and segregation in particulate systems have been studied extensively over the past two decades because multi-component granular systems are important to many industrial processes such as fluidized bed gasification, food processing, and pharmaceutical production. Product quality is often identified by the homogeneity (or lack thereof) of the mixture. Several methods have been developed to assess the mixing or segregation quality in granular systems and various segregation measures have been proposed. This paper presents a method to quantify mixing/segregation in collapsed fluidized beds by using 3D X-ray computed tomography. Additionally, a new characterization measure, the particle segregation number (PSN), is introduced. The PSN expresses the level of "mixedness" in a collapsed fluidized bed (or any granular system) with a single number, thus providing a single measure to quantify mixing or segregation in 3D granular systems.

See more of this Session: Mixing and Segregation of Particulates I
See more of this Group/Topical: Particle Technology Forum

(238g) Mixing of Solid Particles In Coarse Particle Fluidized Beds

Haihong Fan, Yanxin Chen, Hui Li and Jinhe Fan, Xi'an University of Architecture &Technology, Institute of Powder Engineering, Xi'an, China

The mixing of solid particles in fluidized bed reactions has been recognized as an important factor determining the heat and mass transfer rates as well as the overall reaction rates. Many studies have been made in the past in order to understand the underlying mechanisms and predict the behaviour of mixing and segregation, including the investigation of factors affecting the mixing/segregation. Almost all the studies in this area are concerned with monosized particles,and Few studies on mixing of single-size particles in Coarse Particle were reported. This paper extended the work of these work and aimed to the mixing behaviors of solid single-size particles and influencing factors by means of a discrete particle simulation.

In this simulation, the gas phase was taken as continuum medium and the particle phase was discrete phase. The distinct element method (DEM) based on Lagrangian approach was applied to calculate the movement of particles. By this approach, the particle trajectories, particle-particle and particle-wall interactions were considered distinctly for each particle existing within the studied system.The continuity and momentum equations was used for the fluid phase. The motion of the particles was calculated by solving Newton's equation of motion for each individual particle by taking into account the gravity force, the drag force, the fluid pressure and the contact force.

The value of Ashton index was calculated to elucidate the dynamic mixing process.

The solid flow patterns were examined first to generate some visual understanding of the mixing process. The minimal mixing time was defined as the time to reach a macroscopically stable state.

In this simulation, the tracer particles were monitored momentarily to acquire their locations and the mixing index. This paper focused on five factors affecting the mixing extent, the gas velocities, the bed height, the gas temperature, the particle density and the particle size. It was showed that there is a maximum value of mixing index and a relevant minimum mixing time for a given operation condition. These five factors have different effects on particles mixing.

See more of this Session: Mixing and Segregation of Particulates I
See more of this Group/Topical: Particle Technology Forum

(247a) Nanoparticle Synthesis Using Reversible Ionic Liquids

Amy Rohan[1], Emily Nixon[2], Rani Jha[2], Manish Talreja[1], Pamela Pollet[2], Charles Liotta[2] and Charles Eckert[1], (1)School of Chemical & Biomolecular Engineering, Georgia Institute of Technology, Atlanta, GA, (2)School of Chemistry and Biochemistry, Georgia Institute of Technology, Atlanta, GA

Applications of metal nanoparticles have attracted significant attention due to their shape and size dependent properties. Current methods for nanoparticle synthesis remain cost-, energy- and time-intensive. This work will present Au nanoparticle synthesis with reversible ionic liquids, formed upon the addition of CO_2 to a silylated amine precursor. The resulting ion pair allows for reverse micelle formation. Heating the solution to a moderate temperature results in the reversal of the ionic liquid and the break-up of the reverse micelles, for subsequent characterization or deposition of the Au nanoparticles. Formation of metal nanoparticles without the requirement of capping agents or surfactants, sometimes in large excess, could result in significant energy and waste reduction. We have tuned a number of process parameters to demonstrate the formation of monodisperse Au nanoparticles.

See more of this Session: Nanostructured Particles for Catalysis
See more of this Group/Topical: Particle Technology Forum

(247b) Synthesis, Characterization and Catalytic Application of Dendrimer-Mediated Nickel and Iron Alloy Nanoparticles Containing An Average of 40 Atoms

Vanessa A. Castillo and John Kuhn, Chemical and Biomedical Engineering, University of South Florida, Tampa, FL

Bimetallic nanoparticles (NPs), composed of two differentmetals, have recently drawn great interest in the field of catalysis because bimetallization makes it possible to obtain catalysts withimproved catalytic activity that may not be achieved with monometallicnanoparticles [1]. In addition, the size of the metal particles plays animportant role in the activity and selectivity of catalysts,and an enhancement in the catalytic properties has been previously shown formonodisperse particles with ~1nm of diameter [2]. Here we report the synthesis and characterizationof different ratios of Fe-Ni bimetallic nanoparticles of < 2nm diametersupported on SiO_2 for the study of the catalytic activity in theethylene hydrogenation reaction.

Thenanoparticles were synthesized within a hydroxyl-terminated fourth generationpoly-aminoamide (PAMAM) Dendrimer (G4-OH), a hyperbranchedpolymer, in aqueous solution and then immobilized by depositing onto aCAB-O-SIL M-5 Silicon dioxide support. UV-Visabsorption spectrophotometry was used to monitor the coordination processbetween $Fe^{(+3)}$ and $Ni^{(2+)}$metals ions and the internal tertiary amine groups of the dendrimers. TransmissionElectron Microscopy and X-ray absorption spectroscopy were used to measure thesize and distribution of the zerovalent metals afterreduction of the coordinated metal ions. Catalytic activity of the supported Fe-Ninanoparticles was studied with ethylene hydrogenation at different temperaturesafter hydrogen reduction pretreatment at 523 K.

Reference

1. J.H. Sinfelt, Acc. Chem. Res., Vol. 20 p. 134, 1987.

2. Huang,W. et al., Nano Letters. Vol. 8, No. 7 p. 2027, 2008.

See more of this Session: Nanostructured Particles for Catalysis
See more of this Group/Topical: Particle Technology Forum

(247c) Scalable Synthesis of Palladium Nanoparticle Catalysts by Atomic Layer Deposition

Xinhua Liang[1], Lauren Blinn[2], Brittany Michael[1] and Alan W. Weimer[1], (1) Chemical and Biological Engineering, University of Colorado at Boulder, Boulder, CO, (2)Chemical Engineering, University of Florida, Gainesville, FL

Atomic layer deposition (ALD) was used to produce Pd/Al_2O_3 catalysts using sequential exposures of Pd (II) hexafluoroacetylacetonate and formalin at 200 °C in a fluidized bed reactor. The ALD prepared Pd/alumina catalysts were characterized by various methods including hydrogen chemisorption, XPS and TEM, and compared with a commercially available 1 wt.% Pd/alumina catalyst, which was also characterized. The content of Pd on alumina support and the size of Pd nanoparticles can be controlled by the number of ALD coating cycles and the dose time of the Pd precursor. One layer of organic component from the Pd precursor remained on the Pd particle surface. The oxidized/reduced ALD 0.9 wt.% Pd/alumina had greater active metal surface area and percent metal dispersion than the commercial 1 wt.% Pd/alumina catalyst. The ALD and commercial catalysts were subjected to catalytic testing to determine their relative activities for glucose oxidation to gluconic acid in aqueous solution. The ALD 0.9 wt.% Pd/alumina catalyst had comparable activity as compared to the commercial 1 wt.% Pd catalyst. No noticeable amount of Pd leaching was observed during the vigorously stirred reaction. ALD can be an efficient method for the preparation of highly dispersed noble metal nanoparticles.

See more of this Session: Nanostructured Particles for Catalysis
See more of this Group/Topical: Particle Technology Forum

(247d) A First-Principle Study of Pt Clusters On Anatase TiO2 (101) Surfaces

Yun Zhou[1], Christopher L. Muhich[2], Charles B. Musgrave[2] and Alan W. Weimer[2], (1)Department of Chemical and Biological Engineering, University of Colorado, Boulder, Boulder, CO, (2)Department of Chemical and Biological Engineering, University of Colorado, Boulder, Bouler, CO

The absorption properties of Pt_n (n is up to 13) clusters on defect free TiO_2 anatase (101) surfaces have been studied using *ab initio* pseudopotential calculations based on density functional theory. The (111) crystal plane was chosen for the Pt atom arrangement. The change of the Pt cohesive energy and the electronic structure were investigated with the change of the Pt cluster size. The Pt cluster properties were also compared with a full monolayer Pt thin film to understand thoroughly the Pt island growth on an anatase TiO_2 (101) surface. The results indicate that Pt prefers to absorb on existed Pt clusters rather than on the TiO_2 surface. For the same size of the Pt cluster, different absorption positions, orientation and cluster configurations were examined. The binding energy difference of the 1 layer configuration and several 2 layers configurations for the same number of the Pt atoms (eg. 10 Pt atoms) were compared. Results showed that the Pt energetically prefers the formation of three-dimensional particles. Different cycles of Pt nanoparticles were also experimentally deposited on TiO_2 nanopowders by use of the atomic layer deposition method. These Pt/TiO_2 nanoparticles were explored via high resolution transmission electron microscopy and the results validated the theoretical study.

See more of this Session: Nanostructured Particles for Catalysis
See more of this Group/Topical: Particle Technology Forum

(247e) Size Dependence of Sulfur Deactivation Using Silica Supported Platinum Nanoparticles

Lyndsey M. Baldyga, University of South Florida, Tampa, FL and J. N. Kuhn, Chemical & Biomedical Engineering, University of South Florida, Tampa, FL

Sulfur poisoning is of large concern in the fossil fuel and the renewable energy industries. A working hypothesis is that an optimal particle size exists because small particles, with their unsaturated coordinated surfaces, will bind sulfur strongly as compared to the surfaces of larger particles and large particles have a low surface to bulk ratio as compared to small particles.

For this project, the sulfur tolerance of four different sizes of silica – supported Pt nanoparticles are tested using ethylene hydrogenation as a probe reaction. These particles were synthesized using colloidal synthesis and immobilized on silica. The catalyts were characterized using transmission electron microscopy, X – ray diffraction, and temperature programmed experiments using a mass spectrometer. The sizes determined as 2.0, 3.4, 4.3, and 6.8 nm. Using mass spectrometry and gas chromatography for analysis, the sulfur tolerance was tested by performing an ethylene hydrogenation reaction with and without sulfur treatment.

See more of this Session: Nanostructured Particles for Catalysis
See more of this Group/Topical: Particle Technology Forum

(247f) Silica-Supported Ruthenium Nanoparticles for Hydrogenation of Pyruvic Acid In the Aqueous Phase

Xianfeng Ma[1], Rui Lin[1], James E. Jackson[2] and Robert Y. Ofoli[1], (1)Chemical Engineering and Materials Science, Michigan State University, East Lansing, MI, (2)Chemistry Department, Michigan State University, East Lansing, MI

Heterogeneous catalysis is an important technology in the chemical industry. Incorporation of nanoscience into catalysis provides one of the most powerful approaches to understanding reaction mechanisms and designing new-generation catalysts more efficienctly[1]. In this work, we present a strategy for design of monodisperse Ruthenium nanoparticles (RuNPs), immobilization of the RuNPs onto ordered mesoporous silica (MSUF), and reactivity assessment of the supported catalyst (RuNPs/MSUF) by aqueous phase hydrogenation of pyruvic acid to lactic acid. The RuNPs were prepared by a polyol reduction method[2] using ruthenium(III) acetylacetonate and poly(vinylpyrrolidone) (PVP) as precursor and stabilizer, respectively. Transmission electron microscopy (TEM) characterization showed that the particles were spherical with a fairly narrow size distribution and an average diameter of 3.5±0.5 nm. RuNPs/MSUF was prepared by immobilizing the RuNPs onto ordered mesoporous silica support MSUF using sonication method. RuNPs/MSUF was prepared by immobilizing the RuNPs onto ordered mesoporous silica support MSUF using sonication. TEM characterization showed well-dispersed particles on the surface of MSUF with no agglomeration and no damage to the framework of the support. Both inductively coupled plasma-atomic emission spectroscopy (ICP-AES) and X-ray photoelectron spectroscopy (XPS) characterization confirmed the attachment of RuNPs to MSUF. The reactivity of RuNPs/MSUF was evaluated by aqueous phase hydrogenation of pyruvic acid. All the reactions were conducted in a multi-batch reactor system (Parr Instrument Co., Model 5000), and the samples were analyzed by HPLC (Waters Co.) with an RI detector. A differential approach was used to compute the initial turnover frequency (TOF) for each reaction[3,4]. Results showed that, even under much milder reaction conditions than reported to-date[5], the initial TOF of the reaction (685.2 h^{-1}) catalyzed by RuNPs/MSUF is more than 25 times higher than values reported in the literature (24.2 h^{-1})[5]. It is also noteworthy that RuNPs/MSUF shows a significant improvement in TOF compared to commercial Ru catalysts, which is an indication of a higher utility efficiency of Ru in catalysis.

References:

1. Gabor A. Somorjai, Feng Tao, *et al. Top Catal* (2008) 47:1–14
2. Sang Hoon J, Jeong Y. Park, *et al. Nano Lett.*, 2010, 10 (7), pp 2709–2713
3. Zhang, Z.; Jackson, J. E.; Miller, D. J. *Ind. Eng. Chem. Res.*, 2002, 41, 691-696.
4. Mehmet Zahmakýran, *et al. J. Am. Chem. Soc.*, 2010, 132 (18), pp 6541–6549
5. R.. Luque, J.H. Clark. *Catalysis Communications* 11 (2010) 928–931

See more of this Session: Nanostructured Particles for Catalysis
See more of this Group/Topical: Particle Technology Forum

(247g) Single-Step, Plasma-Based Synthesis of Supported Nanoparticle Catalysts

Travis Koh and Michael Gordon, Chemical Engineering, University of California, Santa Barbara, Santa Barbara, CA

Many established and future technologies rely on the unique chemical and opto-electronic properties of nanometer-scale particles. For example, metal, metal oxide, and semiconductor nanoparticles find application in heterogeneous catalysis, photocatalysis, and photovoltaics. In light of such needs, a robust synthesis route for size and composition-controlled nanoparticles would be highly desirable.

In this work, we present a single-step, plasma-based synthesis route to produce a wide range of surfactant-free nanoparticles (*e.g.* metals, alloys, metal oxides, oxide-supported metals, and semiconductors) in the 1-20 nm size range that can be used in catalytic applications. A hydrodynamically-stabilized, atmospheric pressure, micro-hollow cathode plasma discharge with high density and temperature is used to pyrolyze volatile organometallics and semiconductor precursors to nucleate small crystalline nanoparticles. Particle growth is quenched at small, uniform size because the residence time in the plasma is very short (~microseconds). The end result is an aerosol of surfactant-free nanoparticles that can be spray-deposited on a variety of surfaces.

The talk will focus on the synthesis and characterization of metallic (Pd, Fe, Ni), alloy (Pd-Ag, Fe-Ni), and oxide-supported metallic nanoparticles (Pd/SiO_2) realized via microplasma deposition. *In-situ* FTIR and catalytic testing of Pd and Pd/SiO_2 for selective hydrogenation and CO oxidation were performed to probe activity and metal-support interactions. With respect to hydrogenation, both Pd catalysts exhibited low temperature light-off (< 70 deg C) and high ethylene selectivity at high conversion. Pd catalysts spray-deposited on ITO and graphitic supports were also tested for electrochemical oxidation of methanol and formic acid. Finally, we will highlight how microplasmas can be used to realize a variety of nanostructured materials for various technological applications.

See more of this Session: Nanostructured Particles for Catalysis

(260a) Challenges That the FCC Faces Today - Cyclone Reliability

Ye-Mon Chen, Fluid Catalytic Cracking, Shell Global Solution, Houston, TX

Although the fluid catalytic cracking (FCC) process has been in commercial operation for over 60 years, the technology continues to evolve in order to meet new challenges. This paper presents selected snap-shots of a few challenges, high temperature erosion and corrosion, that the FCC process faces today.

See more of this Session: Special Session: To Celebrate Ted Knowlton's Career Long Accomplishments
See more of this Group/Topical: Particle Technology Forum

Decoupling pressure fluctuation signals for the monitoring of bubble behaviour in dense fluidized beds

X. Tony Bi and John Grace
Fluidization Research Centre
Department of Chemical & Biological Engineering
University of British Columbia
Vancouver, British Columbia, Canada

AIChE Annual Meeting, Chicago, October 2011

Outline

1. Measurements of pressure fluctuations and their applications
2. Coupling of pressure fluctuations
3. Decoupling of pressure fluctuations
4. Conclusions and future work

2

1. Measurement of pressure fluctuations and its applications

Pressure fluctuations in fluidized beds

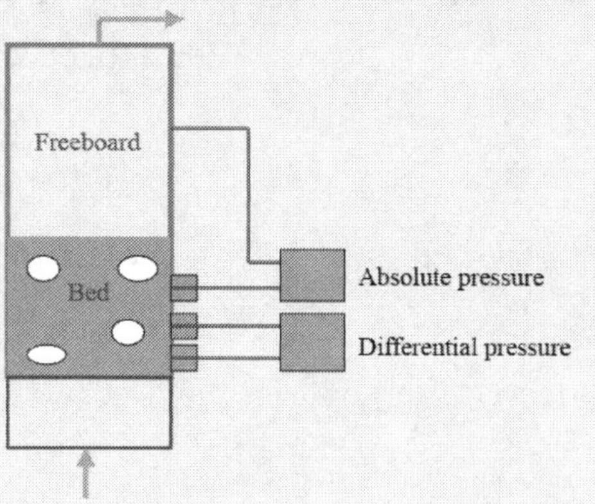

4

Pressure induced by bubbles

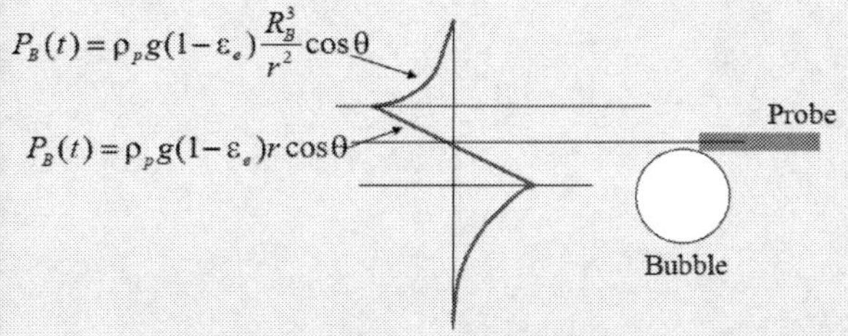

$$s_P \propto \rho_p (1-\varepsilon) g D_B$$

5

2. Coupling of pressure fluctuations from different sources

6

Complex pressure fluctuation phenomena

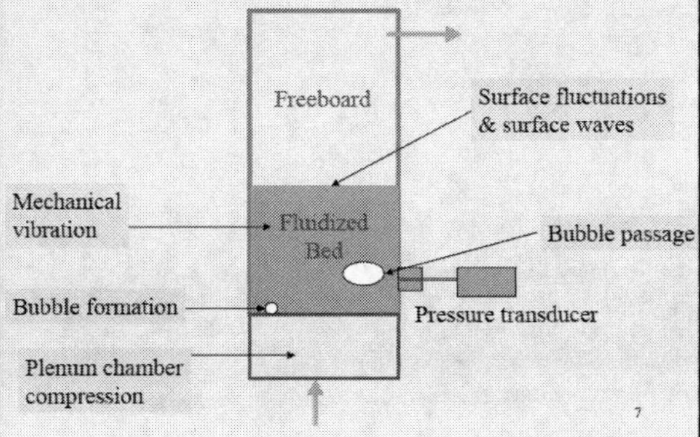

7

Local absolute vs. differential pressure fluctuations

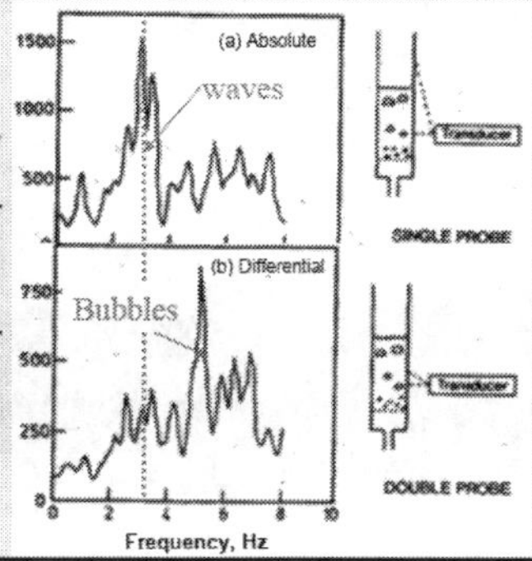

Roy and Davidson (1994)

8

Local voidage vs. local pressure fluctuations

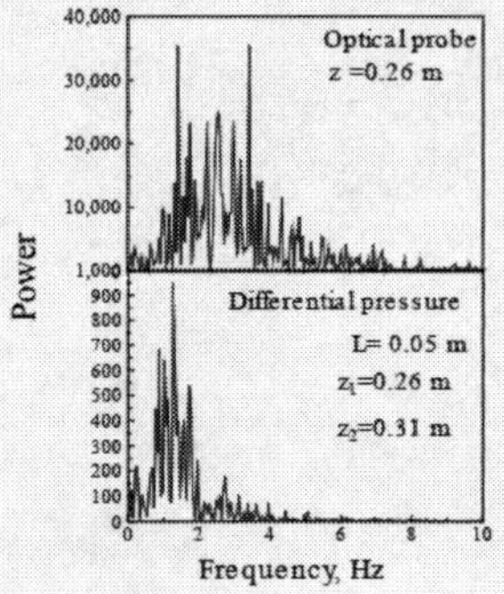

Bi (1994)

9

4. Decoupling of pressure fluctuations

10

Pressure fluctuations at different axial locations

(Liu et al., 2010)

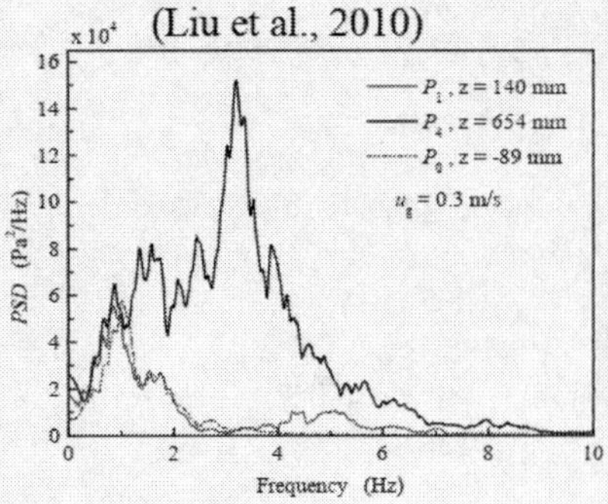

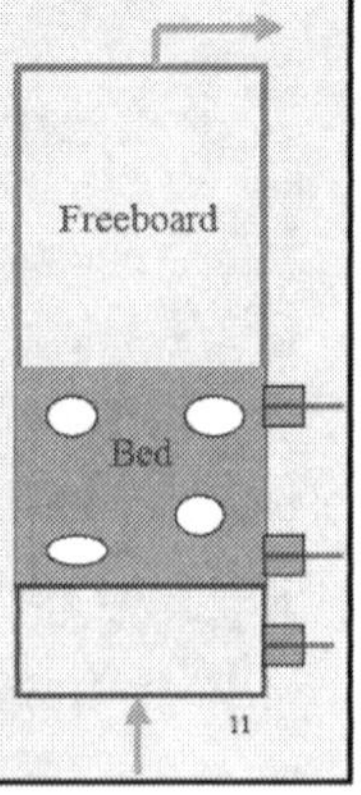

11

Method 1. Coherent and incoherent components

$$\gamma^2_{xy}(f) = \frac{\phi_{xy}(f)\phi^*_{xy}(f)}{\phi_{xx}(f)\phi_{yy}(f)}$$

$$COP(f) = \gamma^2_{xy}\phi_{yy}(f)$$

$$IOP(f) = (1-\gamma^2_{xy})\phi_{yy}(f)$$

The coherent component thus represents propagating pressure waves or global bed oscillations, while the incoherent component will correspond to the local fluctuations induced by bubble motions or local turbulence.

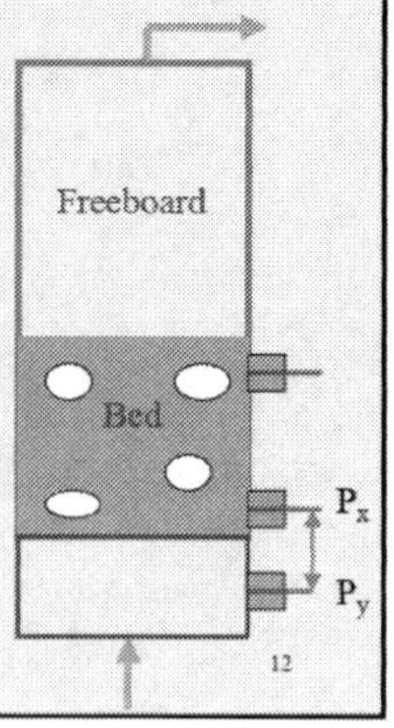

12

Method 2. Differential pressure fluctuations

Roy and Davidson (1994)

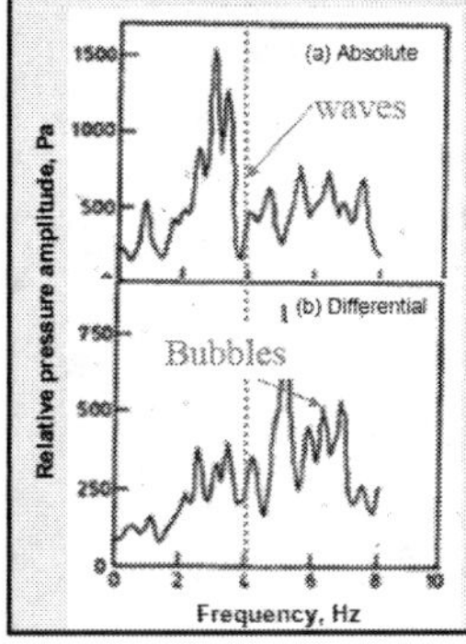

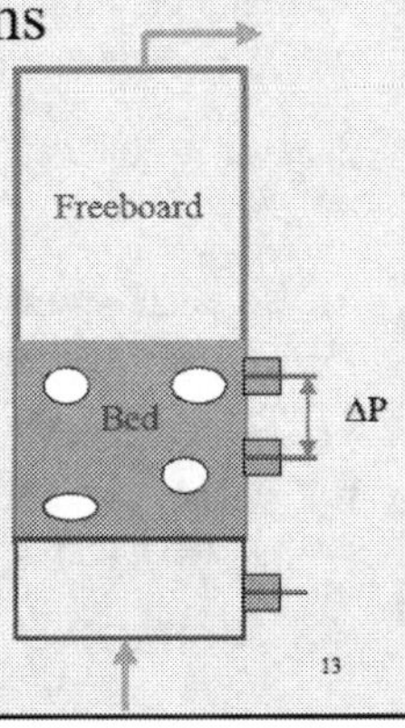

13

Bubble properties

$$d_{b,DP} \propto \frac{\sigma_{DP}}{\rho_p g(1-\varepsilon_{mf})}$$ Differential pressures

$$d_{b,IOP} \propto \frac{\sigma_{xy,i}}{\rho_p g(1-\varepsilon_{mf})}$$ Incoherent components

14

Power spectrum of AP, DP and OP

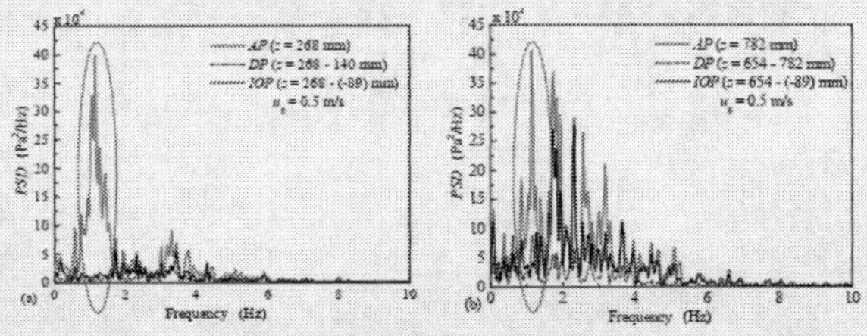

Liu et al. (2010)

15

Mean frequencies for DP and IOP

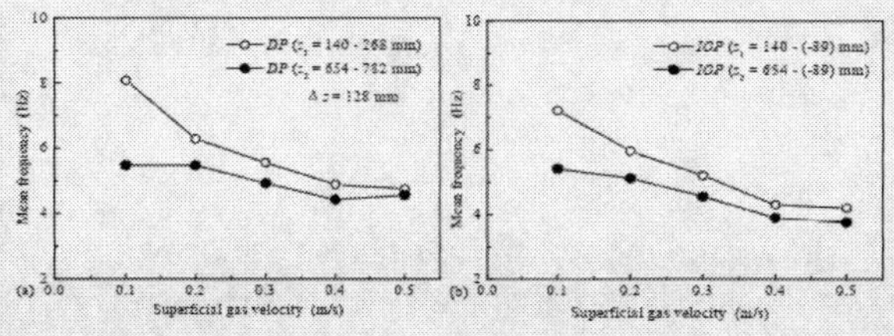

Liu et al. (2010)

16

Mean bubble diameters based on DP, IOP and OP signals

$$d_{b,DP} = \frac{\sigma_{DP}}{\rho_p g(1-\varepsilon_{mf})} \qquad d_{b,IOP} = \frac{\sigma_{xy,i}}{\rho_p g(1-\varepsilon_{mf})}$$

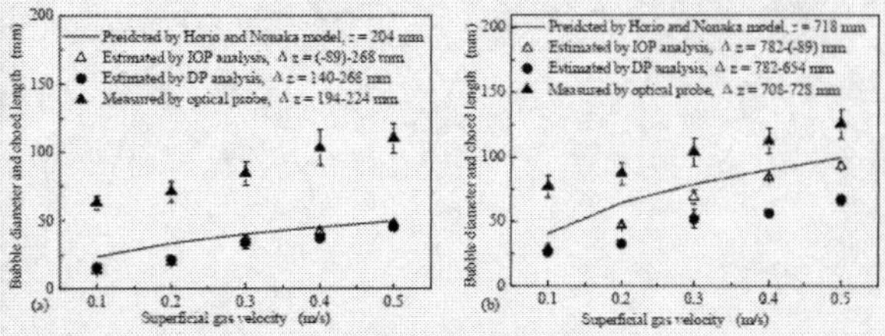

Liu et al. (2010)

17

Conclusions

- Pressure fluctuations in gas-solids fluidized beds are composed of several sources, including local bubble induced fluctuations, global fluctuations and propagating waves.
- Coupling of signals from bubble motion, bed oscillations and propagating waves creates the complexity of local pressure fluctuations.
- Decoupling local pressure fluctuation signals is needed in order to extract useful signals associated with local bubble movement.
- Both differential pressure signals and the incoherence components of absolute pressure signals are mainly contributed from local bubble movement.

18

Future work

- Further development in method (e.g. wavelet analysis) to decouple fluidized bed pressure fluctuations, so as to use the extracted signals for reliable performance diagnosis, on-line monitoring and early-warning, and fluidization quality assessment.
- Further validation of the usefulness of decoupled signals based on more reliable measurement of local bubble properties.

19

Acknowledgement

- Natural Sciences & Engineering Research Council of Canada (NSERC) for a Discovery grant.
- Dr. Mengxi Liu, Dr. Yongmin Zhang and Dr. Aihua Chen

20

(260c) Pulsating Gas-Solid Fluidized Beds for More Efficient Drying

Marc-Olivier Coppens, Chemical & Biological Engineering, Rensselaer Polytechnic Institute, Troy, NY and J. Ruud Van Ommen, Department of Chemical Engineering, Delft University of Technology, Delft, Netherlands

Structuring multiphase reactors introduces extra degrees of freedom, allowing to decouple conflicting design objectives, such as high mass transfer versus low pressure drop in packed beds. For fluidized beds, we have studied four structuring approaches over the past years: pulsating the gas flow, imposing an electric field to induce interparticle forces, distributing the gas injection, and optimizing distributed particle properties such as particle size. We have shown that structuring fluidized-bed reactors can facilitate scale-up, and increase conversion and selectivity by controlling the size and the spatial distribution of the bubbles.

The advantages of structuring are not limited to fluidized-bed *reactors*, but can also be beneficial to other fluidized-bed operations. In this paper, we will focus on fluidized-bed drying. In drying operations homogeneous mixing is important, e.g., to avoid hot spots and deterioration of the product quality. Avoiding gas channelling, and achieving overall homogeneity is especially challenging when the particles are cohesive, as is common in many applications. We will show that pulsating the gas flow is a good way to enhance the mixing of the bed material, while increasing the bubble size. First, this will be demonstrated for beds of dry material. We will also show that the influence of internals is limited; both with and without horizontal heat exchange tubes, pulsation strongly increases the bubble size. We found that there is an optimal pulsation frequency, which was 3.0 Hz for a powder of cohesive, organic material with Sauter diameter of about 0.2 mm. For batch drying of a bed of wet, pharmaceutical granules, we find that pulsation has a beneficial effect as well. Once again, mixing is much better, and drying time is reduced by 15-20% in this case.

These results show that pulsating the gas flow is an attractive way to increase homogeneity and reduce power consumption in fluidized-bed drying. In addition to visual observations, pressure fluctuation analysis and optical probe measurements allow us to offer scientific interpretations, providing the tools that can be used to structure and further optimize pulsed fluidized bed drying operations.

(260d) Fine-Grid Simulations of Gas-Solids Flow In a Circulating Fluidized Bed

Sofiane Benyahia, National Energy Technology Laboratory, Department of Energy, Morgantown, WV

Abstract

Ithas been known for more than a decade that fine heterogeneous structures in theform of clusters and streamers increase the slip velocity in gas-driven flowsof small particles [1]. We have shown the necessity to include sub-grid scaleclosures with coarse-grid simulations of gas-solids flows in order toaccurately predict solids segregation and flux profiles [2] as well as axialpressure gradient profiles [3] in circulating fluidized beds.

Fine-gridsimulations of a small-scale bubbling fluidized bed of Geldart type A materialshowed that a grid size in the order of few particle-diameters is essential topredict the correct bed expansion due to the formation of small bubbles thatwere not resolved with coarse-grid simulations [4]. In the present study, gridrefined computations are conducted on a much larger riser section of acirculating fluidized bed to demonstrate accurate predictions of continuum(two-fluid) models with respect to gas pressure drop and solids hold-upmeasurements. A major hurdle in conducting these transient simulations is therequirement to obtain long time-averaged numerical data to compare withexperimental measurements. Simulations with a maximum of about one millioncomputational cells, translating into a grid size of 1 mm with a grid-sizeratio of one, were conducted in this study and showed clear improvement in thetwo-fluid model (using a standard homogeneous drag law) predictions of the gaspressure drop vertical profiles.

References

1. AgrawalK, Loezos PN, Syamlal M, Sundaresan S. The role of meso-scale structures inrapid gas-solid flows. *J. Fluid Mech.* 2001; 445: 151-185.

2. BenyahiaS. On the Effect of Subgrid Drag Closures. *Ind. Eng. Chem. Res.* 2010;49: 5122-5131.

3. BenyahiaS. Analysis of Model Parameters Affecting the Pressure profile in a CirculatingFluidized Bed. *AIChE Journal*. Accepted for publication, doi:10.1002/aic.12603.

4. WangJ, van der Hoef MA, Kuipers JAM. Why the two-fluid model fails to predict thebed expansion characteristics of Geldart A particles in gas-fluidized beds: Atentative answer. *Chem. Eng. Sci*. 2009; 64: 622-625.

See more of this Session: Special Session: To Celebrate Ted Knowlton's Career Long Accomplishments
See more of this Group/Topical: Particle Technology Forum

(260e) Characterization of Particle Flow In Inclined Standpipes

Lilian de Martn and J. Ruud van Ommen, Department of Chemical Engineering, Delft University of Technology, Delft, Netherlands

Characterization of particle flow in inclinedstandpipes

Lilian deMartn and J. Ruud van Ommen

Delft University of Technology, Department of Chemical Engineering,Product & Process Engineering, Julianalaan 136, 2628 BL Delft, TheNetherlands

Inseveral chemical processes, such as fluid catalytic cracking (FCC), thepresence of several coupled elements (risers, standpipes, cyclones)complicates to have a steady operation. Two main limitations are encountered inthese systems that are associated with the standpipes.

The firstone is commonly found operating at high catalyst recirculation rates. In this condition,an excess of gas can be transferred down the standpipe with the solids. Whenthis occurs, the regime in the standpipe changes and there is no longer awell-defined dense phase [1]. At this point, the apparent density andconsequently the pressure build-up in the standpipe decreases, leading in turnto the air go up the standpipe and a decrease of the collector efficiency [2]. This undesired situation can be avoided increasing the pressure drop across thestandpipe by increasing the apparent density of the solid.

Thesecond one comes from the bubbles formed in the standpipe due to the aerationrequired to maintain the solid circulating. The maximum catalyst flow rate thatthe standpipe can handle is limited by the size of these bubbles [3]. Thislimitation is especially a problem when operating with inclined standpipesbecause the rising bubbles become large at low solids flow rate and low aeration

In thiswork we have characterized the catalyst circulation features in a circulatingfluidized bed (CFB) equipped with an inclined standpipe by means of visualobservation, pressure drop and pressure fluctuations. We have specially focusedon inclined standpipes because they are quite often used in industry, butseldom studied in academia. Moreover, the hydrodynamics behavior in inclinedstandpipes is quite different from that in vertical standpipes or horizontaltransport lines.

The testswere carried out in a CFB consisting in a riser 0.265 m diameter and 5 m tall.The standpipe has an inclination of 45°, 0.09 m diameter and it ismade of Perspex. The solid fluidized is FCC catalyst with an average particlesize of 60 μ m.

[1] J. R. Grace, A. A. Avidan andT. M. Knowlton. Circulating fluidized beds. Springer 1996.

[2] D. Geldart, N. Broodryk and A.Kerdoncuff. Studies on the flow of solids down cyclone diplegs. PowderTechnology, 76 (1993) 175-183.

[3] W.-C Yang. Handbook offluidization and fluid-particle system. CRC Press 2003.

See more of this Session: Special Session: To Celebrate Ted Knowlton's Career Long Accomplishments
See more of this Group/Topical: Particle Technology Forum

(260f) Quantification of Riser Hydrodynamics

Ray A. Cocco[1], Franklin Shaffer[2], Balaji Gopalan[2], S. B. Reddy Karri[1], Jia Wei Chew[3], Christine M. Hrenya[3] and Ted M. Knowlton[1], (1)Particulate Solid Research, Inc., Chicago, IL, (2)National Energy Technology Laboratory, U.S. Department of Energy, Pittsburgh, PA, (3)Chemical and Biological Engineering, University of Colorado at Boulder, Boulder, CO

High speed video analysis was done for riser hydrodynamics of Geldart Group A and B material in a 12-inch (0.3-m) diameter by 60-ft (18-m) tall riser, 12-inch (0.3-m) diameter by 49-ft (15-m) tall riser, and 8-inch (0.2-m) diameter by 72-feet (22-m) tall riser. Analysis of video results clearly showed the core annulus profile with both upflow and down flow at the wall, the cluster formation and stability resulting from transient gas jets and particle rotational velocities. Detailed particle tracking analysis (HSPIV) further revealed shear near the wall plus the radial granular temperature profile.

See more of this Session: Special Session: To Celebrate Ted Knowlton's Career Long Accomplishments
See more of this Group/Topical: Particle Technology Forum

(260g) Cluster Characteristics of Polydisperse Group B Particles In a CFB Riser

Jia Wei Chew[1], Roy C. Hays[2], Drew M. Parker[1], John G. Findlay[2], Ted M. Knowlton[2], S. B. Reddy Karri[2], Ray A. Cocco[2] and **Christine M. Hrenya**[1], (1) Chemical and Biological Engineering, University of Colorado at Boulder, Boulder, CO, (2)Particulate Solid Research, Inc., Chicago, IL

Cluster characteristics of PolydisperseGroup B particles in a CFB riser

Jia Wei Chew[a], Drew M. Parker[a],Roy Hays[b], John G. Findlay[b], Ted M. Knowlton[b],S.B. Reddy Karri[b], Ray A. Cocco[b], Christine M. Hrenya[a,*]

[a]Department of Chemical and Biological Engineering,University of Colorado at Boulder, Boulder, CO 80309

[b]Particulate Solid Research Incorporated, Chicago, IL 60632

The important role of clustering inhigh-velocity, gas-solid flow has been well documented in experiments andsimulations alike, though previous efforts have focused on monodispersesystems. In this work, experiments with a focus on understanding the impact ofpolydispersity on clustering characteristics (appearance probability, durationand frequency) of Geldart Group B particles in a circulating fluidized bed(CFB) riser have been carried out. In a moderately dense CFB riser, threemixtures were investigated: (i) a density-difference binary mixture, withparticles of different material density (ρ_s)but same particle sizes (d_{ave}),(ii) a size-difference binary mixture, with particles of different d_{ave} but same materialdensity ρ_s, and (iii)a continuous particle size distribution (PSD). In addition, in a dilute CFBriser, varying widths of continuous PSDs and varying compositions of binarymixtures were investigated. Afiber optic probe was used for detection of solids, and cluster identificationwas accomplished via wavelet decomposition. Collectively, resultsshow that: (i) local riserposition has a significant influence on all three cluster characteristics inall cases, while effects of operating condition and material type (i.e.,material property or effect of polydispersity) are secondary, (ii) among thethree cluster characteristics, appearanceprobability appears to be largely a function of riser position only, and isinsensitive to changes in operating condition or material type, and (iii)cluster duration and frequency are influenced by operating condition andmaterial type, though only at some axial positions. Theresults presented provide a comprehensive picture of factors affecting clustertrends.

*Corresponding author:hrenya@colorado.edu

See more of this Session: Special Session: To Celebrate Ted Knowlton's Career Long Accomplishments
See more of this Group/Topical: Particle Technology Forum

(291a) Study of Granular Material Multscale Character Based On Complex Networks Perspective

Wu Jintao, Chemical Engineering Machine, Dalian University of Techenology, Dalian, China
See more of this Session: Dynamics and Modeling of Particles, Crystals and Agglomerate Formation
See more of this Group/Topical: Particle Technology Forum

Paper Unavailable

(291b) Large Scale Coupled Eulerian-Lagrangian Simulation of Fluidized Bed

Pradeep Gopalakrishnan and Danesh Tafti, Mechanical Engineering, Virginia Tech, Blacksburgh, VA

Coupled Eulerian-Lagrangian simulation of multiphase flow has been used widely to analyze dense particulate flows. In this approach, the fluid is treated as a continuum while the solid phase is modeled using the Discrete Element Method (DEM). In DEM particle-particle and particle-wall interactions are resolved and the time integration is carried out using Newton's second law of motion. This approach offers better accuracy than the traditional Eulerian-Eulerian Two-Fluid method used for modeling multiphase flows. The computational expense of DEM is very high and hence previous studies have been limited to small scale (mostly 2D) problems. In this study a parallel algorithm for DEM was developed for the open source code, Multiphase Flow with Interphase Exchanges (MFIX). MFIX, developed at NETL has been widely used to simulate hydrodynamics, heat transfer and chemical reactions occurring in bubbling and circulating fluidized beds.

Most previous DEM parallelization efforts available in the literature are based on the domain decomposition framework for fluid and mirror domain method for DEM. In this approach each processor holds the information of all the particles in the system but carries out computations only on a subset. Synchronization of the dataset is carried out via interchange between the processors after every particle time step. The disadvantages of this method are large memory requirements for holding all particle information and high communication cost associated with the synchronization of the data. An alternative approach[1] is to employ the domain decomposition technique for fluid and graph partitioning algorithm based on contact between particles for DEM. In this approach, the fluid variables required by the DEM have to be mirrored. The main advantage of this system is that load balancing is ensured for all configurations. However the communication cost associated with the mirroring of flow variables and the global summation required for particle-fluid interaction terms is high, which affects the scalability for large scale simulations with large processor counts.

In the present study, a hybrid MPI/OpenMP parallelization architecture is employed. Both fluid and DEM are solved using the standard domain decomposition framework along with selective OpenMP parallelization directives for critical routines. Use of OpenMP offers flexibility in decomposing the flow based on the number of computational cells and DEM based on particle numbers. Each processor holds only information related to its domain and hence the memory requirement is low. Further the communication cost is low as only the information near the interface between processors has to be exchanged and no mirroring is required for either fluid or particle information.

For validation, the pseudo-2D fluidized bed experiment by Muller[2] was simulated at two superficial velocities (0.6 and 0.9 m/secs). The average voidage and velocity profiles at different axial locations were compared against the experimental results. The scalability of the code was studied using a fluidized bed with 4.5 million particles with the number of processors ranging from 16 to 128. Both MPI and hybrid MPI/OpenMP simulations were carried out along with TAU profiling. The study shows that the scaling of the MPI simulation drops from 32 to 64 processors due to increased MPI communications required by the flow solver, while the hybrid simulation shows strong scalability up to 128 processors due to reduced communication overhead.

See more of this Session: Dynamics and Modeling of Particles, Crystals and Agglomerate Formation
See more of this Group/Topical: Particle Technology Forum

2D-FRACTAL BASED APPROACH FOR FINE PARTICULATE CHARACTERIZATION: AN APPLICATION TO LIMESTONE FILLERS

Prof. Giuseppe Bonifazi Dr. Nicoletta Picone Dr. Silvia Serranti
Sapienza – Università di Roma, Rome, Italy

Abstract

A full target oriented fine and ultrafine particles systems characterization represents an hard task to reach. Such a difficulty is mainly linked to the problems arising when, starting from a full analysis of single particles properties (i.e. size, shape, surface characteristics and density) a correlation with their behaviour in bulk has to be established.

In this paper an innovative approach is proposed in order to detect and quantify the overall attributes of fine and ultrafine particles resulting from different processes and characterized by different particle characteristics. The proposed approach is based on: a 2D-fractal analysis of fine and ultrafine particles population acquired as digital images. This approach allows to perform a full measurement and control of fine particles structure and texture by evaluating statistical fractal parameters on the base of the topological assessment of the particles on a flat surface. The work was developed with the aim to develop and implement a reliable and simple tool to use for particulate products characterization and recognition able to correlate the characteristics of fine particles detected by the proposed logics and those usually extracted by using classic methods.

Introduction

Fractal geometry concerns the study of non-Euclidean geometrical figures generated by a recursive sequence of mathematical operations. These figures show self-similar features in the sense that their shape at a certain scale is equal to or at least "similar" to the same shape of the figure at a different scale or resolution. This property of scale invariance often occurs also in some natural events. Examples include organisms and real-world phenomena: clouds, river networks, mountain ranges, coastlines, craters, crystals, cauliflower or broccoli, systems of blood vessels and DNA can be analyzed as fractals. All these objects can be modeled on a computer by using a recursive algorithm. An example of this recursive nature is a branch from a river or a tree, they can be considered, in fact, a miniature replica of the whole: not identical, but similar in nature.

Starting from the previous assumptions, the present study was addressed to utilize the fractal approach to perform an "alternative" fine and ultrafine bulk particles systems characterization, whose results fully embed all those physical-chemical properties affecting the behavior of the particles systems considered as a whole. This latter aspect is preeminent in many industrial sectors producing and/or utilizing fine and ultrafine particles.

Usually size class distribution and rheological tests are carried out when a characterization of particles systems, considered as a whole, have to be achieved. Such an approach becomes more difficult dealing with particles characterized by an average size ranging from few microns to hundreds of nanometers.

Surface characteristics and size class distribution strongly influence both sampling and measures, affecting this way the reliability of the results.

Starting from the assumptions that the fractal dimension can not be derived exactly but must be estimated, the proposed approach [1] is based on the study of the structure of the particle distribution on a surface and their agglomerating modalities. From the quantitative analysis of these attributes, depending on their physical surface properties [2], it is thus possible to perform a characterization of particle size distribution and "behavior" in bulk.

The basis of the 2D-Fractal analysis consists in the comparison of the space covering of a geometrical construction, in the case of study the Sierpinski Carpet, and the particles populations acquired as digital images, by evaluating their statistical fractal parameters related to the particles image structure and texture [3]. The choice of the 'Sierpinski Carpet', as compared geometrical construction, is due to its power (ability) showed to be a useful tool in studying bounded two dimensional space filling problems. The work was carried out to develop a simple and reliable tool able to investigate the potential correlations existing among the attributes of fine and ultrafine particles, detected by a 2D-fractal approach, and those usually extracted by adopting classic method (i.e. class size distribution analysis, morphological and morphometrical analyses).

Literature survey

Fundamentals of fractal geometry

Roots of mathematical interest on fractals can be traced back to the late 19th century, the term was coined by Benoit B. Mandelbrot in 1975 [4] and was derived from the Latin adjective *fractus*. The corresponding Latin verb *frangere* means "to break": to create irregular fragment. A fractal is defined by Mandelbrot as "*a rough or fragmented geometric shape that can be split into parts, each of which is (at least approximately) a reduced-size copy of the whole*" [4]: this property is called self-similarity. A fractal is usually characterized by the following features [5]:

- it has a fine structure at arbitrarily small scales,
- it is too irregular to be easily described in traditional Euclidean geometric language,
- it is self-similar (at least approximately or stochastically),
- it has a Hausdorff dimension which is greater than its topological dimension,
- it has a simple and recursive definition.

This characterization introduces two basic concepts of the fractal theory: the fractal dimension and the self-similarity features of shapes. Both concepts are strictly related to each other. A structure is said to be self-similar if it can be broken down into arbitrarily small pieces, each of which is a small replica of the entire structure. The Fractal Dimension (FD) gives a measure of the irregularity of the examined pattern. It will vary depending on the degree of roughness of the pattern. FD will also have a different value for each pattern type, with the fractal dimension being specific for that pattern [6].

The fractal dimension of a rugged line is an indication of its ability to fill space: as is shown in Figure 1, a rugged boundary has a FD between 1.00, a straight line which fill no space, and 2.00, an area in which all the available space is used. Mandelbrot suggests that such curves can have their ruggedness described by allocating a fractional number between one and two which will describe their space filling ability. The difference between the fractional dimension of the lines and their topological dimension describes their ruggedness.

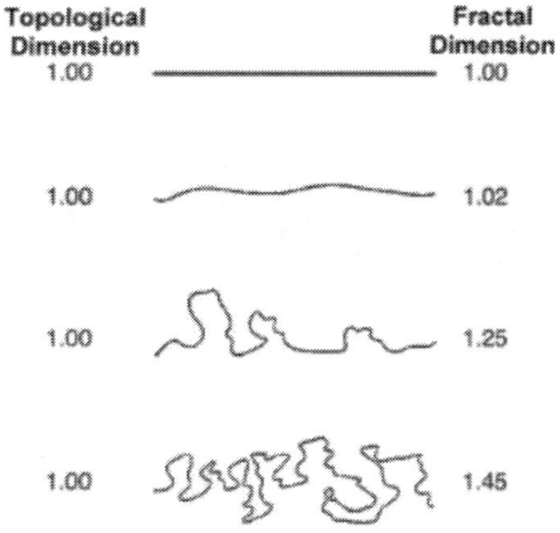

Figure 1: Topological and Fractal Dimension (FD) of a rugged boundary [6].

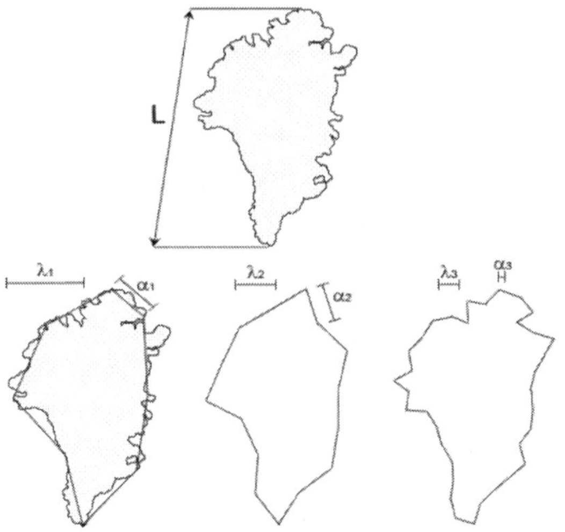

2a

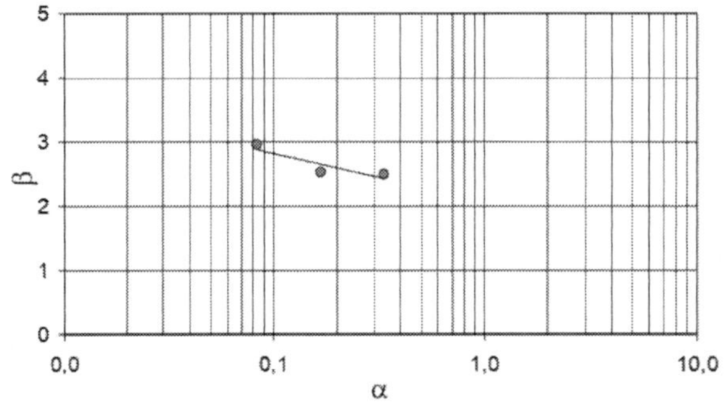

2b

Figure 2: Procedure for characterizing the fractal dimension of the rugged coastline of Groenland (2a). 2β: α is the size of polygon and β is the polygon perimeter, both normalized with respect to L, being L the maximum projected length (Feret diameter) of the Groenland [4].

Let us consider a coast line, just us Mandelbrot did [7]: he posed the question of how to measure the length of a coastline. The obvious answer for the mathematician is to approximate the curve with a polygonal path, each side of which is of length λ. By evaluating the length of the polygonal path as $\lambda \to \infty$, the approximated length $P(\lambda) \to \infty$ also. Mandelbrot discussed the experimental work on this problem which was undertaken by Lewis Fry Richardson. Richardson pointed out that the length of a fractal line (i.e. coastline), $P(\lambda)$, can be defined as:

$$P(\lambda) = n\lambda^{1-D} \tag{1}$$

where $P(\lambda)$ is the length of the fractal line based on unit measurement length λ; n is a proportionally constant, equal to the actual and indeterminate length of the line; and D is the fractal dimension of the curve which is a quantitative descriptor of roughness.

Taking the logarithm of both sides of Equation (1) yields a linear relationship between $P(\lambda)$ and λ with D related to the linear slope coefficient, m, by:

$$D = 1 - m \tag{2}$$

The value of the length of fractal line increases using an increasingly smaller step length

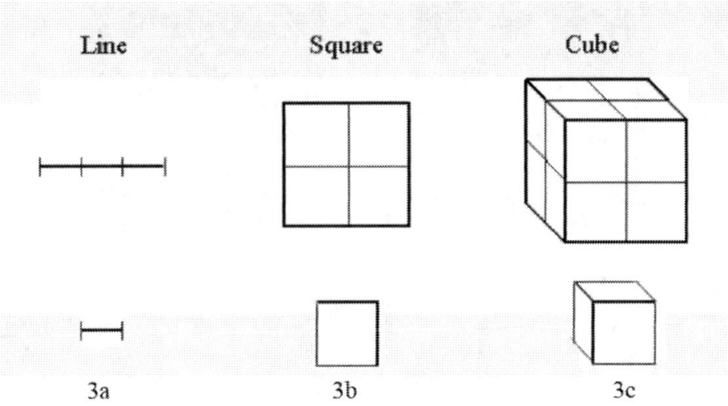

Figure 3: Sets of topological dimension of 1, line segment of length one (3a), 2, square of area one (3b) and 3, cube of volume one (3c) [6].

λ: the estimates of the profile boundary, for each step length, are plotted in a log-log graph. In Figure 2 is shown the structured walk procedure for characterizing the fractal dimension of the coastline of Greenland: it yields a value of FD=1.6754.

The Sierpinski Carpet approach

The main characteristic of fractals is the self-similarity features, so they can be broken down into arbitrarily small pieces, each of which is a small replica of the entire structure object. It is important that the small pieces can be obtained from the entire structure by a similarity transformation. It would be a mistake to believe that if a structure is self-similar, then is also fractal. Take, for example, a line segment, or a square, or a cube: each can be broken into small copies which are obtained by similarity transformation (Figure 3). These structure, however, are not fractals. The common characteristic to all strictly self-similar structure, fractal or not, is that there is a relation between the reduction factor (i.e. scaling factor) and the number of the scaled down pieces into which the structure is divided. To begin with, one consider the set in Figure 3 which is topologically one-, two- and three-dimensional. For the line segment, square and cube there is a power law relation between the number of pieces N and the reduction factor r. By considering the line segment (Figure 3a), dividing it into N equal parts, each part will be scaled by a factor of $r = 1/N$; thus, if $N \cdot r^1 = 1$. For a square (Figure 3b), dividing it into N equal parts, then each part is scaled by a factor of $r = 1/(N)^{1/2}$; thus $N \cdot r^2 = 1$. Finally, by considering a cube (Figure 3c) and dividing it into N parts, each part will be scaled by a factor of $r = 1/(N)^{1/3}$; thus $N \cdot r^3 = 1$. In general, this is the law:

$$N = \frac{1}{r^{D_S}} \qquad (3)$$

where D=1 for the line, D=2 for the square, and D=3 for the cube. The exponent in the power law corresponds with those numbers which are familiar as (topological) dimension of the figures. Equivalently:

$$D_S = \frac{\log N}{\log 1/r} \qquad (4)$$

This formulation constitutes the fundamental equation, adopted in this work, to calculate the mathematical fractals. This novel approach, based on the acquisition, processing and analysis

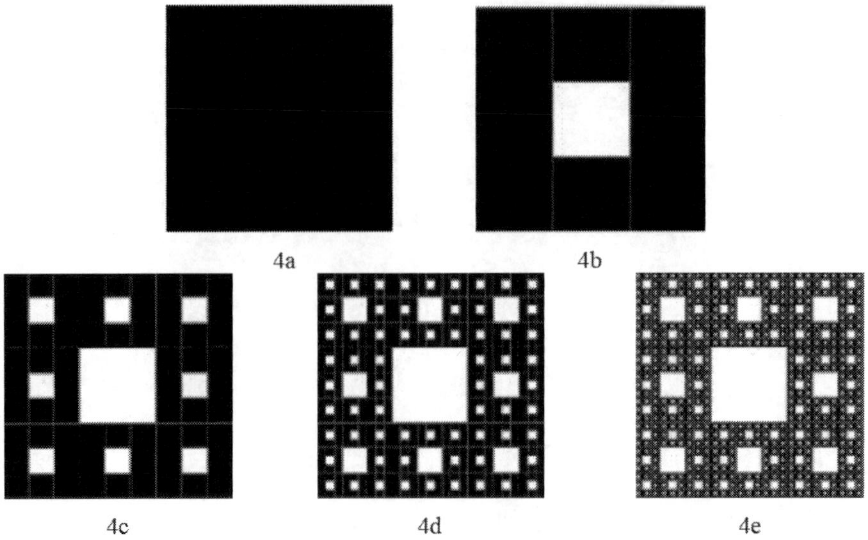

Figure 4: Initiator (4a), first (4b), second (4c), third (4d) and fourth (4e) order iterations of a Sierpinski carpet with a fractal dimension of 1.8928 [6].

of digital images of fine and ultrafine particles, considers fine particle image as a whole and not as a sum of single objects (i.e. particles). The focus on image characteristics can thus provide information on particles spatial disposition and on the structure of the disposition itself, both these aspects being linked to particles system physical-chemical attributes. The approach is based on the Sierpinski Carpet theory. The constructor algorithm of the Sierpinski Carpet along with the 1st , 2nd , 3rd and 4th order iterations, are shown in Figure 4. The algorithm is composed of an initiator, a filled square (Figure 4a), and a generator which dictates the exact form of the subsequent development, governing the division of the initiator into nine equal squares from which the central one has been removed (Figure 4b). Thus, the side of the initiator has been divided into segments of equal interval length according to the ratio of similarity, r, which in this case is 1/3. The total number, N, of the remaining un-removed squares in the generator is eight (Figure 4b). The Sierpinski fractal dimension, D, of the system can be determined using the Equation 4: it yields a value of FD=1.8929. The generator component is then imposed upon each of the remaining eight squares, thus defining a new working length (Figure 4c) and so on ad infinitum, causing the carpet's area to vanish, whilst the total perimeter of its holes becomes infinite. Obviously, it is not possible to describe the natural systems by a symmetrical carpet, such as the type shown in Figure 4. However, if one constructs a self similar randomized carpet (Figure 5a), then a carpet system, which mimics many naturally occurring system, can be generated. Imagine to look at a Sierpinski carpet (Figure 5) with several image magnification and to utilize a preliminarily low image magnification to resolve only the largest hole visible in Figure 5a, we measure the un-removed area β_1 of the carpet left when we could see holes of area α_1. Now, let us use another image magnification to see the next size of holes α_2 and estimate the un-removed area β_2 of the surface of the carpet, following this approach the graph in Figure 5b is plotted and a straight line trough the data can be drawn. The slope of this line is a measure of how fast the carpet is disappearing: in the same way as the step length λ is used at which a perimeter goes to infinity as a measure of the fractal structure of the boundary, the rate at which the carpet disappears can be utilized, with increased resolution of inspection, as a measure of the fractal structure of the system. The FD of this particular Sierpinski carpet is:

$$D = 2 - m \tag{5}$$

where, *2* is the topological dimension of carpet and *m* is the slope of the data line.

5a

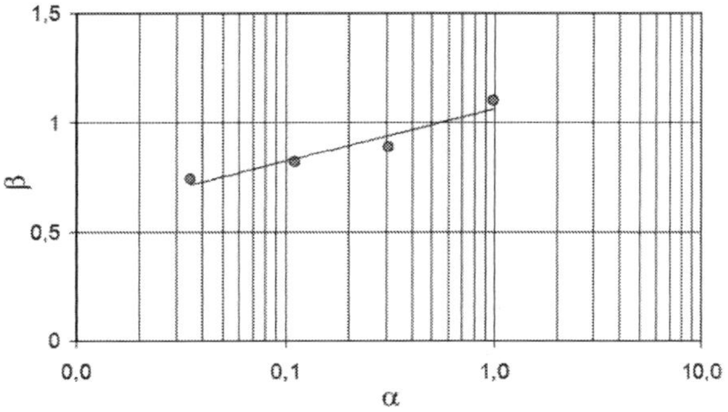

5b

Figure 5: A statistically self-similar version of the third order of iteration (5a).
5b: normalized unremoved area (β: Residual Area)
and normalized ***Just* R*esolved* T*rema*** (JRT) (α: hole) plot for the Sierpinski carpet system (5a) [6].

It follows that, in general, if it is measured the rate at which the background of a dispersed system disappears, at a series of increasing image magnifications, and discovered a linear relationship on a log-log graph, such as shown in Figure 5, then the structure of the system can be described by a Sierpinski fractal, the magnitude of which can be deduced from the slope of the data line.

To measure the Sierpinski Fractal of a dispersed system in two-dimensional space, the data have to be normalized: in particular, it has to be normalized the magnitude of each entity which constitute the analyzed system, such a goal is realized by dividing by the area of the largest entity and the area of the un-removed carpet with the total area of the field of view. Following the Sierpinsky Carpet approach, a low image magnification was thus preliminary utilized, it can only resolve the largest entity (hole) of size α_1 visible in the two-dimensional system. This clearly visible size is described in the fractal language of Mandelbrot's book as the ***Just* R*esolved* T*rema*** (JRT). The fraction of the space (carpet) bigger than JRT, would be β_1: increasing the resolution the whole of the data are transformed.

The developed 2D Fractal algorithm

The Sierpinski Carpet approach for fine particulate characterization

The Sierpinski Carpet approach has been shown to be a useful tool in studying bounded two dimensional space filling problems. In fact, the physical significance of the Sierpinski fractal is that the smaller its magnitude the greater the rate of disappearance of the background, as the resolution is increased. It follows then that it gives a measure of the efficiency with which the entities are covering the available space. In this study, the entities of the system are fine and ultra-fine particles and the available space is the microscope sample holder.

The implemented algorithm was developed in such a way to allow an automatic procedure for the extraction of the fractal dimension and other related parameters (Figure 6) [8].

The algorithm could perform the analysis of processed images with several values of *Area Threshold*, which work as an area filters on the images, to analyze separately several dimensional classes of particles (Figure 7). Once the image has been processed, the implemented algorithm

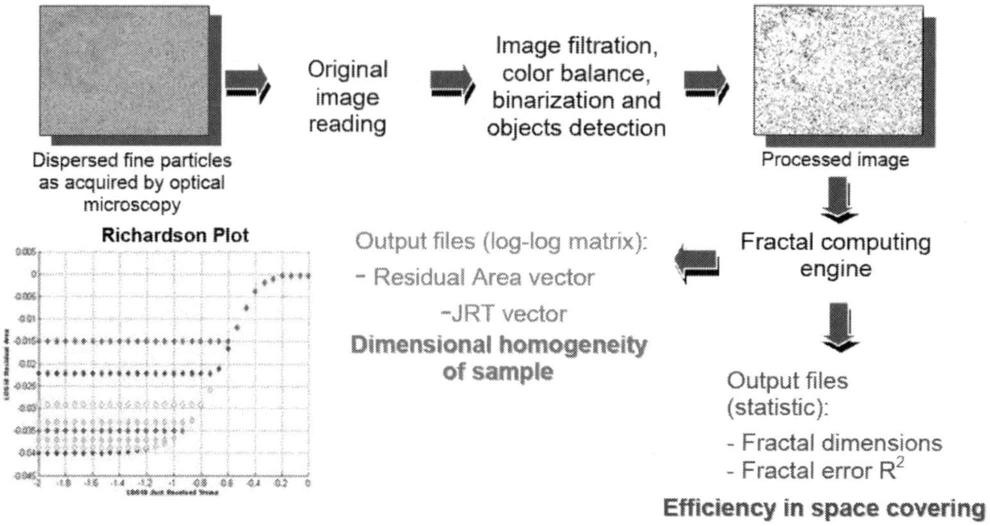

Figure 6: Flow sheet of 2D-fractal analyzer extractor algorithm.

allows to apply the automatic procedure for the extraction of:

- **F***ractal* **D***imension* (FD), which is characteristic of the *Efficiency in Space Covering*, and the precision of the computed parameters calculating their standard deviation with the Maximization of the Pearson coefficient (R^2);
- **R***esidual* **A***rea* (RA) and **J***ust* **R***esolved* **T***rema* vector (JRT), which allow to build the Richardson Plot, representative of *dimensional homogeneity* of samples.

The output files of Fractal Dimension permit to analyze two fractal ranges:

- **L***ow* **F***ractal* **D***imension* (Low FD): occurs at high resolution analysis and shows low values; it can be related to the basic structure of the image and it is indicative of the dimensional characteristics of the matrix of particles;
- **H***igh* **F***ractal* **D***imension* (High FD): occurs when lager particles are considered; these classes of particles or agglomerates seem to represent a type of disturb over the normal disposition structure and in fact, show an higher value which represents a lower efficiency in space covering.

Data were reported in a Richardson plot, on a log-log plane: for each value of Area Threshold, it was drawn a curve in which each point represents the residual background area of the microscope sample holder, so the area not covered by visible particles assuming that the only particles that can be seen are bigger than the relative values of JRT (i.e. applying an area threshold of 80 pixels, it is possible to process only the particles having an area higher than 80 pixels).

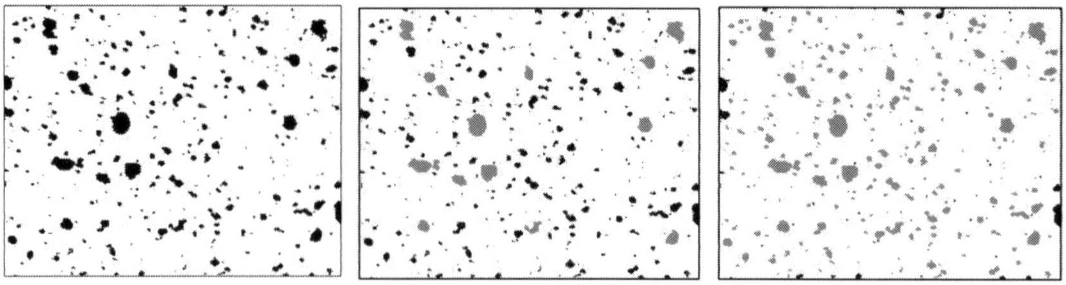

7a 7b 7c

Figure 7: Original image (7a); image filtered adopting a threshold area of 80 (7b) e 0 (7c) pixels, respectively Only the red particles were subjected to the processing.

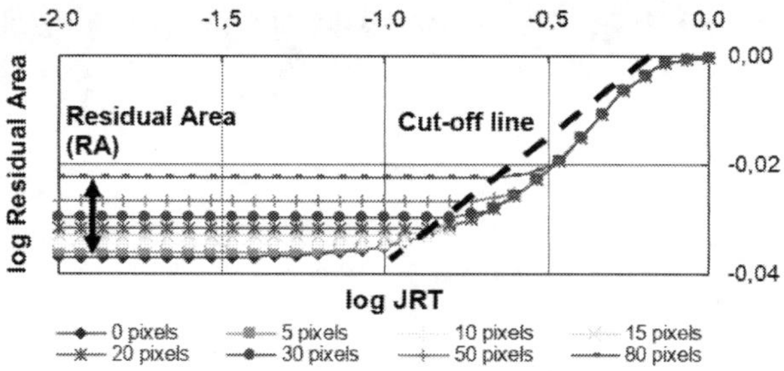

Figure 8: Example of Richardson Plot.

For high values of threshold area it is possible to analyze only the coarse particles and this is the reason because the RA increases, by moving on the right part of graph, the JRT increases (the resolution decreases) and the area not covered by particles increases.

It means that in the Richardson plot there are two linear relationship divided by a line, described in the fractal language of Mandelbrot's theory as the cut-off line (Figure 8): the leftmost part of graph can be related to the high resolution inspection and describes the textural fractal dimension of the sample. Instead, the rightmost part of graph can be related to the low resolution inspection and the structural fractal dimension, usually describing the aerodynamic behaviour of the agglomerates of fine and ultrafine particles. By using several values of threshold area, it has been possible to investigate dimensional homogeneity of samples, evaluating the difference in modulus of RA between the higher and the lower area threshold value (usually, 80 and 0 pixels respectively). The higher is this value, the greater the size distribution: so, the digital acquired image contains an assorted dimensional distribution of fine and coarse particles, thus resulting in a better surface coverage efficiency (lower value of Low FD).

The 2D-fractal approach allows to perform a good characterization of particles size distribution and their efficiency in space covering. The correlation between fractal parameters and Richardson plot shows as the smaller is the RA, calculated to the minimal resolution adopted (intercept in modulus at log JRT=-2.0), the higher is the Low FD. This means that the efficiency by which the particles are distributed over the surface (microscopy glass) is low: there are more big particles or aggregates that cover the space respect to the smaller ones.

It has been possible to investigate the dimensional homogeneity of samples, evaluating the difference in modulus of RA between the higher and lower area threshold value (in the case of study, respectively 80 and 5 pixels). The higher is this value, the greater the size distribution: digital image of particles contains an assorted dimensional distribution of small and big particles, thus resulting in a better surface coverage efficiency (lower value of Low FD).

Experimental

Materials

Different commercially available carbonate fillers have been utilized: two produced by Umbria Filler S.p.A.: TecnoFiller® and RioChim® and two by Omya S.p.A.: Omyalite 90T® and Omyalite 95T®. Their main characteristics are reported in the following.

TecnoFiller®: is a coated hyper-micronized $CaCO_3$ treated on the surface with a monomolecular film of stearic acid. It is a water-repellent powder, very light and white. The main utilizations is in rubber and PVC manufactured goods production.

RioChim®: is a technical hyper-micronized calcium carbonate, treated on the surface with calcium stearic acid. It is commonly utilized for polymers based manufactured products characterized by good mechanical characteristics and a glossy external aspect.

Omyalite 90T® and Omyalite 95T®: they are both fine coated limestone resulting from micronization processes, which differ for their size class distribution. They are usually employed in the PVC based products sector.

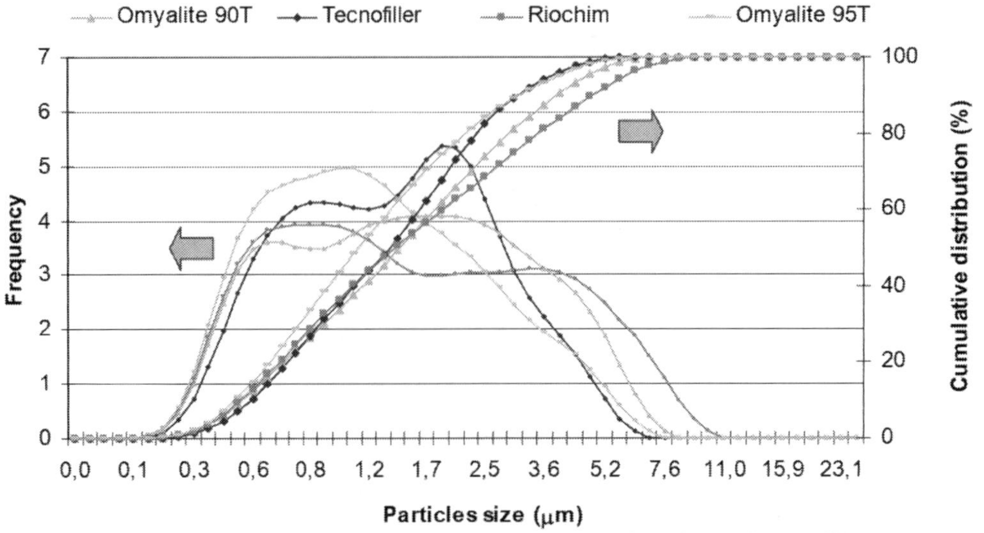

Figure 9: Size class and particles frequency distribution of calcium carbonate fillers.

On all samples morphological, morphometrical, rheological and fractal analyses have been carried out.

Morphological and morphometrical analysis

Particles size distribution (PSD) was analyzed by laser diffraction using a Bettersize BT-2003 Laser Particle Sizer (LDS). Powders were dispersed using ethylic alcohol in an ultrasonic cell for 30 min. From each filler three different samples where collected to perform the analysis and the results averaged. Calculations were based on the Mie theory. Different powders characteristics have been thus detected: the diameter corresponding to x% of the cumulative distribution (d_x); the average of the size distribution (DAD) equal to $d_{16}+d_{50}+d_{84}/3$; the uniformity coefficient (Cu), defined as d_{60}/d_{10}; curvature coefficient (Cc), defined as $(d_{30})^2/d_{10}*d_{60}$ and the fineness moduli (FM), defined as the summary of cumulative passing percentage of selected sieve sizes divided by 100. Three different measurements have been carried out for each sample.

Figure 9 displays the size class distributions of calcium carbonate fillers and their number fraction distribution (frequency). From the exam of frequency plots, it is possible to see as all the samples are characterized by only one particles population (Table 1), they show a very fine particles size distribution, concentrated in the size interval ranging from 1 to 3 μm for TecnoFiller®, Omyalite 90T® e Omyalite 95T® and from 1 to 8 μm for RioChim®.

Rheological characterization

Particles flowability was evaluated using a Manumit Powder Rheometer™ (Stable Micro System Ltd.) integrated with Texture Expert Exceed (TEE) processing software for texture analysis. The output of those tests was a series of flow indices, such as Attrition Coefficient (AC), describing the attrition properties: all the samples show very cohesive properties.

Table 1: Characteristic dimensional parameters of samples extracted by using a Laser Particles Sizer.

Sample	d_{10} μm	d_{50} μm	d_{98} μm	DAD	Cu	Cc	SPAN
Tecnofiller	0,55	1,34	4,46	1,52	2,97	0,84	1,87
Riochim	0,52	1,40	7,07	1,98	3,66	0,71	3,07
Omyalite 90T	0,52	1,44	5,48	1,76	3,47	0,84	2,34
Omyalite 95T	0,49	1,11	4,74	1,38	2,76	0,86	2,32

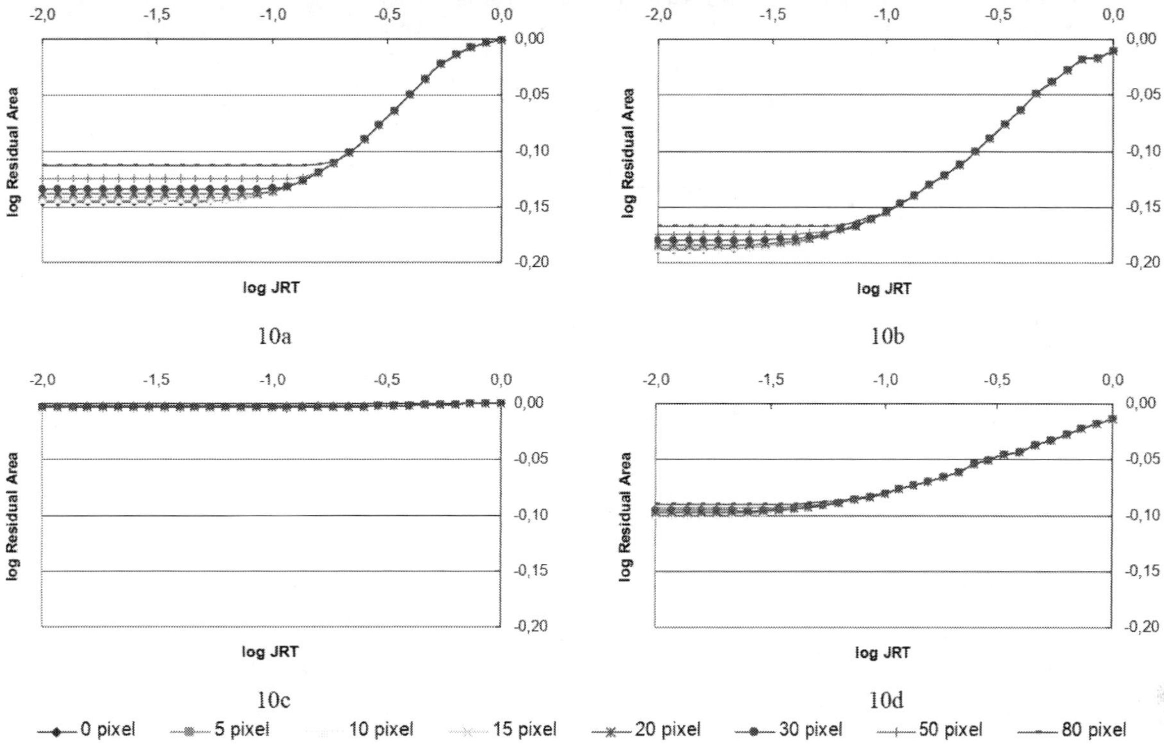

Figure 10: Richardson plots of Tecnofiller (10a), Riochim (10b), Omyalite 90T (10c) and Omyalite 95T (10d).

Fractal Analysis

The analysis of particulate samples were performed arranging them on a glass support and acquiring the images adopting a classical architecture composed by a Leica MZ75 stereomicroscope equipped with a CCD camera, directly interfaced to a PC. The acquisition of images was performed in transmitted light, with an image magnification of 160x. For each samples 20 images have been acquired, processed and analyzed. The procedures implementing the previous described 2D Fractal algorithm were developed utilizing the *MatLab®* environment, utilizing some functions implemented by the *SDC® Mathematical Morphology Toolbox*.

The acquired images were filtered, color balanced and binarized using routine functions implemented inside *MatLab®* environment. This pre-processing has been required by the diffraction effects caused by the microscopic acquisition. Therefore, the threshold function were used to create a boundary between the particle and the background (the threshold value can be changed). The images were decomposed into the constituent part (segmentation) using simple functions implemented under *MatLab®* environment. *Microsoft Excel®* data sheets implemented with visual basic macros, to collect and work out Richardson Plot charts and statistics.

The results of 2D-fractal analysis are reported in Figure 10, where the average Richardson plot for the different fillers samples considered in this study are reported. The two data vectors (log JRT and log RA) constituting the plot, have been calculated as the average over a set of 10 images. The values of the Low FD and High FD and the Low R^2 and High R^2 for each value of threshold areas, for each investigated samples, are reported in Table 2.

Results

Results have been analyzed and discussed with reference to: i) 2D-fractal analysis, ii) particles morphological, morphometrical characterization and iii) rheological characterization.

Table 2: Characteristic fractal parameters of samples:
high (High FD) and low (Low FD) fractal dimensions and their precisions (Low R^2 and HighR^2).

	Tecnofiller				Riochim			
	Low FD	Low R^2	High FD	High R^2	Low FD	Low R^2	High FD	High R^2
0 pixel	1,8792	0,9204			1,8973	0,9744	1,8406	0,8993
5 pixel	1,8793	0,9198			1,8933	0,9760	1,8439	0,8993
10 pixel	1,8797	0,9172			1,8973	0,9744	1,8406	0,8993
15 pixel	1,8818	0,9074			1,8974	0,9737	1,8406	0,8993
20 pixel	1,8830	0,8983			1,8893	0,9717	1,8486	0,9251
30 pixel	1,8882	0,8804			1,8765	0,9618	1,8365	0,8932
50 pixel	1,8977	0,8450			1,8696	0,8991	1,8644	0,8901
80 pixel	1,9103	0,7999			1,8733	0,9430	1,8738	0,9251

	Omyalite 90T				Omyalite 95T			
	Low FD	Low R^2	High FD	High R^2	Low FD	Low R^2	High FD	High R^2
0 pixel	1,9977	0,8782	1,9972	0,9812	1,9503	0,9882	1,9260	0,9355
5 pixel	1,9973	0,7526	1,9981	0,9359	1,9503	0,9882	1,9260	0,9355
10 pixel	1,9973	0,7693	1,9969	0,9479	1,9503	0,9882	1,9260	0,9355
15 pixel	1,9975	0,6938	1,9981	0,9520	1,9503	0,9879	1,9260	0,9355
20 pixel	1,9977	0,6757	1,9971	0,9762	1,9504	0,9871	1,9260	0,9355
30 pixel	1,9981	0,5505	1,9965	0,8941	1,9506	0,9830	1,9259	0,9354
50 pixel	1,9992	0,4813	1,9955	0,8388	1,9489	0,9725	1,9171	0,9433
80 pixel	1,9998	0,3253	1,9966	0,7340	1,9492	0,9576	1,9192	0,9030

Finally in Table 3 the values of the parameters of the different filler samples related to their size class distribution characteristics, rheological properties and those based on the innovative 2D-fractal approach are reported.

Results

All the Results have been analyzed and discussed comparing the results obtained by the 2D fractal approach and the classical ones based on morphological, morphometrical and rheological characterization. The first considerations are related to the relationship existing between particle size distribution and their efficiency in space covering. The correlation between fractal parameters and Richardson plot can be summarized as follow: the smaller is the RA, calculated to the minimal resolution adopted (intercept in modulus at log JRT = -2.0), the higher is the Low FD. This means that the efficiency of particles distribution over the surface (microscope sample holder) is high, because particles are characterized by a high rugged boundary, described by the fractional number, which influences the space filling ability. On the contrary, the higher is the RA

Table 3: Characteristic parameters of samples extracted by using: 2D-fractal based approach (fractal parameters are referred to threshold value of the area equal to 5 square pixels), size class distribution and rheological analysis.

	2D-Fractal Based Approach		Size Class Distribution				Rheological Analysis	
Sample	Residual Area (80-0 pixel)	Low FD (0 pixel)	d_{10} µm	d_{50} µm	DAD	Cu	r_b kg/m^3	Attrition Coefficient
Tecnofiller	$3{,}34x10^{-2}$	1,8792	0,55	1,34	1,52	2,97	618	2,43
Riochim	$2{,}07x10^{-2}$	1,8973	0,52	1,40	1,98	3,66	817	3,65
Omyalite 90T	$1{,}85x10^{-3}$	1,9977	0,52	1,44	1,76	3,47	711	2,45
Omyalite 95T	$9{,}57x10^{-3}$	1,9503	0,49	1,11	1,38	2,76	583	0,96

on the Richardson plot charts, the lower is the Low FD: this means that to the minimal resolution, the performance in space covering is low. The results of analysis showed the best covering efficiency for Omyalite 90T®, followed to Omyalite 95T®, RioChim® and TecnoFiller®.

Using several values of threshold area, it has been possible to investigate dimensional homogeneity of samples, evaluating the difference in modulus of RA between the higher and the lower values of threshold area (usually, 80 and 0 pixels respectively). The higher is this value, the wider is the size distribution. With reference to TecnoFiller® and Omyalite 90T® samples, the first shows a multimodal frequency, according to its assorted particles distribution, on the contrary, the second one shows a unimodal frequency according to its higher degree of homogeneity.

With regard to rheological analysis, and the corresponding values assumed by the Attrition Coefficient (AC), the results showed that Omyalite 95T® has the best flow properties, followed to TecnoFiller®, Omyalite 90T® and RioChim®. The same behavior could be noted examining the dimensional homogeneity of samples: it is evident that Omyalite 95T® having a high grade of homogeneity (low Cu), is characterized by the presence of air voids that improve the flowability properties. On the contrary, whit an assorted dimensional distribution (high Cu), such as TecnoFiller®, Omyalite 90T® and RioChim®, the entrained air decreases and AC increases. The same behavior could be noted examining the bulk density (ρ_b). All these characteristics are well evidenced by the results obtained following the proposed 2D Fractal approach and outlined in the resulting Richardson plots (Figure 10).

Conclusions

An innovative method based on 2D Fractal analysis is proposed to detect and to quantify the overall attributes of fines and ultra-fine particles. The developed set of procedures allows to perform the study of fine particles topological assessment on a flat surface, evaluating fines population structure and texture by computing statistical fractal parameters.

The results obtained with this innovative approach have been linked to those provided by classical morphological, morphometrical and rheological analysis in order to develop and implement a reliable and simple tool to use for particulate products characterization and recognition in order to correlate the characteristics of fine particles detected by the proposed logics and those usually extracted by using classic method (e.g. flowability, attrition, size class distribution, surface characteristic, etc.).

References

1. Feder, Jans (1998), Fractals. Plenum Press, New York and London, pp.285.
2. Adler, P.M. (1989), Flow in porous media. *The Fractal Approach to Heterogeneous Chemistry*, John Wiley & Sons Ltd. Chichester, pp. 341-359.
3. Edgar, Gerald A. (1990), Measure, Topology, and Fractal Geometry, Springer-Verlag, New York, pp.268.
4. Mandelbrot, Benoit B. (1982), The Fractal Geometry of Nature, W.H. Freeman and Company, San Francisco, pp.471.
5. Falconer, Kenneth (2003), Fractal Geometry: Mathematical Foundations and Applications, John Wiley & Sons, Ltd., pp. 328.
6. Kaye, Brian H. (1989), A Random Walk Trough Fractal Dimensions, VCH New York, pp. 453.
7. Mandelbrot, Benoit B. (1976), How long is the Coast of Great Britain, Statistical Self Similarity and Fractional Dimension, *Science*, vol. 155, pp. 636-638.
8. Bonifazi, Giuseppe (2007), 2D-Fractal BasedAlgorithms for Fine and Ultrafine Particulate Solids Systems Characterization. *Particle and Particle Systems Characterization*, vol. 24, pp. 313-321.

(291d) Wet Agglomerate Formation DEM Simulation In a Bladed Mixer

Charles Radeke[1], Daniele Suzzi[1] and Johannes G. Khinast[2], (1)Research Center Pharmaceutical Engineering (RCPE), Graz, Austria, (2)Institute for Process and Particle Engineering, Graz University of Technology, Graz, Austria

Granular flows are extremely important for the pharmaceutical and chemical industry, as well as for other scientific branches dealing with granular media. Thus, the understanding of the impact of particle size and related effects on the mean as well as on the fluctuating flow field in granular flows is critical for design and optimization of powder processing.

We use a specialized simulation tool written in C and CUDA, a massive parallelization technique which runs on the Graphics Processing Unit (GPU). It was chosen to be able to simulate significantly higher amounts of particles in acceptable runtime.

We focus on flow fields and wet-mixing properties obtained in the lower million particle range. In our talk, we present simulation results of a pitched blade mixer on a laboratory scale. Particles with a medium particle size of 365 micrometers are blended at 30 rpm for several turns (10 revolutions). The GPU implementation enable the DEM simulation of powders in high resolution. In order to investigate the mixing process during spraying of a liquid, the DEM code was extended suitable. Liquid films and bridges implement the spread of the fluid in the aggregates. Depending on the wetness level on particle surfaces, capillary forces account for cohesion and agglomeration. The simulations will shed light on the optimum location and the type of fluid entry, and allow predictions about the time until a homogeneous moisture distribution in a blended bulk can be achieved. Agglomerate formation is studied on different cohesion levels. Their spatial distribution, growth and breakage is analyzed in order to study the granulation process.

See more of this Session: Dynamics and Modeling of Particles, Crystals and Agglomerate Formation
See more of this Group/Topical: Particle Technology Forum

(291e) Aggregate Sintering Dynamics of Crystalline Ceramic and Metal Nanoparticles

Max L. Eggersdorfer and Sotiris E. Pratsinis, Particle Technology Laboratory, Institute of Process Engineering, Department of Mechanical and Process Engineering, ETH Zurich, Zurich, Switzerland

AggregateSintering Dynamics of Crystalline Ceramic and Metal Nanoparticles

Fractal-like aggregates consist of multiple particles that are connected by chemical (e.g.sintering) bonds. Such aggregates form by natural and man-made processes,typically at high temperatures, like fly ash from volcano eruption and coalcombustion as well as aerosol synthesis of ceramics (titania,fumed silica, alumina) and metals (Ni, Fe, Ag etc.). The morphology of suchparticles has critical implications in their performance. In design of nanoparticle synthesis by aerosol processes the variation ofaggregate structure during particle formation hardly affects the primaryparticle diameter. In contrast, it profoundly affects the collision andmobility diameter that determines the transport, mechanical and opticalproperties of such aggregates.

Here a recently developed model for viscous sintering ofamorphous aggregates[1] is extended tosimulate solid state sinteringmechanisms that describe coalescence of two differently sizedparticles and multiparticle aggregates of crystalline particles (e.g. TiO_2 or Ag).This model reproduces the initial neck growth and evolution of particles center-to-center distance forequally sized pairs of particles and compared to the classic models[2]and characteristic sintering times[3]. So the evolution of thedetailed morphology, radius of gyration and effective fractal dimension of ensemblesof irregular particles is presented as they asymptotically approach fullcompactness by coalescence or sintering and compared to experimental data. Figure 1 shows the evolution of aggregate morphologyduring grain boundary diffusion sintering with normalized time t/t_0 for initially 512 primary particles. At the beginningthe driving force for sintering is largest with the highest curvatures in theparticle necks (curvature = 1/radius: large = blue, small = red).

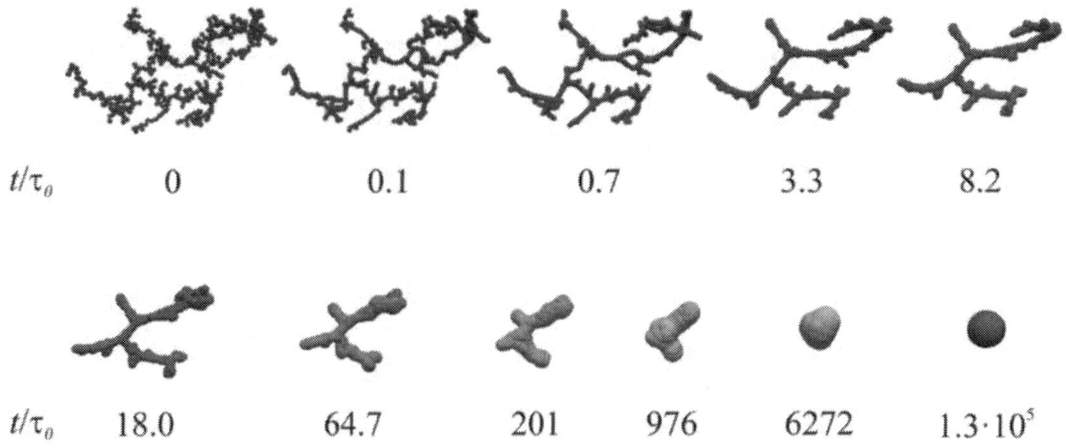

Figure1: Snapshots of aggregates undergoing grain boundary diffusion sintering andconsisting of initially 512 monodisperse primaryparticles generated by diffusion limited cluster-cluster (DLCA, D_f=1.79). The colors correspond to the aggregate curvature (= 1/radius): large =blue, small = red.

1. Eggersdorfer, M.L., Kadau,D., Herrmann, H.J., Pratsinis, S.E., Multi-ParticleSintering Dynamics: from Fractal-like Aggregates to Compact Structures. Langmuir in press (2011).

2. Coblenz, W.S., Dynys J.M., Cannon R.M., Coble, R.L., *Initial Stage Solid State Sintering Models: a Critical Analysis andAssessment*. Mat. Sci. Res. 13 (1980) 141-157.

3. Seto,T., Shimada, M., Okuyama, K., *Evaluation of Sintering of Nanometer-Sized TitaniaUsing Aerosol Method*. Aerosol Sci. Technol. 23 (1995) 183-200.

See more of this Session: Dynamics and Modeling of Particles, Crystals and Agglomerate Formation
See more of this Group/Topical: Particle Technology Forum

(291f) Application of a 2-D Population Balance Model to a Pharmaceutical Crystallization Process

Anwesha Chaudhury, Chemical and Biochemical Engineering, Rutgers, The State University of New Jersey, Piscataway, NJ, Jose E. Tabora, Chemical Development, Bristol-Myers Squibb Company, New Brunswick, NJ, Saravanababu Murugesan, Process Research & Development, Bristol-Myers Squibb Co, New Brunswick, NJ, Brenda Remy, Process Research and Development, Bristol-Myers Squibb Co., New Brunswick, NJ and Rohit Ramachandran, Department of Chemical and Biochemical Engineering, Rutgers, The State University of New Jersey, Piscataway, NJ

Population balance models (PBMs) have acquired increased popularity in the modeling of pharmaceutical crystallization processes. In this study, we develop and evaluate a two dimensional PBM as a tool to predict the crystal habit and the crystal size distribution obtained in the cooling crystallization of an active pharmaceutical ingredient. The crystallization system was found to exhibit minimal agglomeration, and a PBM with growth and secondary nucleation parameters was found to be an adequate descriptor for the system. Laboratory and plant (50 L scale) data was used to fit the model parameters and evaluate the model predictions. Two distinct crystal planes with independent secondary nucleation and growth properties were used to describe the evolution and distribution of the crystal properties. The model successfully captures the transition from rod-like particles with high aspect ratio to prisms-like structures. Also discussed are the strategies to model situations in which the seed acts as a source of heterogeneous nucleation.

See more of this Session: Dynamics and Modeling of Particles, Crystals and Agglomerate Formation
See more of this Group/Topical: Particle Technology Forum

(291g) Resolved Simulation of Calcium Carbonate Crystallization Using the One-Dimensional Turbulence Model

Derek D. Harris and David O. Lingell, Chemical Engineering, BYU, Provo, UT

Resolved simulation of calcium carbonate crystallization using the one-dimensional turbulence model

Derek D. Harris, David O. Lignell

Abstract –AIChE 2011 Annual Meeting

Mineralization of carbonate species is an important process with wide ranging use and application. There has been recent interest in technologies that produce carbonate precipitates from CO_2 in combustion flue gas, either for cement production, or as a stable means for CO_2 sequestration. More generally, carbonate precipitation is a specific application of aqueous crystallization processes.

The modeling and simulation of these processes are a significant challenge. These systems often involve turbulent mixing among two or more input streams, and require the accurate representation of a crystal population balance. Often, only rough details of this balance are of interest, such as product conversion or mean particle size, but the accurate determination of such quantities depends on the detailed physical processes occurring. The final total conversion of reactants may include dependence on particle size, shape factors, crystal structure, as well as the liquid composition. In turbulent environments, turbulent mixing may strongly impact the precipitation and crystal growth processes due to dependence on local liquid state. Disparate time and length scales are present, with scales ranging from fast liquid ionic chemistry, to particle nucleation, to slower particle growth processes. Additional complexities arise due to strong differential diffusion between particles (with essentially no diffusivity) and the mixing liquid streams (so that the solids will transport through the liquid composition space as bulk convection occurs).

Full resolution of these processes is commonly achieved using direct numerical simulation (DNS), where computational costs are extremely high, and configurations are limited to relatively low Reynolds numbers in canonical flows. Reynolds averaged Navier-Stokes (RANS) and large eddy simulation (LES) are effectively employed, but rely on subgrid models for unresolved fine structure, and cannot capture the detailed chemical and mixing processes that occur. The one-dimensional turbulence (ODT) model is an approach that fully resolves all turbulent flow structures in a single dimension, which may be regarded as a notional line-of-sight through a flow. This model has been successfully applied to a variety of turbulent flows, e.g., jets, wakes, mixing layers, buoyant stratified flows, and turbulent combustion, to name a few applications. ODT is computationally affordable since it solves only unsteady diffusion reaction equations in one dimension. Turbulent advection is modeled through stochastic eddy events implemented through domain remapping processes in a manner that is consistent with turbulent scaling laws.

We have applied the ODT model to the problem of aqueous calcium carbonate precipitation via the mixing of two reactant streams. The population balance is treated using direct quadrature method of moments (DQMOM), currently in terms of particle size, and polymorphic composition. Equilibrium is assumed for the nonideal aqueous streams, and carbonate is nucleated and grown kinetically. The yields of calcium carbonate and particle distribution properties will be presented, which depend on mixing rates, stream compositions, and temperature. These results are being applied for model development and validation for LES simulation of practical carbonate precipitation processes.

See more of this Session: Dynamics and Modeling of Particles, Crystals and Agglomerate Formation
See more of this Group/Topical: Particle Technology Forum

(291h) Control of Crystallization Using Polymeric Additives

Michael Tomasini and M. Silvina Tomassone, Chemical and Biochemical Engineering, Rutgers University, Piscataway, NJ

The control of crystallization from solution is important in a variety of fields, particularly in the pharmaceutical industry. Dependent on the particular application at hand, it may be desirous to either induce or inhibit drug precipitation. For example, many drugs currently in development are poorly water soluble, resulting in reduced bioavailability in the aqueous environment of the GI tract following oral administration. One method for increasing drug uptake, and thus bioavailability, is to create a transient supersaturated solution in the GI tract so as to increase the driving force for drug uptake. However, the natural tendency of a supersaturated solution of poorly soluble drug is to form a crystalline precipitate. It is possible to delay this onset of precipitation through the use of bioactive polymers in the drug formulation which influence drug crystallization and keep the supersaturated solution in a metastable state. Conversely, polymeric additives could also aid in promoting crystallization of a desired drug polymorph. The addition of polymeric additives to drug solutions is a topic that has been studied in the literature, but most studies are concerned with the control of crystal growth rather than nucleation. In this work, we aim to use molecular simulation techniques to further understand what general interactions influence crystal nucleation in a ternary drug-polymer-solvent system.

Our work initially focuses on a qualitative understanding of the system followed by extension to real systems. We perform atomistic molecular dynamics (MD) simulations to obtain relative interaction ratios between the components of our desired drug-polymer-solvent system. We then use this data to parameterize a coarse-grained Monte Carlo (MC) system which we utilize to study nucleation. The MC simulations are done using the gauge-cell MC technique selected because it allows for the simulation of metastable and unstable states such as crystal nuclei below their critical size. Given the resulting isotherms produced from the gauge-cell MC technique, we calculate the energetic barrier to crystal nucleation and how this barrier is influences by the presence of polymer.

In simulating a model drug-polymer-solvent system we have found that the nucleation behavior is dependent on the interaction affinities of the polymer with the drug and with the solvent in addition to the relative molecular volumes of the polymer and drug. When the polymeric additive interacts moderately with both the drug and solvent, and has a molecular volume equivalent to that of the drug, the result is for the polymer to promote nucleation. For inhibition of drug nucleation, we have found that the polymer should have a strong interaction with the drug and have a molecular volume incommensurate with that of the drug. This results in the polymer interrupting the emerging lattice structure of a newly forming drug crystal.

To validate our simulation results, we have extended our general system to a few specific systems. We performed simulations on the systems of the drug felodipine with either the polymers hydroxypropyl methylcellulose (HPMC) – a case in which nucleation is inhibited or Polyvinylpyrrolidone (PVP) – in which the polymer does not seem to influence nucleation. In addition, we have performed simulations on naringenin with either PVP or HPMC, an example of a system in which polymer promotes nucleation.

See more of this Session: Dynamics and Modeling of Particles, Crystals and Agglomerate Formation
See more of this Group/Topical: Particle Technology Forum

(310a) Anomalous Diffusion On Segregating Mono-Layers of Granular Materials: Experimental, Computational and Theoretical Observations

David C. Parra and Watson L. Vargas, Chemical Engineering, Universidad de los Andes, Bogot, Colombia

Despite recent efforts to better understand the so called "Four State of Matter", there is still no complete comprehension of the phenomena that surrounds this kind of materials. For example, segregation of granular materials due to size or density differences under shear or externally energized conditions is a very well known phenomenon but poorly understood. Such a lag in understanding makes mixing of granular materials notoriously difficult. Here we present an experimental, computational and theoretical work on segregation/mixing dynamics in a very simple system such as a bi-dimensional diffusion cell one particle deep for a wide range of filling fractions. The system considered in this work comprises a set of mono-layer round particles of low-density polyethylene of approximately 6 mm in diameter which are put under thermal agitation and the dynamics of mono-dispersed and bi-dispersed mixtures in the system are followed in time. Experimentally a piezoelectric and a wave magnitude amplifier were used in order to generate the agitation needed; a square wave with constant amplitude and frequency was implemented. A digital camera was used to record the motion of the particles and the trajectories of such particles where analyzed statistically. Granular Brownian motion, caging dynamics and other relevant behaviors and parameters are obtained from the experimental observations. A computational implementation using DEM that mimics the experimental setup is also considered. Values for mean square displacement (MSD) show the deviation from the Fick's law of diffusion as the filling fraction increases. The well-known caging effect also arises in both, experimental and computational results. Finally a theoretical model for the anomalous diffusion in such a dissipative systems is considered and the three approximations are compared.

See more of this Session: Mixing and Segregation of Particulates II
See more of this Group/Topical: Particle Technology Forum

(310b) Segregation and the Solid-Solid Drag Term

Wei Bai, Department of Chemical and Biological Engineering, Iowa State University, Ames, IA, Theodore J. Heindel, Department of Mechanical Engineering, Iowa State University, Ames, IA and Rodney O. Fox, Department of Chemical & Biological Engineering, Iowa State University, Ames, IA

A solid-
solid drag model is used to simulate segregation rates in binary mixtures of Geldart D-type particles differing in size. Three cases with different mixture compositions are simul ated and compared to available experimental data. The superficial gas velocity in each ca se is adopted to match the experimental conditions. The performance of the proposed mo del is then compared with the solid-
solid drag model available in the literature. The comparison shows that the solid-
solid drag model in the present study predicts the segregation rate better than the one fou nd in the literature. Furthermore, a modification to the literature model is made by introdu cing an energy scaling factor. The simulation results show that the modified model also p rovides a better prediction of the segregation rate.

See more of this Session: Mixing and Segregation of Particulates II
See more of this Group/Topical: Particle Technology Forum

(310c) Influence of Particle Size In a Bladed Mixer Using the Discrete Element Method (DEM)

Avik Sarkar, School of Mechanical Engineering, Purdue University, West Lafayette, IN and Carl Wassgren, Mechanical engineering, Purdue University, West Lafayette, IN

Present computational hardware limits the number of particles that can be realistically simulated using DEM to $O(10^6)$ or less, whereas typical granular systems may contain more than $O(10^9)$ particles. Most DEM simulations of granular mixers model particles that are several orders of magnitude larger than the particle size expected in real systems. The goal of this work is to characterize the influence of particle size on flow and mixing in a mechanically agitated granular system. A periodic section of a horizontal cylindrical mixer with a pair of impeller blades is modeled using the discrete element method (DEM). The ratios of mixer diameter to particle diameter investigated range from 12.5 to 75. Comparisons of flow, microstructure, and mixing are presented using measurements of flow fields, velocity distributions, coordination number, and granular diffusion coefficients. Results suggest that particle level measurements such as coordination number and mixing rates are a strong function of particle size, but the macroscopic flow velocities are not affected significantly when using larger particles. The choice of particle size in DEM simulations should be decided based on the measurements that are of interest, along with the available computational resources.

See more of this Session: Mixing and Segregation of Particulates II
See more of this Group/Topical: Particle Technology Forum

(310d) Granular Segregation In Rotating Cylinders

Suman K. Hajra, TBA, Pittsburgh, PA and Devang V. Khakhar, Department of Chemical Engineering, Indian Institute of Technology, Bombay, Mumbai - 400076, India

Segregation is a phenomenon that has long been recognized as the cause of several problems in materials handling and processing industries. The segregation of particles during the handling of bulk solids can have a detrimental effect on the quality of products in industrial processes. Rotating cylinders are commonly used in the chemical and process industries as kilns, mixers, dryers, reactors, and granulators to process granular materials. Good cross-sectional mixing in the cylinders is important for product quality and efficient operation in terms of energy usage for continuous flow systems. Granular mixtures of particles with different properties (e.g., size, density or both) tend to spontaneously demix and form segregated regions. Neither the experiments nor the models have yet proved universal enough to describe all the mechanisms that cause segregation in shear flow. The flow of granular materials in most industrial situations is very complicated. There is a great need for experiments, simulation and mathematical modeling that give quantitative results of the evolution of segregation. The dynamics of granular segregation are not well understood yet, hence the basis of design of solid mixers is in ad hoc. Experimental technique and results of granular segregation for binary mixtures will be discussed. A mathematical model for segregation will be presented. Results of granular segregation for ternary mixtures will be discussed.

See more of this Session: Mixing and Segregation of Particulates II
See more of this Group/Topical: Particle Technology Forum

(310e) Rheology and Segregation of Granular Mixtures In Dense Flow

Devang V. Khakhar, Department of Chemical Engineering, Indian Institute of Technology, Bombay, Mumbai - 400076, India and Anurag Tripathi, Chemical Engineering, Indian Institute of Technology, Bombay, Mumbai - 400076, India

We investigate the rheology of granularmixtures, comprising particles of different size and density, in aflow on a rough inclined plane by means of Discrete Element Method(DEM) based simulations. Steady flows at relative high particlevolume fractions (> 0.5)are considered for different density ratios, size ratios andinclination angles. A new model based on the viscoplastic model ofJop et al. (*Nature*, **441**,727, 2006) is shown to describe the rheology of the mixtures quitewell. Segregation of particles is analyzed by considering thesedimentation of single particles in the flow. We find that modifiedversions of Stokes Law and Archimedes principle may be used to obtainthe drag force and buoyancy experienced by the sedimenting particle.A continuum model is developed to predict the equilibrium segregationin the flow and compared to simulation results.

See more of this Session: Mixing and Segregation of Particulates II
See more of this Group/Topical: Particle Technology Forum

(310f) Predicting Segregation In Handling Processes Using Hybrid Discrete and Continuum Approaches

Kerry D. Johanson, Material Flow Solutions Inc., Gainesville, FL

One of the main causes for unscheduled plant shut-down is segregation of the mixture within the system. Individual components can segregate due to sifting of fines through a matrix of course particles. Differences in repose angle can induce material separation during pile formation. Fines may be carried by air currents and deposit in the process equipment where gas velocities decrease sufficiently to drop the particles out of the gas stream, causing separation as material fills a process vessel. Each of these mechanisms has a segregation pattern that is characteristic of the segregation profile. Many theories describing segregation are limited to two components. Multiple components often complicate the interpretation of segregation mechanism. In addition, prediction of segregation in process vessels is a complicated process especially when dealing with multi-component mixture or multiple mechanisms. To understand the influence of segregation in process vessels one must be able to compute the velocity profiles in process equipment during flow coupled with a mathematical model describing the cascade action of the free pile surface during both filling and emptying cycles. The effect of multiple processes in series can complicate the segregation issue, making prediction using simple rules ineffective. This paper presents a methodology to predict segregation in real processes using results from a novel segregation test method, standard flow properties, process geometry, and operation parameters. The net result of this approach is the ability to predict the segregation expected as material passes a prescribed location in the process. This approach should be valuable to process engineers to aid reliable process design.

See more of this Session: Mixing and Segregation of Particulates II
See more of this Group/Topical: Particle Technology Forum

(310g) Size Induced Segregation Inside the Feed Frame and During the Filling the Dies Using DEM Simulations

Daniel Mateo, Chemical Engineering, University of Puerto Rico Mayaguez Campus, San Juan, PR, Luis Obregon, Chemical Engineering, University of Puerto Rico Mayaguez, Mayaguez, PR, Rafael Mendez, Chemical Engineering, University of Puerto Rico, Mayaguez, PR, Carlos Velazquez, Chemical Engineering, ERC for Structured Organic Particulate Systems, University of Puerto Rico, Mayaguez, PR and Fernando. J Muzzio, Chemical and Biochemical Engineering, Rutgers University, Piscataway, NJ

Feed frame is used to fill powder into the tablet press dies. Die filling is one of the key steps to control final properties of tablets. Granular materials may segregate upon processing and handling based on differences in particle properties such as size, density, or shape. The Discrete Element Method (DEM) is a computational method for simulating the dynamics of particle processes. DEM was used to simulate the equipment used in the experimental part**to** understand powder phenomena inside the feed frame. Segregation behavior of a single material with a particle size distribution was investigated using this method. The computational setup was a fully 3D model of a Manesty Betapress feed frame with real boundaries, which represents die-filling process without the compression step. A cohesion model was used for some simulations. The DEM simulation component includes a 2 feed frame speed (24and 72 RPM) and 2 die disc speeds (29 and 57 RPM). It was found that the lower the die disc speed, the lower the relative standard deviation (RSD) of the die weight. Simulations with and without the cohesion linear model show basically the same hold-up and die weight variability. Results demonstrated size segregation. This effect was more noticeably for low feed frame speed. Velocity profile shows higher speed when powder enters to the feed frame and for powder closer to the paddles but separated to the center of the wheels. The most complicated flow pattern occurs between the two paddles and over the exit in the second stage. Particle velocity profile was lower at low feed frame speed and suggesting the posibility of a more significant particle rearrengement when going into the feed frame partially explaining a significant size segregation.The cumulative size distribution for 50 random dies demonstrate a more significant die size segregation at lower feed frame speed demonstrating size segregation inside the feed frame and dies.

(335a) Chemical Looping Technology for Fossil Energy Conversion - Solids Flow Issues

L. - S. Fan, Chemical and Biomolecular Engineering, The Ohio State University, Columbus, OH

Chemical Looping Technology for Fossil Energy Conversion - Solids FlowIssues*

by

L.- S. Fan

Distinguished University Professor

C John Easton Professor in Engineering

Professor of Chemical and Biomolecular Engineering

The Ohio State University

*to be presented at the Fluidization Session of the 2011 AIChE Annual Meeting in honor of Ted Knowlton

Abstract

The concept ofchemical looping reactions has been widely applied in chemical industries, e.g.,the production of hydrogen peroxide (H_2O_2) from hydrogenand oxygen using 9,10-anthraquinone as the loopingintermediate. Fundamental research on chemical looping reactions has also beenapplied to energy systems, e. g., the splitting of water (H_2O) toproduce oxygen and hydrogen using ZnO as the loopingintermediate. Fossil fuel chemical looping applications had been usedcommercially with the steam-iron process for coal from the 1900s to the1940s andhad been demonstrated at a pilot scale with the carbon dioxide acceptor processin the 1960s and 1970s. There are presently no chemical looping processes usingfossil fuels in commercial operation. A key factor that hampered the continueduse of these earlier processes for fossil energy operation was the inadequacy ofthe reactivity and recyclability of the looping particles. This factor led tohigher product costs for using the chemical looping processes, compared to theother processes that were petroleum or natural gas based. With CO_2 emission control now beingconsidered as a requirement, interest in chemical looping technology hasresurfaced. In particular, chemical looping processes are appealing due totheir unique ability to generate a sequestration-ready CO_2 stream whileyielding high energy conversion efficiency. Renewed fundamental and appliedresearch since the early 1980s has emphasized improvement over the earliershortcomings. New techniques have been developed for direct processing of coalor other solid carbonaceous feedstock in chemical looping reactors. Significantprogress is underway in particle design, reactor development, and loopingsystem integration, as demonstrated by the operation of several pilot orsub-pilot scale units worldwide, making it possible that chemical loopingtechnology may be commercially viable in the future for processing carbonaceousfuels.

Thispresentation will describe the fundamental and applied aspects of modernchemical looping technology that utilizes fossil and biomass as feedstock. Thepresentation will discuss solids flow issues associated with this technology. Specifically, it will highlight solids-gas contact modes with respect to theoptimum feedstock conversion and relationship among the metal oxide conversion,solids flux and reactor size. Opportunities and challenges for chemical loopingprocess scale-up and commercialization will also be illustrated from the solidsflow standpoint.

See more of this Session: Special Session: To Celebrate Ted Knowlton's Career Long Accomplishments II
See more of this Group/Topical: Particle Technology Forum

(335b) Effect of Stokes Number On Dilute Turbulent Liquid-Solid Flow: An Experimental and Numerical Study

Akhil A. Rao[1], Mark Pepple[2], Deepak Rangarajan[2], **Jennifer Sinclair Curtis**[1], Bruno C. Hancock[3] and Carl Wassgren[4], (1)University of Florida, Gainesville, FL, (2)Chemical Engineering, University of Florida, Gainesville, FL, (3)Pfizer Inc., Groton, CT, (4)Mechanical engineering, Perdue University, West Lafayette, IN

This work presents an experimental and numerical study of dilute turbulent liquid-particle flow spanning a wide range of Stokes numbers. LDV/PDPA experiments performed with 0.5, 1 and 1.5 mm glass particles, with solid concentration 0.7-3% by volume and in highly turbulent water with Re ~ $2x10^5$-$5x10^5$ investigate the viscous, transitional and collisional flow regimes in a vertical pipe.

The experimental data are used for validating an Eulerian based two-fluid model from which flow predictions are obtained. The drag force, solid stress, and the fluctuating energy interaction closures (previously published and developed herein) vary depending on whether the suspension is in the viscous, transitional or collisional flow regimes. The numerical results compare favorably to the experimental data.

See more of this Session: Special Session: To Celebrate Ted Knowlton's Career Long Accomplishments II
See more of this Group/Topical: Particle Technology Forum

(335c) Rapid Coal Pyrolysis to Acetylene In Thermal Plasma Reactor

Yi Cheng and Yong Jin, Department of Chemical Engineering, Tsinghua University, Beijing, China

Using thermal plasma for coal pyrolysis to acetylene provides a direct route to make chemicals from coal resources. The possibility of this route was first proved in the 1960s. Since then, rapid pyrolysis of coal to gaseous products was ever paid much attention worldwide. As the coal particles are pyrolyzed in a plasma reactor under extreme conditions (e.g., the ultra-high temperature greater than 3000 K), complicated physical phase changes and chemical reactions occur simultaneously in milliseconds contact time. The rapid heating of coal particles to release volatile matters has significant impact on the overall reactor performance. It has been confirmed that thermal energy is the driving force for the coal devolatilization. Different heating rates and the residence time of particles in the high temperature zone will lead to different yields of light gases. Therefore, the reactor design is directed to the formation of desired temperature field inside the reactor to guarantee the sufficient heating of coal particles in milliseconds.

Plasma reactors with different configurations at the lab scale of ~10 kW have been built to study the millisecond pyrolysis performance for a variety of coal based feedstock. The influences of atmospheric environment, power input, coal injection position and feeding rate on the reactor performances were investigated. A comprehensive computational fluid dynamics with discrete phase model (CFD-DPM) has been established to describe rapid coal pyrolysis process in a reactor under ultra-high temperatures. The simulations based on this model helped to understand the complex gas-particle reaction behavior in the millisecond process of coal pyrolysis. The particle-scale physics such as the heat conduction inside solid materials, diffusion of released volatile gases, coal devolatilization, and tar cracking reactions were incorporated. The improved chemical percolation devolatilization (CPD) model was applied to describe the devolatilization behavior of rapidly heated coal based on the physical and chemical transformations of the coal structure. The CFD-DPM method was validated by comparing the predicted volume fractions of main species and yields of light gases to the experimental data under a set of typical operating conditions. Accordingly, the reactor design can be optimized with the guidance of the aforementioned simulations. In particular, the multi-stage design for coal pyrolysis using thermal plasma will be evaluated by both experiment and simulation results. At last, the recent progress of a 5-MW pilot-plant industrial unit is to be overviewed, showing the perspective of this novel, clean coal conversion technology.

See more of this Session: Special Session: To Celebrate Ted Knowlton's Career Long Accomplishments II
See more of this Group/Topical: Particle Technology Forum

(335d) Numerical Simulation of CO2 Removal Process Using Solid Sorbent In a Fluidized Bed

Emadoddin Abbasi and **Hamid Arastoopour**, Department of Chemical and Biological Engineering, Wanger Institute for Sustainable Energy Research (WISER), Illinois Institute of Technology, Chicago, IL

Carbon capture process based on solid sorbents is being considered as a potential solution to remove CO_2 from power plant flue gas. The process includes a circulating fluidized bed consisting of a riser as carbonator and a bubbling fluidized bed as regenerator to ensure feasibility of continuous operation of the system.

Experimental data in the literature on fresh, carbonated and regenerated sorbent particles have shown that the sorbent particles size distribution remains almost constant during carbonation and regeneration processes. However, the density of sorbent particle is a function of particle porosity and changes during both sorption and desorption processes. The structural changes in the particles due to the reaction between CO_2 and sorbent lead to the changes in porosity of particles. The focus of this study is to couple the Eulerian-Eulerian CFD model (Arastoopour, 1999 and Benyahia et al., 2000) with the population balance equation (PBE), to account for the evolution of particle density distribution. The model is used to simulate the reactive gas-solid flow inside the riser part of a circulating fluidized bed. The effect of change in particle density distribution on the hydrodynamics and the reaction are investigated.

See more of this Session: Special Session: To Celebrate Ted Knowlton's Career Long Accomplishments II
See more of this Group/Topical: Particle Technology Forum

(335e) A Low Pressure Vertical Lift System for Coarse Powders and Granular Solids

Clive E. Davies, School of Engineering and Advanced Technology, Massey University, Palmerston North 4442, New Zealand

This paper outlines the construction, development and operation of a simple feeder inspired by the L-Valve. Non-mechanical seals are attractive in hygienic applications such as food processing because there are no wear surfaces, and the threat of contamination of the product by filings and other wear detritus is eliminated. The lifter is essentially comprised of a vertical tube attached to the central arm of a Tee; an elbow attached to one of the horizontal arms leads into a vertical downcomer which holds the feed material and another elbow attached to the other horizontal arm admits conveying air. Solids flow rate is controlled by aeration air admitted to a point on the centre-line of the Tee, at the base of the downcomer. This device is cheap to make and install, or to retrofit to existing installations, and has found widespread and continuing application in the food industry.

See more of this Session: Special Session: To Celebrate Ted Knowlton's Career Long Accomplishments II
See more of this Group/Topical: Particle Technology Forum

(335f) Investigation of Minimum Stability Factor for Plate Grid Distributors

E. Nicholas Jones, ExxonMobil Research and Engineering Company, Fairfax, VA

Plate grid distributors have been widely used in fluidized bed operation for over fifty years. A critical design parameter for grid distributors is the stability factor defined as the ratio of the grid pressure drop to the pressure drop within the fluidized bed (DP_{grid}/DP_{bed}). Sufficiently large grid pressure drop is required to activate all the grid holes and uniformly fluidize the bed of solid particles. A low value for the stability factor can lead to defluidized regions resulting in large temperature and pressure fluctuations as well as adversely affecting the performance of cyclone diplegs returning entrained particles to the fluidized bed. Several literature studies have investigated the minimum value for the stability factor in order to achieve uniform fluidization.[1, 2, 3, 4] A key focus for these studies was the impact of the bed aspect ratio (diameter:height) on the quality of fluidization. However, these studies exclusively investigated plate grid distributors where only the fluidizing gas passes through the grid holes. Older designs of fluidized catalytic cracking units (FCCUs) employ grids where both gas and catalyst particles flow through the grid holes, and it has been typical practice to use the same minimum stability factor for these designs as for the air-only grids.

This study presents experimental results on the fluidization behavior for both air-only and air-catalyst plate grid designs. The investigation was conducted at Particulate Solid Research, Inc (PSRI) in a scaled-down circulating fluidized bed of FCC catalyst where the stability factor was varied by manipulating grid hole velocity, solids flux, and bed depth. A fluidization index is defined to account for variations in fluidization quality occurring in different regions of the catalyst bed as measured by high speed pressure transducers. Results demonstrate that the fluidization index varies with changes in the stability factor for the different grid designs. The findings from these experiments are shown to be reasonable based on an analysis of the kinetic energy of the gas and catalyst mixture flowing through the grid distributor.

1. Gregory, S. A., *Proc. Int. Symp. Fluidization*, Eindhoven, Neterhlands

University Press, Amsterdam, 1967.

2. Zenz, F. A., *Proc. of Tripartite Chem. Engr. Conf.*, Montreal, Inst. Chem. Engr., 1968

3. Geldart, D. and Baeyens, J., *Powder Tech.*, **42** (1985)

4. Sathiyamoorthy, D. and Horio, M., *Chem. Engr. J.*, **93** (2003)

See more of this Session: Special Session: To Celebrate Ted Knowlton's Career Long Accomplishments II
See more of this Group/Topical: Particle Technology Forum

(335g) Applicability of a Uniflow Cyclone As Third Stage Separator in the FCC-Process

M. Kraxner[1], U. Muschelknautz[1], S. B. Reddy Karri[2], Ray A. Cocco[2] and Ted M. Knowlton[2], (1)MCI Management Center Innsbruck, A-6020 Innsbruck, Austria, (2)Particulate Solid Research, Inc., Chicago, IL

Motivation

Actually the Third Stage Separator (TSS) in a FCC-process is constructed as a standard reverse flow

cyclone which can be designed by approved calculation models, as for example by using the model of

Barth/Muschelknautz

Experimental

The experimental testing was performed on 4 uniflow cyclone configurations differing from each other

regarding the gas outlet tube diameter and the guide vane angle at the cyclone inlet generating the swirl

strength. Figure 1 shows the schematic assembly of the multi uniflow cyclone test facility. The operating

conditions have been scaled down from a typical FCC-process.

84

86

88

90

92

94

96

98

0 1 2 3 4 5 6

Separation Efficiency in %

0

2.000

4.000

6.000

8.000

10.000

12.000

14.000

Pressure Drop in Pa

Separation Efficiency Pressure Drop

Config A

Config B

Config D

+Underflow

Config C

Config D

Figure 1: Schematic assembly of the testing facility (left) and the performance data of the investigated configurations A to D (right)

Results

Figure 2 shows the particle size distribution (PSD)

and the fractional efficiency (

h

0

10

20

30

40

50

60

70

80

90

100

1,0 10,0 100,0 1000,0

Particlediameter d

p

in μ m

Weight percent under size,

h

(d) in %

Feed

Collected

Fines

h

h

Figure 2: PSD and fractional efficiency of configuration D

with 3% underflow

((dd))

(d)) of the multi uniflow

cyclone configuration D with an underflow of 3

percent of the main gas flow. The usage of an

underflow raises the separation efficiency from 94.8

up to approximately 95.5 percent, Figure 1. The PSD

of the used FCC-powder shows an amount of 24

percent fines below 20 microns, Figure 2. Fine

particles, leaving the cyclone with the pure gas, have

a mean diameter of 2 microns. The fractional

efficiency of configuration D allows a complete

separation of particles greater than 8 microns at a

pressure drop of 7.900 Pa (=1,14 psi).

. The FCC-process efficiency mainly depends on the cyclone performance i.e. on

its separation efficiency and its pressure loss. The conical shape of the lower separation body of a

reverse flow cyclone leads to high tangential velocities and therefore to high erosion which strongly

affects its life time. A second cyclone type, which is not yet well investigated, is the uniflow cyclone. Its

essential advantages compared to the reverse flow cyclone are its compact construction and its easy

integration into pipes. A further advantage is its low pressure loss and therefore its low energy

consumption.

See more of this Session: Special Session: To Celebrate Ted Knowlton's Career Long Accomplishments II
See more of this Group/Topical: Particle Technology Forum

Delivery of Cancer Drug Candidates using Core-Shell Environmentally Responsive Nanoparticles

Sriya Sanyal, Huang C. Huang, Kaushal Rege and Lenore L. Dai, Arizona State University, Tempe, AZ, USA

Abstract

It is possible to obtain intracellular drug delivery through some physical methods such as the microinjection/gene gun and hydrodynamics but there is concern that these techniques cause tissue damage. This is why most efforts focus on developing drug delivery vehicles that coat or encapsulate drugs and deliver them efficiently to the desired target. Surfactant free emulsion polymerization is used to synthesize the nanoparticles as this allows control over the size of nanoparticles. As no surfactant is used to stabilize the system, the formed nanoparticles are easily cleaned. This is an excellent process to make pure synthetic nanoparticles for biological applications. We have recently developed core (polymer)-shell (silica) nanoparticles that are promising candidates for the delivery of hydrophobic (e.g. anti-mitochondrial) drugs. A drug has been loaded in the polymeric cores during formation of the nanoparticles. An investigation into drug loading and release properties of the nanoparticles has been carried out. Thermally responsive polymers have been integrated into the cores of the nanoparticles in order to develop smart drug delivery vehicles that can be remotely triggered for releasing the drug at the tumor site. In vitro uptake of nanoparticle-mediated delivery using dye as model drug was also observed using prostate cancer cell lines. Also cytotoxicity studies using the drug loaded nanoparticles was done and showed promising cell death compared to the controls.

Introduction

There has been a lot of recent scientific interest in nanoparticles since they have properties that are significantly different than those of bulk materials [1,2]. Their high surface-to-volume ratio and narrow size distribution makes them a suitable carrier or delivery system for drugs, proteins or genes which can be either delivered locally or targeted specifically [3]. Here we employ the concept of Pickering emulsions to synthesize core-shell composite particles. This idea uses a Pickering emulsion which is an emulsion that is stabilized by solid particles (for example colloidal silica) which adsorb onto the interface between the two phases. This type of emulsion was named after S.U. Pickering, who described the phenomenon in 1907, although the effect was first recognized by Walter Ramsden in 1903. Here the polymer serves as the core and the inorganic particles serve as the shell. Such materials provide a new class of supramolecular building blocks which cannot be found simply by mixing the soft polymer and inorganic material together. Incorporation of a thermally responsive polymer: Poly (N-isopropyacrylamide) (PNIPAAm) in the core-shell nanoparticles makes dramatic changes in the degree of swelling upon small changes in the environmental conditions possible. A temperature-responsive polymer is a polymer which undergoes a physical change under external thermal stimuli. The ability to undergo such changes under easily controllable conditions makes this class of polymers fall into the category of smart materials. These physical changes can be exploited for many analytical techniques, especially for separation chemistry. After numerous investigations of poly(N-isopropylacrylamide) (PNIPAAm), there was a sparked interest in the applications of this and many other stimuli-responsive polymers. There has been extensive research in the applications of intelligent polymers for use as stationary phases, extraction compounds, surface modifiers, drug delivery, and gene delivery.

In this work, we report a synthetic pathway involving Pickering emulsion polymerization which provides adequate stability as evidenced by zeta potential measurements. Cancer drug candidate 17-(Allylamino)-17-demethoxygeldanamycin (17-AAG), an ansamycin antibiotic which binds and inhibits Hsp90 (Heat Shock Protein 90), is incorporated into the core during synthesis and used as a hydrophobic drug in the release experiments. Cellular uptake and cytotoxicity were studied in prostate cancer cells incubated at 37°C, with composite nanoparticles encapsulating 17-AAG and their controls.

Experimental Methods

Materials

IPA-ST (Nissan Chemicals) is a dispersion of 10-15 nm silica nanoparticles in 2-isopropanol at a concentration of 30-31 wt %. Nonionic azo initiator VA-086 (98%, 2,2-azobis(2-methyl-*N*-(2-hydroxyethyl)propionamide), Wako Chemicals), styrene monomer (99.9%, Fisher), N-isopropylacrylamide monomer (NIPAAm, 97%, Aldrich), and water (HPLC grade, Acro Organics) were used in the polymerization without further purification. The nonpolar dye BODIPY (493/503) (4,4-difluoro-1,3,5,7,8-pentamethyl-4-bora-3a,4a-diaza-s-indacene) was obtained from Invitrogen, Molecular Probes. The serum-free medium contains RPMI-1640 medium, 25 mM HEPE, L-Glutamine, and 1% Penicillin/streptomycin (HyClone). The serum-containing medium has 10% FBS (heat inactivated fetal bovine serum, HyClone) in addition to the serum-free medium. PC3-PSMA prostate cancer cells were generous gifts from Dr. Michael Sadelain at the Memorial Sloan Kettering Cancer Center, New York and were used as provided; PC3-PSMA cells are a sub-clone of PC3 cells retrovirally transduced to stably express the PSMA receptor [4]. PC3 cells were purchased from ATCC. The drug candidate, 17-(Allylamino)-17-demethoxygeldanamycin (17AAG) was obtained from LC Laboratories, MA. 3-(4, 5- dimethylthiazol-2-yl)-2, 5-diphenyl tetrazolium bromide (MTT) was the assay (ATCC Inc.) used in cytotoxicity experiments.

Composite Nanoparticle Synthesis

Water, IPA-ST, NIPAAm and styrene were agitated mechanically with an IKA Ultra Turrax T25 homogenizer at 10,800 rpm for 2 min in an ice bath to emulsify. Then, the emulsion was degassed with nitrogen and kept in nitrogen atmosphere under magnetic stirring. When the temperature was raised to 70°C, the initiator aqueous solution was added to start the polymerization. The dye BODIPY493/503, when used for cellular uptake experiments, was added prior to adding the initiator. A typical formulation of the thermo-responsive nanoparticle includes 0.66 g NIPAAm, 3.76 g styrene, 32 mL water, 4.1 g IPA-ST silica nanoparticle dispersion, 0.037 g initiator VA-086, and 1 µg BODIPY 493/503. For the release experiments, 17AAG was added to the emulsion before the polymerization took place, prior to adding the initiator at a concentration of 600 g/ml. Before characterization, release and uptake experiments, the samples were washed twice by centrifuging-redispersing cycles using an Eppendorf 5810R centrifuge.

Characterization Methods

Particle size distribution and zeta potential of composite particles were measured using a Brookhaven 90Plus Particle Size Analyzer with the dynamic light scattering (DLS) technique. The washed composite particles were further dispersed to appropriate concentrations with water before measurements. Specimens for electron microscope were prepared by placing a droplet of polymer samples onto mica substrates and dried in air. The specimens were then sputter coated with gold and viewed by SEM-XL30 (FEI). Fourier transform infrared spectra (FTIR) were scanned over the range of 400-4000 cm^{-1} with potassium bromide pellet on a Bruker IFS 66V/S FTIR spectrometer. To remove the silica shell, hydrofluoric acid (HF) etching procedure described by Han and coworkers [5] was adopted.

Cellular Uptake Experiments

Composite core-shell nanoparticles loaded with BODIPY 493/503 dye in the core for imaging and as a model drug were used in cellular uptake experiments. First, prostate cancer cells (PC3 and PC3-PSMA) were seeded in a 24-well plate at a density of 50,000/well followed by overnight incubation at 37°C under 5% CO_2. Dye-containing nanoparticles were added to cancer cells in absence of serum for 5 h, following which, nanoparticles and serum-free medium were removed and cells were washed three times with 1X phosphate buffered saline (PBS). Serum-containing medium was then added to the cells for 20 min prior to phase contrast and fluorescence imaging using a confocal microscope.

Drug Loading and Release Experiments

Composite core-shell nanoparticles loaded with 17AAG were used for the drug release experiments. The amount of drug released was determined by UV analysis. Absorbance of a number of standard solutions of the reference substance at concentrations encompassing the sample concentrations was measured using a Biotek Microplate Reader and the calibration graph was constructed.

Cytotoxicity Experiments

The cytotoxicity of nanoparticles (NPs) was determined using 3-(4, 5- dimethylthiazol-2-yl)-2, 5-diphenyl tetrazolium bromide (MTT) assay. Prostate cancer cells were cultured in medium (RPMI 1640 plus 10 % fetal bovine serum and 1% penicillin/streptomycin) under 5% CO_2 at 37°C. Cells were seeded in a 96-well plate at a cell density of 5,000 cells per well and allowed to attach overnight prior to NP treatment (0.01-1 μg/ml). At 72 h after treatment with different doses of NPs, 10 μl of MTT reagent was added to each well and incubated under 5% CO_2 at 37°C for 4 h, followed by addition of detergent. Absorbance at 570 nm was determined with microplate reader (Biotek Synergy 2). Studies were performed with at least four individual samples and repeated three times. Any interference from NPs alone (i.e. without cells) was normalized for all samples. Two-tailed Student's t-test was employed for statistical analysis.

Results and Discussion

The synthesized polystyrene/PNIPAAm-silica core-shell nanoparticles are responsive to thermal stimuli. Figure 1(a) shows the dependence of average diameter of the composite particles on temperature. The average particle size at 25°C is approximately 92 nm. The size decreases sharply as the temperature reaches 32°C, around the LCST for homopolymer PNIPAAm and size change is nearly reversible upon cooling. Control experiments of polystyrene-silica nanoparticles did not show a size transition over a temperature range of 25-45°C (data not shown). The transition temperature is not shifted by the silica nanoparticle encapsulation. This is consistent with the recently reported composite microspheres with a PNIPAAm core and a silica shell which also show a volume transition starting at 32°C [6]. It is likely due to the fact that silica particles are physically adsorbed on the surfaces of PNIPAAm microspheres thus no chemical bond formation with silica occurs which might change the transition temperature.

These temperature sensitive particles were next synthesized to contain a hydrophobic drug as a model. The drug candidate chosen was 17-(Allylamino)-17-demethoxygeldanamycin (17AAG). The main obstacle to the delivery of 17-AAG is its poor water solubility and it requires complicated formulations with Cremophor EL (CrEL), DMSO, or EtOH before parenteral administration [7]. A safer administration of 17 AAG can be made by a surfactant free delivery system rather than using harmful surfactants to solubilize the drug. . Therefore, it requires a viable drug carrier for its time-controlled release. The release study of 17-AAG from the nanoparticles was performed in deionized water. The composite nanoparticles encapsulated with 17-AAG to be tested were divided into two parts. The first portion was kept at room temperature and the second was in a temperature-controlled water bath at 40°C during the duration of the experiment. At regular time intervals, the samples were taken out and then centrifuged at 10,000 rpm for 3 min in a Beckman Coulter tabletop centrifuge and 100 L of the supernatant was drawn out from release system for analysis. Cumulative drug release measurements were performed. Figure 1(b) depicts the cumulative fractional drug release-time plot of 17-AAG from drug loaded nanoparticles at various time intervals at room temperature and 40°C. At room temperature there was no significant release of the drug. However, at a higher temperature of 40°C, the drug release from the nanoparticles reached a maximum after 7 h.

It was hypothesized that polystyrene/PNIPAAm-silica core-shell nanoparticles are sufficiently small to be taken up by cancer cells. Cell uptake experiments were performed using human prostate cancer cells (50,000/well in 24 well plates). The uptake of the composite nanoparticles by PC3-PSMA cells following incubation for 5 h was visualized using confocal microscopy.

We next compared the ability of 17-AAG-containing polystyrene/PNIPAAm-silica core-shell nanoparticles (17-AAG-PS/PNIPAAm) to induce death in prostate cancer cells, and compared it with cell death induced by polystyrene-silica nanoparticles (PS), 17-AAG-laoded polystyrene-silica nanoparticles (17-AAG-PS) and polystyrene/PNIPAAm-silica nanoparticles (PS/PNIPAAm). Two-tailed student's t-test was employed to compare the two groups, 17-AAG-PS/PNIPAAm and PS/PNIPAAm. As seen in Figure 2a, thermoresponsive nanoparticles loaded with 17-AAG (17-AAG-PS/PNIPAAm) induced the greatest dose-dependent death in PC3 cells at nanoparticle dosages from 0.01 to 1 µg/ml compared to other nanoparticle formulations. Drug-loaded nanoparticles are minimally toxic at lower doses (0.01 µg/ml), but induce death in 35 – 90% of the PC3 population at doses from 0.03-0.3 µg/ml. This is statistically significant from other nanoparticle formulations tested under similar conditions (Figure 5a), indicating that uptake and drug release of the 17-AAG containing nanoparticles resulted in death of PC3 cells. The nanoparticles were not as efficacious in inducing death in PC3-PSMA cells compared to PC3 cells (Figure 2b). While 17-AAG containing nanoparticles demonstrated higher average losses in cell viability in PC3-PSMA cells compared to other nanoparticle formulations, these differences were not statistically significant except at the highest concentration (0.3 µg/ml). This can be explained in part by the susceptibility of this cell line to nanoparticles without the drug; while 'bare' nanoparticles do not result in loss of viability of PC3 cells, they induce death in a large population of PC3-PSMA cells in a dose-dependent fashion. In addition, previous results in our laboratory have demonstrated differential therapeutic efficacy in these cell lines due to differential intracellular trafficking and localization of nanoparticles [8,9]. These differences in closely related cell lines underscore the role of the cancer cell phenotype in determining efficacy of delivered nanoparticle therapeutics and are currently under investigation in our laboratories. In addition, it is hypothesized that conjugation of prostate cancer cell specific biomolecules (e.g. antibodies to the Prostate-Specific Membrane Antigen [10]) will help receptor-mediated uptake of these nanoparticles leading to increased efficacies.

Conclusions

The focus of this work was to employ a unique drug delivery vehicle which can be taken up by cancer cells and can release the loaded drug. Polystyrene/PNIPAAm-silica core-shell nanoparticles were successfully synthesized with N-isopropylacrylamide incorporated into the core of the nanocomposite as a co-monomer. The work has demonstrated the temperature sensitivity, controlled drug release properties of the synthesized core-shell nanoparticles, and their effectiveness for inducing death of human prostate cancer cells.

(a)

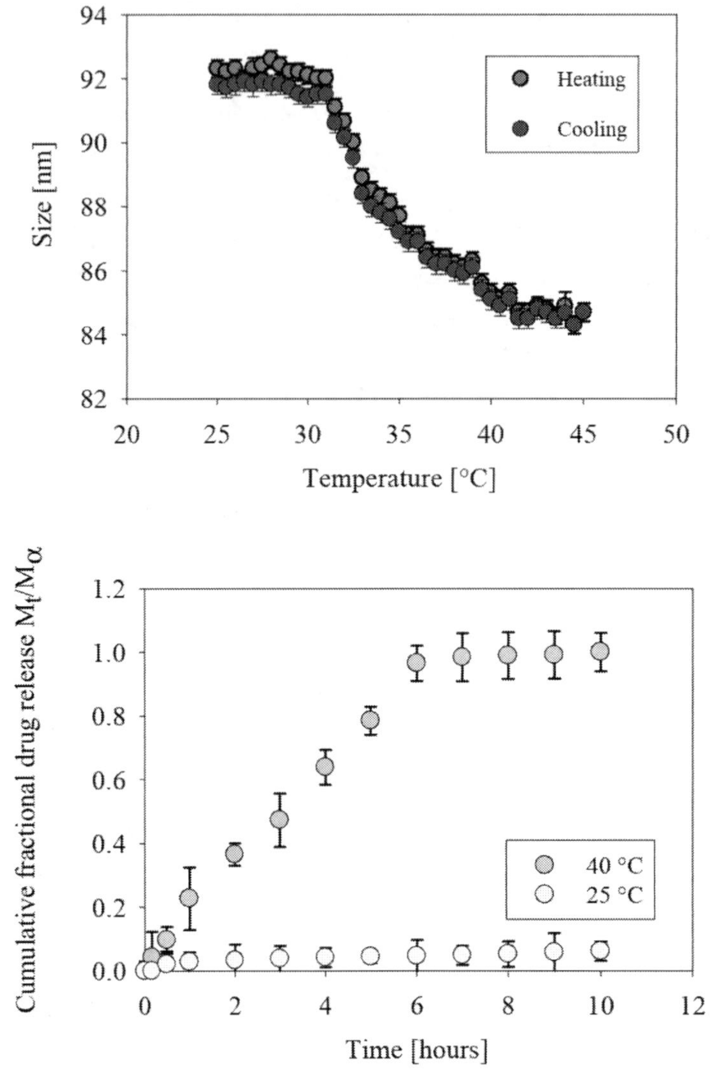

Figure 1

Figure 1. (a) The dependence of average diameter of composite nanoparticles on temperature. The error bars show standard deviations of particles made in three different batches. The transition temperature is around 32°C. There size transition is nearly reversible. (b) Cumulative fractional drug release versus time curve indicating release at room temperature and at 40°C. There is no release at 25°C which is below the transition temperature of the composite particle.

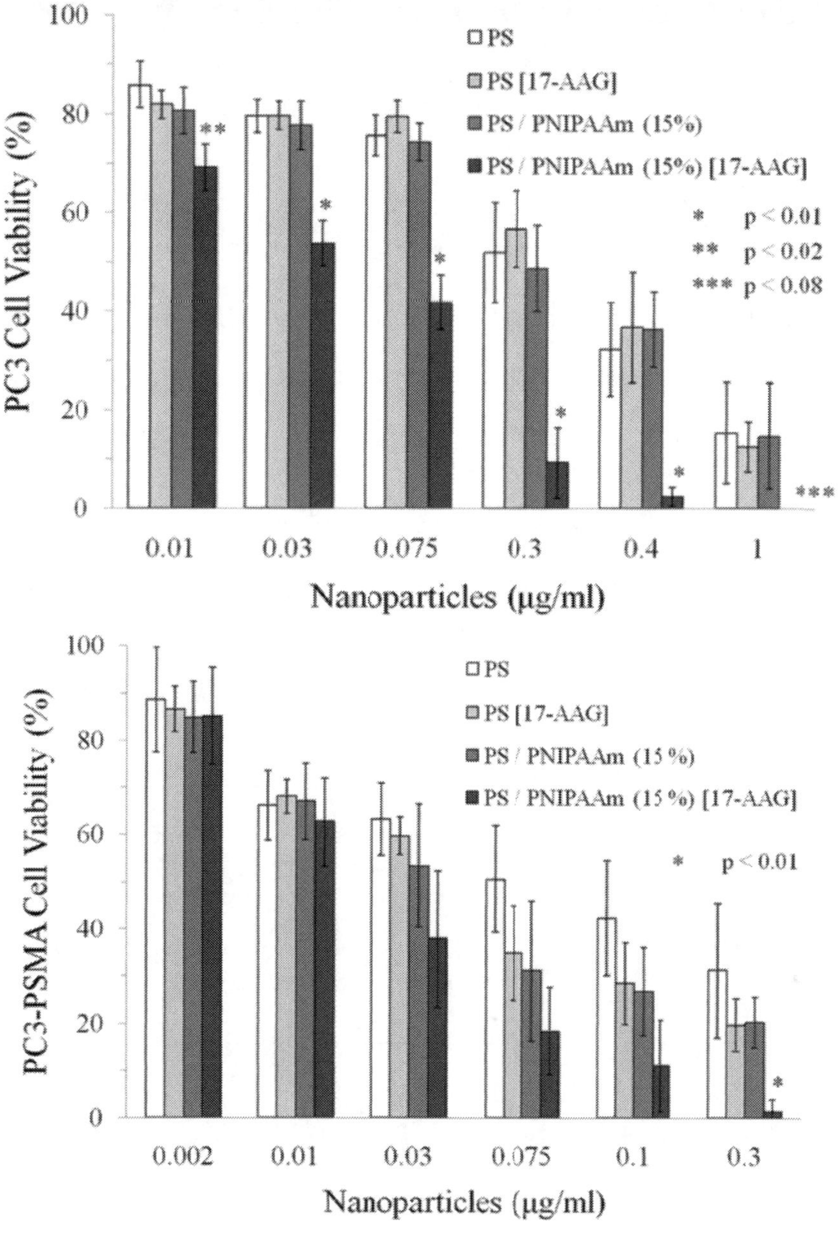

Figure 2

Figure 2. Cytotoxicity of 17-AAG loaded particles against prostate cancer cell lines (a) PC3 (b) PC3-PSMA Two-tailed student's t-test was employed to compare between 17-AAG-PS/PNIPAAm and PS/PNIPAAm treatment.

References

1. Jain P.K., Lee K.S., El-Sayed I.H., El-Sayed M.A. (2006), *Journal of Physical Chemistry, 110,* 7238.

2. Liu C. and Zhang Z. J., (2001), *Chemical Materials, 13,* 2092.

3. Ding B.S., Dziubla T., Shuvaev V.V., Muro S., Muzykantov V. R., (2006), *Molecular Interventions, 6,* 98.

4. Gong M.C., Latouche J., Krause A., Heston W.D.W., Bander N.H., Sadelain M., (1999) *Neoplasia*, *1,* 123.

5. Han, Armes S.P., (2003) *J.Colloid Interface Sci. 262,* 418.

6. Duan L., Chen M., Zhou S., Wu L., (2009), *Langmuir, 25*, 3467.

7. Shin H. C., Alani A. W., Rao D. A., Rockich N. C., Kwon G. S., (2009), *J Control Release,140,* 294.

8. Barua S., Rege K., (2010) *Biomaterials. 22,* 894.

9. Barua S., Rege K., (2009) *Small*, *5*, 370.

10. Rege K., Patel S. J., Megeed Z., Yarmush M. L., (2007) *Cancer Res. 67,* 6368.

(341b) Coupling Molecular Dynamics with Computational Fluid Dynamics Via Parameter Passing for the Simulation of Polymer Nanoparticle Precipitation

Nicodemo Di Pasquale[1], Paola Carbone[2], Daniele Marchisio[1] and Antonello Barresi[1], (1)Dismic, Politecnico di Torino, Torino, Italy, (2)School of Chemical Engineering and Analytical Science, The University of Manchester, Manchester, United Kingdom

Interest in polymer nanoparticles has grown in recent years due to their unique features as nanocarriers for controlled drug delivery. A very common technique for nanoparticle production is solvent-displacement that consists in mixing a solvent (e.g., acetone) with an anti-solvent (e.g., water). The polymer (polycaprolactone, PCL) is dissolved in the solvent, which is completely miscible with the anti-solvent. The polymer is however sparingly soluble in the anti-solvent therefore as soon as solvent and anti-solvent are mixed particles are formed. Since the process is highly sensitive to mixing, very efficient micro-devices, such as confined impinging jets reactors (CIJR), are generally employed for performing this operation.

Mathematical models are very useful in this context for reactor design and optimization. One interesting approach to describe and simulate the process is based on computational fluid dynamics (CFD) via the so-called Reynolds-averaged Navier-Stokes approach (RANS). According to this approach turbulence is modeled and mixing at the molecular level (i.e., micro-mixing) is described by using a presumed probability density function approach (i.e., direct quadrature method of moments, DQMOM, coupled with the exchange and interaction with the mean, IEM). Nucleation, growth and aggregation of the nanoparticles is modeled solving the corresponding population balance equation (PBE) through the quadrature method of moments (QMOM) [1]. One very important issue is related to the identification of all the kinetic parameters and properties needed in these models. Some of these can be obtained from independent experimental measurements (e.g., equilibrium concentration and interface tension) but for some others this is not possible. One important parameter that highly affects the nucleation rate predictions (and in turn the final predictions for a key particulate characteristic such as the particle size distribution) is the molecular volume of the polymer molecules. This changes continuously from pure acetone to pure water and therefore is not costant but changes from point to point in the CIJR.

The aim of this work is to estimate this parameter for PCL by using atomistic Molecular Dynamics (MD) simulations under different conditions (i.e., PCL in pure water, pure acetone and different intermediate concentrations). This parameter is estimated from the accessible surface area (SAS) from MD simulations, allowing to obtain the molecular volume as a function of the composition of the solvent/anti-solvent mixture. The whole model is validated by using experimental data [2]. Preliminary results show that the approach is indeed interesting, since the main trends are well described and acceptable agreement with experiments is observed.

[1] E. Gavi, D.L. Marchisio, A.A. Barresi, Olsen M.G., and Fox R.O., Turbulent precipitation in micromixers: CFD simulation and flow field validation, Chemical Engineering Research & Design, 12:1182-1193, 2010.

[2] F. Lince, D.L. Marchisio, and A.A. Barresi, Strategies to control the particle size distribution of poly-caprolactone nanoparticles for pharmaceutical application, Journal of Colloid and Interface Science, 322:505-515, 2008.

See more of this Session: Synthesis, Characterization and Modelling of Nanoparticle Systems with Pharmaceutical Applications
See more of this Group/Topical: Particle Technology Forum

(341c) Effects of Block Copolymer Properties On Nanoparticle Surface Coverage, Complement Activation, and In Vivo Circulation Half-Times

Suzanne M. D'Addio[1], Walid S. Saad[1], Stephanie J. Budijono[1], Christine Vauthier[2] and Robert Prud'homme[1], (1)Chemical Engineering, Princeton University, Princeton, NJ, (2)Universite Paris-Sud, Paris, France

There has been a dedicated effort in recent decades towards the development of efficient nano carriers that deliver drugs to the majority of tissues. For passive and active targeting, long circulation times are required, which necessitates that the drug carrier not trigger mechanisms involved in the immune system that are dedicated to the elimination of foreign particles. While it has been established that a poly(ethylene glycol) (PEG) surface coating extends circulation times, the details of polymer surface coverage and chemistry must be investigated to optimize circulation times, permitting passive and active targeting mechanisms. Rapid precipitation and assembly of nanoparticles and stabilizing block copolymers enables the formation of a PEG corona which is denser than an equilibrium adsorption process permits. We have compared data from *in vivo* nanoparticle circulation studies to *in vitro* complement activation to gain insight on the effects of hydrophobic block chemistry and molecular weight, finding that large crystalline blocks destabilize nanoparticles in the presence of serum proteins and short, amorphous blocks enable enhanced protection and longer circulation times.

See more of this Session: Synthesis, Characterization and Modelling of Nanoparticle Systems with Pharmaceutical Applications
See more of this Group/Topical: Particle Technology Forum

(341e) Facile and Robust Fabrication of Sub-100nm Poly(lactic-co-glycolic acid) Nanoparticle-Based Drug Delivery Vehicles Via Self-Assembly

Gang Ruan, Department of Chemical and Biomolecular Engineering, The Ohio State University, Columbus, OH, Kalpesh Mahajan, Chemical and Biomolecular Engineering, The Ohio State University and Jessica Winter, Chemical and Biomolecular Engineering, The Ohio State University, Columbus, OH

Controlled drug release using polymer (most commonly poly(lactic-co-glycolic acid), or PLGA) nanoparticles has achieved great success including clinical validations. However, it has been challenging to produce PLGA-based nanoparticles smaller than 200 nm. It has been well established that smaller particle size leads to superior performance in biodistribution and tumor penetration. Although the "nanoprecipitation method" has been shown to result in smaller particles, in practice this process is difficult to control. Here we report a novel approach, which is based on a bottom-up rather than top-down process, to produce PLGA nanoparticles that have particle sizes well below 100 nm.

This novel method involves using micelles as a template for self-assembly of PLGA. Amphiphilic poly(styrene-co-ethylene glycol) molecules were mixed with PLGA to form micelles with particle size approximately 50 nm. Because this method is based on self-assembly, the process is highly robust and no external energy input is required. Dexamethasone was used as a model drug to examine the potential of PLGA-micelles for controlled drug release. It was found that, compared with PLGA nanoparticles fabricated by the conventional solvent extraction method, PLGA-micelles led to significantly lower initial drug release burst. Noting that PLGA acts as an encapsulant within the fixed micelle nano-container, this technology offers a simple way to modulate drug release without changing particle size and surface chemistry, which determine the transport behavior of nanoparticles in biological environments.

See more of this Session: Synthesis, Characterization and Modelling of Nanoparticle Systems with Pharmaceutical Applications
See more of this Group/Topical: Particle Technology Forum

(341f) Quantitative Correlation for the Stability of Pharmaceutical Nanosuspensions and Their Physicochemical Properties

Wusheng Zhu, Frank Romanski and M. Silvina Tomassone, Chemical and Biochemical Engineering, Rutgers University, Piscataway, NJ

Recent results from high-throughput screening and combinatorial chemistry have reported that a large portion of viable drug candidates will be poorly water soluble, and as a consequence, exhibit a poor *in vivo* bioavailability. As a result, a decade of intense research has focused on the production of these poorly soluble drugs into nanosuspensions to increase dissolution rate, bioavailability, and improve overall drug delivery. Unfortunately, the small size of the nano-scale drug crystals causes many unfortunate and undesired consequences including agglomeration, Oswald ripening, and growth, thus destroying the desired nanosuspension properties. Therefore, in order to maintain the desired properties of the nanosuspension, the addition of surfactants and/or polymers are required to introduce steric and electrostatic stability, yet, the current selection and implementation of these additives remains limited to heuristics. As a result, our work has focused on the understanding of molecular interactions between the surface of drug nanoparticles and additives, where the thermodynamic binding energy between the two has been shown to correlate with suspension stability, as well as the ability to quench crystal growth. In this work, a correlation was established for the interaction of additives with the intrinsic properties of the target drug compounds. Nine common, poorly soluble compounds, exhibiting a wide variety of physicochemical properties, including griseofulvin, fenofibrate, itraconazole, azodicarbonamide, ibuprofen, sulfamethozazole, metochlopramide, pyrimethamine, and piroxicam were simulated and correlated with the stabilizer hydroxypropyl methylcellulose (HPMC). HPMC was selected based on our previous work establishing it as an excellent stabilizer for two experimentally validated case studies: griseofulvin and fenofibrate. The newfound correlation accurately establishes the relationship between the simulated binding energy, and the intrinsic properties including molar aqueous solubility, hydrophobicity, melting point, molecular weight, and the supersaturation ratio, where a strong linear correlation was discovered between binding energy, aqueous solubility, and supersaturation ratio. Additionally, these results allow for the intrinsic drug properties to predict and screen for effective surfactants by establishing optimal values for concentration and hydrophobicity. The ultimate goal of this work is to quickly and accurately predict surfactant interactions with nanosuspensions without the need for experimentation.

See more of this Session: Synthesis, Characterization and Modelling of Nanoparticle Systems with Pharmaceutical Applications
See more of this Group/Topical: Particle Technology Forum

(341g) Relationship Between Material Properties and Structure of Nanocoated Organic Crystals

Kalyana C. Pingali[1], Adrian Mann[2], Troy Shinbrot[3] and Fernando Muzzio[1], (1) Chemical and Biochemical Engineering, Rutgers University, Piscataway, NJ, (2) Materials Science and Engineering, Rutgers University, Piscataway, NJ, (3) Biomedical Engineering, Rutgers University, Piscataway, NJ

Anisotropy between orientations of organic crystals and their different faces have been studied with respect to the material properties of bulk pharmaceutical powders. In the first stage, experimental measurements were adopted to study various techniques to grow single pure crystals of acetaminophen. The influence of operating conditions on the geometry of crystal habit was investigated in order to control the growth rate of crystals. Surface properties of crystals were characterized by surface wetting, contact angle measurements, atomic force microscope (AFM) measurements of force separation curves, lateral force microscope (LFM) measurements of frictional forces and electron microscopy of crystal structure and morphology. The mechanical response of the crystal surface with respect to the functionalized groups was obtained. Influence of humidity on the mechanical response of both empty and functionalized AFM tips was obtained on crystal surfaces. In the second stage, crystal surfaces were modified by coating with nanoparticles of colloidal silica. The particle size of nanocoating was controlled to obtain a uniform layer deposition. Atomic Layer Deposition (ALD) was performed to deposit uniform thin films of nanosilica with atomic precision on a carefully mounted crystal surface. Control of the atomic scale deposition based on self-limited ALD film growth through chemisorption and sequential surface reaction was performed. In addition, another method of Pulse Layer Deposition (PLD) was also used to deposit the thin film on the substrate (crystal). The thin film of nanosilica was grown on crystal surface due to the direct flow of ablation particles. Nucleation and growth of the thin film on the substrate surface resulted in a sequential monolayer deposition of nanosilica. The mechanical response of thin nanosilica layers on crystal surfaces was estimated by functionalizing AFM tips with both hydrophilic and hydrophobic terminal groups. In the next stage, binding energies or molecular forces of nanosilica layers as a function of extent of coverage of nanolayer was reported.

See more of this Session: Synthesis, Characterization and Modelling of

(341h) Fabrication of Monodisperse Polyanhydride Nanoparticles by Microemulsion Polymerization

Chelsea K. Sackett[1], Heidi Culver[2], **Timothy Brenza**[1], Drew Sikkema[1], Balaji Narasimhan[3] and Jennifer M. O'Donnell[1], (1)Chemical and Biological Engineering, Iowa State University, Ames, IA, (2)Materials Science and Engineering, Johns Hopkins, Baltimore, MD, (3)Department of Chemical and Biological Engineering, Iowa State University, Ames, IA

Biodegradable polyanhydrides have been conventionally synthesized with solution polymerization and used to fabricate 200-800 nm nanoparticles.[1] Particle size and size distribution is an important determinant of immune cell activation[2], and small (< 50 nm), uniform particles could be effectively used as drug or vaccine delivery devices and adjuvants. Microemulsion polymerization is a technique with which sub-50 nm polymer particles can be synthesized from monomer in one step via a free radical mechanism.[3] In this work, the monomer sebacic acid dimethacrylate (MSA) has been used to synthesize crosslinked polyanhydrides using free radical polymerization.[4] An inverse microemulsion system composed of MSA- and chloroform-swollen micelles stabilized by Span 80 in hexane was studied. The phase behavior of the system over a range of temperatures and compositions was determined visually. PolyMSA nanoparticles were then synthesized by microemulsion polymerization. The particle diameter was controlled by varying reaction times and monomer concentration, resulting in particles with diameters ranging from 20 nm to 300 nm. The particles were characterized by scanning electron microscopy and dynamic light scattering. Nuclear magnetic resonance (NMR) and monomer mass loss were used to determine polymerization rate. NMR was also used to determine distance between crosslinks. Hydrolytic degradation rates of polyMSA particles were compared to conventional polySA particles by measuring the diameters of the surface-degrading particles. The microemulsion polymerization technique offers a new approach to fabricating monodisperse particles with controlled sizes for applications in drug and vaccine delivery.

[1] B. Ulery, Y. Phanse, A. Sinha, M. Wannemuehler, B. Narasimhan, B. Bellaire, *Pharm. Res.* **2008**, *26*, 683-690.

[2] S. Reddy et al., *Nat. Biotechnol.* **2007**, *25*, 1159-1164.

[3] M. Antonietti, T. Nestl, *Macromol. Rapid Comm.* **1994**, *15*, 111-116.

[4] A. Weiner, E. Bock, M. Gipson, V. Shastri, *Biomat.* **2008**, *29*, 2400-2407.

See more of this Session: Synthesis, Characterization and Modelling of Nanoparticle Systems with Pharmaceutical Applications
See more of this Group/Topical: Particle Technology Forum

(357a) Flowability of Binary Mixtures: Effect of Fines Concentration

Ben Freireich[1], Karl Jacob[2] and Madhusudhan Kodam[1], (1)Core R&D - Solids Processing, The Dow Chemical Company, Midland, MI, (2)Core R&D, Dow Chemical, Midland, MI

At the previous two year's conferences our group presented work (Jacob and Vasquez, 2009, "The Role of Fines in Controlling the Flow of Bulk Solids", AIChE Conference, Session 56, Talk a; and Freireich, Jacob, and Kodam, 2010, "Flowability of Binary Mixtures: Effects of Fines Concentration", AIChE Conference, Session 544, Talk c) demonstrating that, contrary to common knowledge, the coarse fraction of a bulk solid does impact its flowability. Since that presentation further studies have been conducted scrutinizing these effects. Specifically, using a Schulze ring shear tester, key parameters such as unconfined yield strength, angle of internal friction, and effective angle of internal friction as function of consolidation stress were determined for narrow size distributions of coarse and fine materials. Experimental data and analysis will be presented.

See more of this Session: Characterization and Measurement In Powder Processing
See more of this Group/Topical: Particle Technology Forum

(357b) Monitoring of Moisture Content and Bulk Density In Rapidly Flowing Powders Using Microwave Spectroscopy

John S. Austin III[1], Pei-Fang Sung[1], Mark Kovacevic[1] and Michael T. Harris[2], (1)Chemical Engineering, Purdue Universtiy, West Lafayette, IN, (2)Chemical Engineering, Purdue University, West Lafayette, IN

In the chemical and pharmaceutical industries, it is very useful to know both the density and the moisture content of bulk powders. Real-time knowledge of these values can help optimize a process. Traditionally, the techniques used to measure these parameters require intrusion into the system, which makes real-time measurements nearly impossible. However, they can be estimated noninvasively from the dielectric properties of the materials using a microwave sensor. While many studies have focused on measuring the dielectric properties of static powders, this research has investigated the feasibility of using a microwave sensor in the presence of rapidly flowing powders.

An experimental apparatus was fabricated to uniformly stir powders above the microwave sensor at high velocities, around 800 RPM, while measurements of the complex dielectric properties were taken in real time. Calibration curves of the calculated real versus imaginary components of the complex permittivity were constructed along lines of constant moisture content. Similar to the results generated for static powders, the slopes of the calibration curves from the flowing powder increased with increasing moisture content. Both the bulk density and the moisture content of unknown samples can be determined by utilizing the calibration curves. An additional experiment was performed to further confirm the accuracy of the calibration curves.

See more of this Session: Characterization and Measurement In Powder Processing
See more of this Group/Topical: Particle Technology Forum

(357c) In-Line Determination of the Bulk Density of Particulate Solids by Direct Weighing Methods

Clive E. Davies, School of Engineering and Advanced Technology, Massey University, Palmerston North 4442, New Zealand, Bob Lankshear, EMAX Engineering Limited, Lower Hutt, New Zealand and Emile S. Webster, Riddet Institute, Massey University, Palmerston North, New Zealand

Bulk density is an important, often critical, property of many manufactured products, but direct determination is challenging to measure in real time during processing, because simultaneous measurements of both the mass and volume of the process material are required. Two approaches to direct measurement are considered here. First, a method previously developed, in which a free-flowing the test material flows continuously through a test cell is reviewed and its potential and limitations outlined. Secondly, a novel system in which a powder is collected and characterised using a moving cup which periodically fills and empties is described. Materials tested using the two methods include bread crumbs, plastic pellets, flour, icing sugar and agricultural peat. Measured densities were reproducible, and consistent with values obtained off-line.

See more of this Session: Characterization and Measurement In Powder Processing
See more of this Group/Topical: Particle Technology Forum

(357d) Flow Properties Characterization for the Analysis of Catalyst Support Impregnation Process Performance

Alisa Vasilenko[1], **Fernando J. Muzzio**[2] and Benjamin J. Glasser[2], (1)Chemical and Biochemical Engineering, Rutgers University, Piscataway, NJ, (2)Chemical and Biochemical Engineering, Rutgers, the State University of New Jersey, Piscataway, NJ

Powders and granular materials can be found at any stage of processing in catalyst manufacturing; they exhibit a variety of flow patterns, and their state and behavior differs from application to application. Since there is a lack of fundamental understanding of powder behavior, multiple problems can be encountered during production, such as jamming of the hoppers, sub-standard blending performance, and weight variability of final products due to segregation and/or agglomeration. Scale-up could also be a challenge, since the lack of constitutive equations for granular materials forces most scale-up efforts to follow the trial-and-error route. The field of powder characterization is employed as both, a distinguishing method for choosing the best-fit material and a predictive tool to analyze the process performance, and therefore, plays a very important role in process and product development. There are numerous methods to characterize the flow properties of granular materials, such as avalanching testers, fluidizers, shear cells, indicizers, density methods, angle of repose, etc.; however, most of them are application-specific, and it is not clear, how they correlate with each other or with process performance. For this reason, the use of most of these testers is restricted to a specific application, for which they were designed, and any attempts to apply the results of such experiments to a different application frequently result in process failure. In this respect, relating the variables and parameters outputted by various testers to the fundamental properties of materials is of great interest for successful implementation of powder characterization as a predictive tool for process optimal performance.

The research efforts of this study concentrated on applying powder characterization methods to optimization of manufacturing processes; in particular, studying the effects of impregnation processing on the flow and shear properties of catalyst support materials. During a dry impregnation process a solution of catalyst metal is typically sprayed onto a dry catalyst support powder. In this case the DI water was used instead of metal suspension to emulate the impregnation process onto a set of blends containing coarse alumina boehmite (~60 μ m) and Benecel as a binder (HPMC, ~70 μ m). Shear cell, compressibility tester, and moisture content measurements were used to characterize the axial uniformity of the impregnated blends within the double-cone impregnation setup; SEM imaging and particle size distribution measurements were used to characterize the changes in the microstructure and connect these changes with the bulk properties measurements. Bulk flow properties were evaluated with the GDR (an avalanching tester) and dilation measurements. Methods for improving impregnation uniformity were evaluated for efficiency.

The study has found a procedure regularly used in industry for impregnation produced mixtures that were very non-uniform; the use of characterization methods have correctly identified the flaw in the blender design that prevented a better water distribution in the blend (a "dead zone" within the blender). With respect to blend uniformity within the impregnator it was also observed that the addition of an intermediate amount of water (50% of the particle pore volume) resulted in greater flow properties variability within the blender when compared to the blends with significant amount of water (95% of particle pore volume). This variability was also more pronounced at higher concentrations of HPMC. This finding suggests that the transfer of water in the impregnator bulk does not occur uniformly, but rather is dependent upon the pore saturation of the particles exposed to the water stream. A particle is more likely to transfer water to the neighboring particles when its pores are mostly saturated with water and no significant absorption can occur at this point.

Two different approaches to enhance mixing performance were tested for efficiency: mixing for an extended period of time after the solution has been added – a method currently in use in industry, and applying shear with an intensifier bar at a very low level (~300rpm), while the solution is being added. The blend uniformity for each case was tested using the methods of the base case study. The testing showed that applying low levels of shear during impregnation results in a much more uniform distribution of solution within the blender, while extensive mixing fails to perform to the same extent for the mixtures containing binders (such as HPMC). These findings may also be helpful in designing wet granulation systems for pharmaceutical applications, as the process for water-binder solution addition is similar.

See more of this Session: Characterization and Measurement In Powder Processing
See more of this Group/Topical: Particle Technology Forum

(357e) Spatially Resolved Characterization of Capping Using Nondestructive Ultrasonics and Finite Element Modeling

Ilgaz Akseli, Xiaorong He and Nadia Ladyzhynsky, R&D Pharmaceutical Development, Boehringer Ingelheim, Ridgefield, CT

Capping is one of the mechanical defects in tableting process in which catastrophic failure of the compact can occur. Understanding what influences capping in terms of process variables, mechanical properties, and density/stress distributions in compacts and developing specialized techniques to quantitatively correlate these variables to capping/lamination tendencies of formulations are practical interests of the pharmaceutical industry. In this talk, we will present a nondestructive ultrasonic method to quantitatively evaluate the capping tendencies of single-component, binary, tertiary mixtures as well as real-life formulations. The speeds of pressure (longitudinal) and shear (transverse) waves in uni-axially compressed tablets were measured in different directions using the developed ultrasonic method. Average elastic (*E*) and shear (*G*) moduli of each tablet in these directions were characterized. The ultrasonic measurements revealed that elastic and shear modulus values vary with different testing orientations which indicate elastic and shear moduli anisotropy in compacts. The extent of capping tendency (*EG*) was quantified by using a dimensionless ratio. It was found that *EG* values increase with increasing anisotropic mechanical properties of the compressed powder. The presented method is an attractive tool for process and formulation scientists since it has the advantages of being nondestructive, fast (measurement time less than 15 microseconds), requiring no calibration and requiring a small amount of material thus it supports material-sparing development (< 1g). This talk also focuses on finite element modeling of powder compaction when approaching this problem using case studies. The results of the current study point out that the nondestructive ultrasonics coupled with finite element analysis provide insight into tablet capping.

See more of this Session: Characterization and Measurement In Powder Processing
See more of this Group/Topical: Particle Technology Forum

(357f) The Effect of Excipient Viscoelastic Properties On Predicting the Tensile Strength of Pharmaceutical Compacts

Dhananjay A. Pai, Agricultural and Biological Engineering, Purdue University, West Lafayette, IN and Martin Okos, Agriculture and Biological Engineering, Purdue University, West Lafayette, IN

Powder compaction is a complex process involving rate dependent effects such as friction, temperature increase, entrapped air removal and viscoelasticity. Currently, elastoplastic and elasto-viscoplastic models are used to explain powder densification. These models predict permanent deformation to occur once the compaction pressure exceeds yield pressure. However, in our experiments, we have observed significant permanent deformation of Microcrystalline Cellulose (MCC) at compaction pressures much lower than the yield pressure. This observation can be explained due to the presence of viscoelastic properties of excipients used in tablet making. Unlike plastic deformation, the presence of a viscous component results in the onset of permanent deformation upon immediate application of a deforming force. Since viscosity is deformation rate dependent, the final deformation in powder compacts is rate dependent as well, as demonstrated by higher densities and tensile strengths for lower compaction speeds (Nokodchi et al. 1996). Hence, it is critical to incorporate viscoelastic properties while developing models that predict the final compact properties.

The contact area between two particles is an important property in wet and dry granulation that influences the final granule properties. Due to compaction, there is an increase in the contact area between two particles which is also concomitant to an increase in tensile strength. The amount of contact area formed is dependent on the particle deformability, which in turn depend on process variables such as contact time and pressure as well as the particle viscoelastic properties. The viscoelastic properties in turn, depend on raw material attributes such as moisture and drug content. The final tensile strength of a pharmaceutical compact has been shown to depend on both raw material properties and process variables (Almaya and Aburb (2008), Narayan and Hancock (2005)). For a given material, it has been found that larger the at-rest contact area, the stronger the bonding (Sebhatu and Alderborn (1999), Hiestand (1997a)). Viscoelastic properties increase the strength because of the effects on the work done during the separation of surfaces and on the radius of curvature of the surfaces in contact (Hiestand, 1997b).

The overall goal of this research is to predict tensile strength as a function of viscoelastic properties. The specific objective was to explore the role of viscoelastic contact area to predict tensile strength of Microcrystalline Cellulose (MCC) and acetaminophen (APAP) mixtures. Powders used for experiments were composed of 0, 25 and 50% APAP (rest MCC PH200). Magnesium stearate (0.5%) and Silicon Dioxide (0.2%) were used as lubricant and flow aide respectively. Moisture content of each drug composition corresponded to 40 and 70% RH. The powders were compressed in a Uniaxial testing machine at a punch velocities of 2 mm/min to final pressures in the range of 12.5 to 85 MPa. The diametral tensile strength was measured based on breakage force at 2 mm/min. The compaction response and stress relaxation of the uniaxial compact so formed were measured on the same machine. The small deformation contact area was calculated using equation by Lum and Duncan-Hewitt (1999) by using the median particle size of MCC, compaction and relaxation properties. The stress relaxation response was found to fit well to Maxwell model with two Maxwell elements and a spring in parallel (R^2>0.99). The compaction curves were modeled using an exponential model (R^2>0.99). It was found that stress relaxation and compaction parameters and hence the contact area, depended on the APAP composition, moisture content and on compact pressure. The viscoelastic contact area and final compact tensile strength were found to increase with increasing compaction pressure, moisture content and MCC. A unified model that expresses the compact tensile strength as a function of viscoelastic contact area was successfully developed. Thus, viscoelastic properties, which are dependent on important raw material properties and process variables, were incorporated in predicting the tensile strength of pharmaceutical compacts.

References:

1. Nokhodchi, A., Ford, J., Rowe, P. and Rubinstein, M. The effects of compression rate and force on the compaction properties of different viscosity grades of hydroxypropylmethylcellulose 2208. International Journal of Pharmaceutics 129 (1996) 21-31
2. Almaya, A. and Aburb, A. Effect of Particle Size on Compaction of Materials with Different Deformation Mechanisms with and without Lubricants. AAPS Pharm Sci Tech, 9 (2), 414-418.
3. Narayan, P and Hancock, B. The influence of particle size on the surface roughness of pharmaceutical excipient compacts. Materials Science and Engineering A 407 (2005) 226–233.

4. Sebhatu, T. and Alderborn, G. Relationships between the effective interparticulate contact area and the tensile strength of tablets of amorphous and crystalline lactose of varying particle size. *European Journal of Pharmaceutical Sciences* 8 (1999) 235 –242
5. Hiestand, E. Principles, tenets and notions of tablet bonding and measurements of strength. European Journal of Pharmaceutics and Biopharmaceutics 44 (1997a) 229- 242
6. Hiestand, E. Mechanical Properties of Compacts and Particles that Control Tableting. Journal of Pharmaceutical Sciences 86(9) (1997b) 985-990.
7. Lum, S. K. and Duncan-Hewitt, W.C. (1999). Powder Densification. 1. Particle-Particle Basis for Incorporation of Viscoelastic Material Properties. Journal of Pharmaceutical Sciences 88(2): 261-276.

See more of this Session: Characterization and Measurement In Powder Processing
See more of this Group/Topical: Particle Technology Forum

(357g) Development of a Database Tool for Understanding Excipient Properties and Variability

Kristine Alston, School of Mechanical Engineering, Purdue University, West Lafayette, IN and Carl R. Wassgren, Mechanical Engineering, Purdue University, West Lafayette, IN

There has been a push within the pharmaceutical manufacturing community to move toward a QbD approach to the development of drug formulations. Excipients, which are the inactive ingredients in these formulations, are frequently derived from natural sources. As a result, excipients can have considerable variation in their properties, depending on their origin and seasonal climate variations. In order to enable formulation design, it is first necessary to gain a clear understanding of excipient properties and their effect on formulations. The objective of this project is to obtain a better understanding of excipient properties, including lot-to-lot and vendor-to-vendor variability, through the development of a collaborative database tool. This publicly available, web-based tool will enable users to view and analyze property measurements for a variety of excipient products. A thorough survey was performed to collect and assess results and test methods published in the literature. Within the data available, there exists considerable scatter in measured properties. In addition, incomplete reporting of testing conditions and methodology makes it difficult to assess the source of discrepancies. To address this issue, measurements of material properties were performed in the laboratory to ensure consistent test methods and reporting of a complete set of test conditions. This report aims to quantify the natural variability of excipients through the analysis of lot-to-lot and vendor-to-vendor variability of a number of excipients. The properties and test methods critical to the characterization of excipients are also discussed. Finally, the implementation and features of a collaborative database tool and its applicability to the understanding of excipient properties are presented.

See more of this Session: Characterization and Measurement In Powder Processing
See more of this Group/Topical: Particle Technology Forum

(374a) Encapsulation and Permeability Characteristics of Amorphous Hydrogenated Carbon Films Formed by Plasma Enhanced Chemical Vapor Deposition Technique

Anaram Shahravan and Themis Matsoukas, Department of Chemical Engineering, The Pennsylvania State University, University Park, PA

We have used plasma enhanced chemical vapor deposition technique to coat the materials and control the rate of their release. This powerful technique coat nanoparticles and nanopowders ranging from 15 nm to 1 micron and we can control the thickness (from 1nm to upwards of 100 nm) of the film by controlling the duration of the deposition time. We present results with two different core materials, silica nanopheres and KCl nanoparticles. By coating silica and etching the core in HF, we form hollow nanoshells whose dimensions depend on the size of the silica particle and the thickness of the shell. Permeation studies are conducted using KCl, whose release profile is measured easily by monitoring the conductivity of the solution. The thickness of the shell is one of the independent variables by which to control the rate of release. It has a strong effect on the rate of release, which increases from a six-hour release (shell thickness is 20 nm) to a long-term release over 30 days (shell thickness is 95 nm). The profile of release for any thickness follows a unique behavior: a fast release (35% of the final materials) during the first five minutes after the beginning of the dissolution, and a slower release till all of the core materials come out. In order to analyze this behavior a model based on an unsteady-state diffusion from sphere is developed. Very good agreement between experimental data and the model after half an hour from the beginning of the release is observed and based on that a diffusion coefficient for the shell is calculated. The diffusion coefficient of $(3.5 \pm 0.5) * 10^{-21}$ is obtained for four different samples with different shell thicknesses (20, 40, 75, 95 nm).

See more of this Session: Functional Nanoparticles and Nanocoatings on Particles I
See more of this Group/Topical: Particle Technology Forum

(374b) From Embedded to Supported Metal/Oxide Nanomaterials: Thermal Behavior and Modelling of Structural Evolution At Elevated Temperatures

Stephanie B. Bubenhofer, Wendelin J. Stark and Robert N. Grass, Institute for Chemical and Bioengineering, ETH Zurich, Zurich, Switzerland

The relation between structure and function is of major importance at material interfaces which often excel in terms of chemical or physical effects. Therefore, multi-phase nanocomposites (e.g. core-shell, janus-shaped and carrier-supported nanoparticles) attract attention due to their possibility to contact two materials at the nanometer scale profiting from their highly increased interaction area.

Especially at metal/oxide interfaces, accounting for a majority of heterogeneous catalysts, the structure-function relationship has been studied intensively. High temperature, dry aerosol processes (flame spray synthesis) are common for building such composites. But the formation of key structural elements in such multi-component systems, depending on coagulation-order and solid-solid diffusion, is not fully understood.

Here, we present the synthesis of silica coated crystalline Pd nanoparticles (Pd in silica) in a single step flame spray pyrolysis process. The nanometer scale transformation of this core-shell material to nanocrystalline Pd supported on the amorphous silica matrix (Pd on silica) was achieved at temperatures exceeding 700°C. A physical, population balance based aggregation/diffusion model gives an understanding of the influence of temperature, matrix viscosity, particle sizes and concentrations on this transformation process. It enables morphological predictions (core-shell vs. carrier-supported particles) in nanocomposite formation processes such as flame spray synthesis. It further allows to understand the performance of multi-phase materials during thermal cycling in catalysis or fuel cells.

Figure 1. Morphological transformation of noble metal encapsulated in oxide (core-shell, left TEM image) nanoparticles to supported on oxide particles (right TEM image).

Reference :

S. Bubenhofer et al., J. Phys. Chem. C, 115(4), 1269-76 (2011).

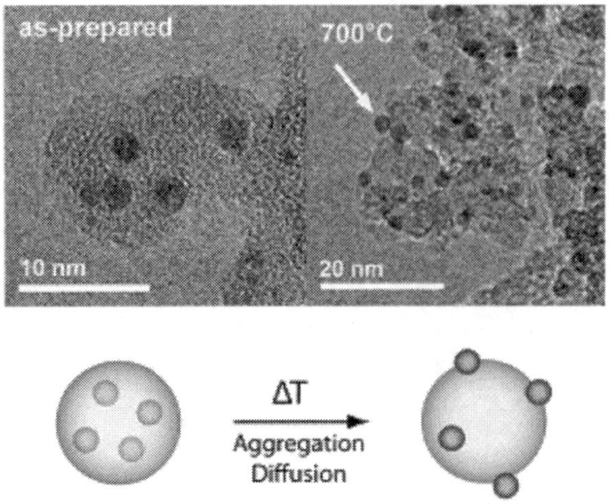

See more of this Session: Functional Nanoparticles and Nanocoatings on Particles I
See more of this Group/Topical: Particle Technology Forum

(374c) Atomic Layer Deposition As a Catalyst Synthesis Technique for Nickel Nanoparticles

Troy D. Gould[1], John L. Falconer[2], J. Will Medlin[3] and Alan W. Weimer[2], (1) Chemical and Biological Engineering, University of Colorado at Boulder, Boulder, CO, (2)Chemical & Biological Engineering, University of Colorado at Boulder, Boulder, CO, (3)Department of Chemical and Biological Engineering, College of Engineering and Applied Science, University of Colorado-Boulder, Colorado Center for Biorefining and Biofuels, Boulder, CO

Efforts to design new catalysts for renewable energy applications have revolved around increasing selectivity & activity by exploring advantages of catalysis at the nanoscale, mainly due to geometric & electronic effects. This work investigates the synthesis of Ni nanoparticle catalysts through the scalable gas phase synthesis route of Atomic Layer Deposition (ALD). Typically ALD deposits Ni as a NiO film, however; Ni nanoparticles were synthesized by using elevated reaction temperatures to produce particles with diameters <5nm. Smaller particles were created to have more catalytically active defects on the surface of the particles. The catalysts were characterized via Temperature Programed Reduction (TPR), chemisorption, Transmission Electron Microscopy (TEM), & Inductively Coupled Plasmon Mass Spectroscopy (ICP-MS) & compared to traditional incipient wetness Ni catalysts. The incipient wetness Ni catalysts were also used as a base comparison of catalytic activity for alkene hydrogenolysis reactions. We will also present the investigations of the controlled preparation of bimetallic catalysts using ALD.

See more of this Session: Functional Nanoparticles and Nanocoatings on Particles I
See more of this Group/Topical: Particle Technology Forum

(374d) Facile Synthesis and Optical Characterization of Hybrid Upconverting and Plasmonic $NaGdF_4$: Yb^{3+}, Er^{3+}/Silica/Gold Nanoparticles

Sha Liu[1], Guanying Chen[2], Tymish Y. Ohulchanskyy[2], Paras N. Prasad[2] and Mark T. Swihart[1], (1)Chemical and Biological Engineering, The University at Buffalo (SUNY), Buffalo, NY, (2)Institute for Lasers, Photonics, and Biophotonics, The University at Buffalo (SUNY), Buffalo, NY

We present a simple process for preparing water dispersable hybrid $NaGdF_4$: Yb^{3+}, Er^{3+}/silica/gold nanoparticles. Monodispersed upconversion nanoparticles of $NaGdF_4$:2% Er, 20% Yb capped by oleic acid were first synthesized via a thermal-decomposition method. The hydrophobic UC NPs were then capped with silica via a reverse-micelle method to make them hydrophilic. Amino-propyltriethoxysilane (APTES) was then used to functionalize the silica-capped UC NPs with amine groups. Finally, gold NPs capped with PVP were mixed with the silica-coated UC NPs. Amine groups on the UC NPs displaced the PVP from the AuNP surface, yielding AuNP-decorated UC NPs that were dispersible in water or ethanol. The emission intensity and color of the upconverting cores are modulated by the plasmonic absorbance and field enhancement of the gold nanoparticles. Energy transfer happened in the hybrid nanoparticles from UC core to the decorating gold nanoparticles, especially at the wavelength of 530 nm at which gold nanoparticles have strong absorption. The lifetime analysis shows that the plasmon resonance effected the emission of upconversion nanoparticles. The applicability of hybrid NPs for multi-modal bioimaging probes was also illustrated with confocal microscopy of living cancer cells in vitro.

See more of this Session: Functional Nanoparticles and Nanocoatings on Particles I
See more of this Group/Topical: Particle Technology Forum

(374e) Single-Walled Aluminosilicate Nanotubes with Organic-Modified Interiors

Dun-Yen Kang, Ji Zang, Christopher W. Jones and Sankar Nair, School of Chemical & Biomolecular Engineering, Georgia Institute of Technology, Atlanta, GA

This talk describes a methodology for modifying the interior of single-walled metal oxide (aluminosilicate) nanotubes by covalently immobilizing organic functional entities on the interior surface of the nanotube structure. Characterization of the modified nanotubes by a range of solid-state characterization techniques - including nitrogen physisorption, thermogravimetric analysis, transmission electron microscopy (TEM), powder X-ray diffraction (XRD), and solid-state NMR - strongly indicates that the organic entities are immobilized on the inner surface of the nanotubes by reaction with the silanol groups on the interior wall of the nanotube. The resulting organic-modified SWNTs show higher hydrophobicity than bare nanotubes based upon water adsorption measurements. Furthermore, a mechanistic understanding of water adsorption in the modified SWNTs is developed, by interpretation of the water adsorption data with a multilayer adsorption model. The degree of interior surface silanol substitution is estimated, with up to 35% of the silanols being substituted through the present modification chemistry. This methodology of immobilizing various functional entities at the inner wall of aluminosilicate nanotubes opens up a range of previously inaccessible "molecular recognition"-based applications for nanotube materials in areas such as catalysis, molecular encapsulation, sensing, and separation.

See more of this Session: Functional Nanoparticles and Nanocoatings on Particles I
See more of this Group/Topical: Particle Technology Forum

(374f) Phase Separation of Mixed Monolayers On Silica Nanoparticles Induced by Hydrogen Bonding

Daniel Sunday, Chemical Engineering, University of Virginia, Charlottesville, VA and **David L. Green**, Department of Chemical Engineering, University of Virginia, Charlottesville, VA

Controlling the self-assembly of mixed monolayers on silica nanoparticles is challenging due to irreversible ligand attachment. In the course of attempting to control the graft density of polystyrene brushes on silica nanoparticles, we synthesized both a surface atom transfer radical polymerization (ATRP) initiator along with an inert analog containing an amide group. The initiator and analog were reacted to the nanoparticles in methylisobutyl ketone to form a mixed monolayer from which polymers could grow. The inert analog reacted preferentially to the nanoparticle surface, which affected the polymerization of styrene with ATRP. In particular, analysis of the initiation efficiency, or the ratio between the ATRP and graft polymer surface coverage provides evidence for phase separation between the ATRP initiator and inert analog in the mixed monolayer. This is most likely due to hydrogen bonding primarily between the inert analog in solution and on the particle surface. Small-angle x-ray scattering (SAXS) and NOESY NMR were used to show that in solution, the analog and initiator form disc-shaped molecular aggregates about 1 nm thick and 1.6 nm in diameter through hydrogen bonding between amide groups. Based on kinetic deposition experiments and FTIR measurements, the molecular aggregates, composed primarily of the inert analog, adsorb strongly to the surface. This adsorption leads to phase separation between the inert analog the ATRP initiator in the monolayer as inferred from FTIR by tracking the non-linear shift in the methylene stretches of the carbons in the ligands, which was a function of the amount of the initiator and analog used to functionalize the nanoparticles. To our knowledge, this reaction mechanism in which the ligand solution behavior affects mixed monolayer structure on silica nanoparticles has not been previously observed.

See more of this Session: Functional Nanoparticles and Nanocoatings on Particles I
See more of this Group/Topical: Particle Technology Forum

(394a) Integrating Nanocellulose Production with Biofuel

Junyong Zhu, USDA Forest Products Laboratory, Madison, WI

This study demonstrated the feasibility of integrating the production of nano-fibrillated cellulose (NFC), a potentially highly valuable biomaterial, with sugar/biofuel (ethanol) from wood fibers. Commercial cellulase enzymes were used to fractionate the less recalcitrant amorphous cellulose from a bleached kraft eucalyptus pulp resulting in a highly crystalline and recalcitrant cellulose (RC). The RC is difficult to hydrolyze to sugars but very suitable for producing biobased nanomaterials through mechanical homogenization. A range of fractionation yields of RC from 10–70% can be achieved by varying fractionation duration and enzyme dosage. The crystallinity of the RC was found to be as much as 24% higher than that of the original bleached pulp. The cellulase fractionation process facilitated mechanical homogenization by significantly reducing the degree of polymerization (DP) to about 400 and the length of the fibers to about 200 mm. The hydrolyzed sugars were found to be easily converted to ethanol through yeast fermentation with excellent efficiency of 91%. Films made from nano-fibrillated cellulose were found to be optically transparent with opacity as low as 12%. The NFC films were mechanically strong and stiff with tensile strengths and moduli approximately 10 and 6 times higher than film made from fibers that had not been nano-fibrillated.

See more of this Session: Nanocellulose-Based Materials and Composites
See more of this Group/Topical: Forest and Plant Bioproducts Division

SURFACE AND AGING CHARACTERIATICS OF PAPER COATED WITH NANO SIZE ZINC OXIDE PIGMENT

Sanjay Tyagi. Renu Tyagi* , B P Thapliyal* R M Mathur* and A K Ray***

**Central Pulp and Paper Research Institute, Saharanpur – 247001*
*** Indian Institute of Technology, (Saharanpur Campus), INDIA*

ABSTRACT

Quality of paper is improved on pigment coating in terms of surface, optical properties and printability characteristics. One of the major parameter for aqueous pigment coating is the type of pigment used and also its particle size and distribution. Zinc oxide is one of the pigment s employed for special coating application. It is offlet nano size particles are playing prominent roles in several areas ranging from medicine to electronics. In paper making there is no exception. Few specialty grades of paper are already using nano size calcium carbonates both as filler and also as coating pigments. Application of water based barrier coating in paper and board in food packaging industry have proven successful. It is also important due to substantial smaller size particles (nano range) compared to macro size solid particles (in micron range) can provide better functional properties of paper in some specialized applications. However, ZnO is present in various forms nano size ZnO particles can enhance the properties significantly to improve optical and surface properties and specially the effect of aging. In this present therefore an attempt has been made to study the potential technological applications of nano zinc oxide particles on paper surface made from hardwood furnish and bamboo using 100% nano ZnO and with admixture of coating grade China clay using laboratory rod coater. All the required properties were evaluated as per ISO standard and compared with the same base paper coated with micro size ZnO pigment particles. From the detailed study it is observed that paper coated with nano ZnO particles imparts better properties as compared to bulk ZnO in terms of brightness, whiteness, gloss, smoothness and printability. Aging properties were also thoroughly studied under artificial condition of varying temperature (50% – 120%)and humidity (10% - 90%). And also UV light, paper coated with nano ZnO showed better resistance to brightness reversion and UV protecting potential.

INTRODUCTION

Paper is pigment coated to improve the quality of paper in terms of surface, optical and printability characteristics. Paper coating ingredients which generally contains pigments, binders and additives is applied to the paper surface to improve the above properties of the paper. One of the major parameter for aqueous pigment coating is the type of pigment used and also its particle size distribution [(1,2,3)]. A commonly used pigment includes china clay, calcium carbonate, silica, zeolite, titanium dioxide, alumina trihydrate and polystyrene [(2)]. Zinc oxide is one of the pigments employed for special coating application [(4)]. Presently all above mentioned conventional pigments are available in the market have their particle size in the range of 1 – 10 micron [(5,6)]. The size of these particles is too big for the preparation of coating pigments to be used efficiently in coating machines. Evenness and uniformity in surface smoothness and thickness of coating layer which is one of the two prime requirements for producing good multi color print, generally lacks with conventional micron sized coating pigments [7,8,9]. Pore structure and contact angle wettability effect print quality by preventing ink spreading. In order to prevent wicking and feathering, it is important that the thickness of coating layer be homogeneous to a scale of few microns in depth which also helps in the absorption of successive droplets of ink at high delivery rates and any water present[9,10].

Application of nano size Zinc oxide in papermaking

While papermaking is normally a statistical process, nano technology makes use of the classic process where the position of every atom is specified and controlled. Yet despite this apparent discrepancy, between two technologies, the paper industry has been practicing nano particles over the conventional pigments long before it was marketed as latest technology trend [5,6].

Nano sized calcium carbonate has been used by paper industry since long as filler due to the advantage of high opacity, high brightness and whiteness [11]. At present, it is also used in special paper products such as baby diapers, sanitary napkins, etc.

Zinc oxide is also one of the pigments finding its use for special coating applications[4]. It is offlet nano size particles are playing prominent role in several areas ranging from medicine to electronics [12]. ZnO is essential for rubber and tier industry and also used as natural rubber, synthetic rubber and latex vulcanization active agents, reinforcing agent and coloring agents. In paint industry, ZnO in addition to color strength and hiding power outside but also a light coating of preservative and agents [13]. In medical and health and food industries, ZnO has a Sida, bleeding, hygienic convergence of functions. In electronics industry, ZnO varistor is the main raw material is magnetic, optical material such as the main additive. Prepared using ZnO varistor, not only has low sintering temperature, but also improve the performance of porous sensitive resistor such as the flux, the non linear coefficient [14,15].

In the present study, an attempt has been made to study the potential technological applications of nano zinc oxide particles on the paper surface made from the pulp of hardwood furnish and bamboo using 100% nano ZnO and with the admixture of coating grade china clay by taking trials on laboratory rod coater. The Zinc oxide soluble starch nano composite was prepared by wet chemical method using Zinc nitrate and sodium hydroxide as precursors and soluble starch as stabilizing and compared with the same base paper coated with micro size ZnO pigment particles.

EXPERIMENTAL

Coating formulation

The coating trials were carried out using Laboratory rod coater. The coated sheets were dried immediately in laboratory oven at the temperature of 105 C for maximum 2 minutes. The following two coating formulations were used for coating trials.

Ingredient	Composition I	Composition II
Pigment		
• Zinc Oxide (Nano or Ordinary)	500 parts	1000 parts
• China clay	500 parts	
Binder		

• Oxidized Starch	90 parts	90 parts
• Styrene Butadiene Latex (SBR Latex)	60 parts	60 parts
Dispersant		
• Anionic polymer	5 parts	5 parts
	Total solids maintained- 65±2%	

Preparation of coating colour

The preparation method for the coating slip is as under-

(a) Starch cooking

The oxidized starch solution was prepared separately by cooking under continuous agitation in hot water at temperature 90-95 ^{0}C, for about 15 minutes, i.e. upto the point when the whole solution became fully gelatinized.

(b) Pigment preparation

Pigment slurry was made in laboratory high-shear coating mixer. The dispersant was firstly added to desired quantity of water. When homogeneous solution was formed, weighed quantities of pigments were added, under constant stirring. A thick paste thus resulted, which became uniform at high shear. Time of agitation was, however kept at about 15 minutes.

(c) Addition of binders to pigments

The freshly prepared starch solution and synthetic binder SBR latex (in required amount) were added to the pigment paste in laboratory high-shear coating mixer, under constant agitation. The mixture was kept under constant agitation at high shear for about 30 minutes to ensure uniform mixing and dispersion of all the ingredients. Finally the slurry was taken out of the mixer.

(d) Screening

The coating slurry was next screened on a 300-mesh screen to eliminate foreign matters, coagulated coating and particles of dried coating slurry etc. Thus a uniform coating slurry; free from dirt, was ready for coating.

Calendering

Coated sheets after dried in laboratory oven were calendered using the laboratory calender with softnip mode, at 40 ^{0}C temperature & 30 bar pressure. The sheets were passed twice through the softnip.

Evaluation of optical characteristics of the pigments

The two pigments- Zinc Oxide (ordinary) and Zinc Oxide (nano particles), which were used in coating trials, were evaluated also for following optical characteristics; as an extension of the detailed study.

- Brightness (%)
- Whiteness (%)
- Yellowness (%)

Evaluation of base paper and coated paper

Base paper as well the coated & calendared papers were conditioned at 27±1 ^{0}C temperature and 65±2 % relative humidity prior to testing of following properties as per standard methods mentioned below.

Property	Standard methods
Grammage (g/m^2)	ISO 534
Brightness (% ISO)	ISO 2470

Gloss (%)	Tappi 480 om-99
Whiteness (%)	ISO 2470
PPS Roughness (micron)	ISO 8791/4
Picking velocity (cm/s)	ISO 3783
Print density (g/cm^3)	Using IGT printability tester Model AIC2-5
Print uniformity index	Using IGT printability tester Model AIC2-5
Stain length (mm), oil absorbency	Using IGT printability tester Model AIC2-5

RESULTS & DISCUSSION

Optical characteristics of base paper

The base paper used in the coating trials, was procured from commercial paper mill. The base paper was made from the pulp containing furnish of hardwood pulp and bamboo. Base paper was evaluated for different optical & printing properties for comparison purpose with coated & calendered paper. The result is tabulated in Table-I.

Optical characteristics of pigments

The optical characteristics of the pigments used in coating are given in Table-II. By comparing the values of the properties obtained for nano Zinc Oxide and ordinary Zinc Oxide, it was found that the starch-encapsulated nano Zinc Oxide particles had slightly lower brightness, whiteness and was relatively more yellowish in colour as compared to ordinary Zinc Oxide particles. This is probably due to the preparation process.

Characteristics of Coated & Calendered Paper samples

The coated paper samples, after softnip calendering, were evaluated for

(a) Optical characteristics
(b) Surface characteristics
(c) Printing characteristics
(d) Aging characteristics

(a) Optical characteristics

The effect of nano size zinc oxide particles on brightness of coated and calendered paper is given in Figure 1. Results show that nano zinc oxide gives better brightness as compared to conventional zinc oxide in both cases. i.e. when used alone and when used in combination with coating grade china clay. The same trend is true with whiteness value of coated paper (Fig. 3).

The value gloss of paper coated with nano ZnO also improved as compared to paper coated with conventional ZnO (Fig. 2). However, the improvement is low as compared to brightness and whiteness values of base paper, there is a fall in their values after coating with ZnO and China clay. The reduction in values is more in case when china clay is used in coating formulation. This is probably due to the lower brightness of china clay and binder additives which results in overall reduction of optical properties of coated paper.

(b) Surface characteristics

PPS Roughness

The surface smoothness of coated paper which is an important property in deciding the print performance of paper, improved with the application of nano ZnO. The PPS roughness in case of nano ZnO is relatively lower than conventional ZnO. The same trend was obtained when nano ZnO is used with china clay. A smooth surface was obtained when china clay is used in coating furnish. This is probably due to particle shape. China clay being platelet shape flattened more during the calendaring operation in compared to ZnO which is spherical in shape (Fig. 4).

(c) Printing characteristics

Picking velocity

Picking velocity of paper coated with nano ZnO was found to be higher as compared to conventional ZnO (Fig. 5) in both cases when used alone and in combination with china clay. This indicates that paper coated with nano ZnO can run easily on high speed printing presses or at the same printing speed, less binder is required in coating formulations with nano ZnO. The excess amount of binder always has negative effect on the printing characteristics.

Print Mottle (Print Non-Uniformity)

Print non uniformity leads to print mottle which is uneven appearance of colour density in a print. This is due to the surface factors to absorb ink evenly due to variation in base paper formation or coat weight. The non uniformity printing index is lower for paper coated with nano ZnO as compared to paper coated with conventional ZnO (Fig. 8).

The variation in sheet matrix formation formed by coating layer of conventional ZnO (due to non uniform size of ZnO particles) probably the reason for this. Paper coated with ZnO alone showed lower number as compared to paper coated with a mixture of china clay and ZnO since particle size of china clay is largely varied. The sheet matrix variation can lead to print mottle non uniform compression during calendering, resulting in variation in coating layer densification. Sheets that contain binder migration, coat weight variation, inferior coating profile and heterogeneous pigment blends can all cause mottle formation on the printed sheet.

Print density

Print density is affected by ink setting/ink holdout. For good density the vehicle of printing ink should stay on the top of paper together with ink pigment. The more the dried ink is retained on the outer surface of the paper, the higher will be the ink density, accordingly, the lower will be the ink demand. Paper with large pores in coating will have more penetration of printing ink leading the lower density. This is observed in case of papers coated with conventional ZnO coating (Fig. 7). The same is true with china clay in the coating formulation. This indicates that more ink will be required to print coated papers with conventional ZnO coatings.

Oil absorbency

Oil absorbency which is an indication of rate of penetration of oil was found to be more in the case of paper coated with nano size ZnO alone and in combination with china clay. (Fig 6)

(d) Effect of Aging

The effect of aging on the optical properties of coated and calendered paper is shown in the graph (Fig. 9-11) and tabulated in table (4-6). The results show that nano zinc oxide gives excellent UV protecting strength in both cases when used alone and when used in combination with coating grade china clay. The reduction in brightness and whiteness is minimum in the case of nano zinc oxide. The lower value of indicated the higher yellowness, which incase after aging maximum when china clay is used.

CONCLUSIONS

The use of nano zinc oxide as a coating pigment improves the paper properties in terms of optical, surface and printability as compared to conventional zinc oxide. This indicates that use of nano ZnO can be a good substitute for the replacement of conventional ZnO in the case of speciality grade papers. This opens a new scope for the use of nano Zinc oxide as a coating pigment for speciality grade of papers. The application of nano ZnO also exhibits excellent UV protection characteristics when tested at different condition of humidity and temperature. Paper represents good aging resistance properties which increases the longevity of paper. This opens a new scope of manufacturing paper with much longer life which lasts for longer years paper manufactured under connectional methods

The use of nano Zinc oxide as a coating pigment results in the paper with better properties as compared to conventional ZnO in terms of brightness, whiteness, gloss, smoothness, UV protection and printability characteristics i.e. picking velocity, print uniformity, print density, and oil absorbency .

Refrences:

1. 'Y.V.Sood, S.Tyagi. R.Tyagi, P.C.Pande, R.Tandon & T.K.Roy', 'Quality of indigenous coated paper and ways to improve it', *'J.IPPTA, 2008, 20 (4), pp'*
2. 'Coating pigments: from size and shape to nanotech', *'www.thefreelibrary.com/coating+pigments:+from+size+and+shape+to+nanotech.....'*
3. 'Surface characteristics of coating layers formed by coating pigments with different particle sizes and size distribution' *'J. Coat. Tech. and Res'*
4. 'N. Vatansever & S. Polat', 'Effect of zinc oxide type on aging properties of styrene butadiene rubber compounds', *'www.sciencedirect.com, 2009'*
5. Roxare, 'Nanotechnology in paper making', *'http://pulpandpapercircle.com/2008/07/nanotechnology –in-paper-making'*
6. 'J.Shen, Z.Song, X.Ojan, F.Yang & F.Kong', 'Nano fillers for papermaking wet end applications', *'www.bioresources.com, 2010, 5(3)'*
7. G. Chinga, 'Detailed characterization of paper surface structure for gloss assessment', *'JPPS, 2004, 30 (8), pp 222-227'*
8. G. Ström, S. E. Research, A. Englund & M. Karathanasis, 'Effect of coating structure on print gloss after sheet fed offset printing', *'J. Nord. Pulp & Paper, 2003, 18 (1), pp108-115'*
9. 'G. Chinga-Carrasco, M. Axelsson, Ø. Eriksen & S. Svensson', 'Structural characteristics of pore networks affecting print through', *'JPPS, 2008, 34 (1), pp 13-22'*
10. 'D.W. Donigian, J. N. Ishley & K.J. Wise, 'Coating pore structure and offset printed gloss', *'J.TAPPI, 1997, 80 (5), pp 163-172'*
11. T. Nypelö, M. Österberg & J. Laine, 'Ultra thin coating of paper by tailor made nanoparticles', *'TAPPI International Conf. on Nanotechnology for the Forest Product Industry, 2010'*
12. 'V. Bianchard & P. Blanchet', 'Color stability for wood products during use: Effects of inorganic nanoparticles', *'www.bioresources.com, 2011, 6(2)'*
13. 'B. Wang, N. Han, D. Meng, R. Yue, J. Yan & Y. Chen', 'Filter paper- templated preparation of ZnO thin films and examination of their gas sensing properties', *'www.sciencedirect.com, 2011'*
14. 'V. Prasad, A. J. Shaikh, A. A. Kathe, D. K. Bisoyi, A.K. Verma, N. Vigneshwaran', 'Functional behavior of paper coated with zinc oxide-soluble starch nanocomposites', *'J. Mat. Proc. Tech., 2010, 210 (14), pp 1962-1967'*
15. 'Surface characteristics of coating layers formed by coating pigments with different particle sizes and size distribution' *'J. Coat. Tech. and Res.'*

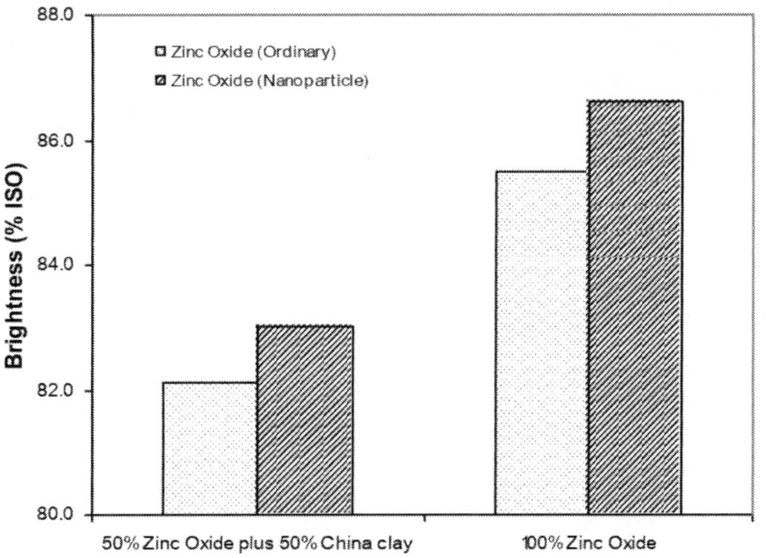

Fig. 1: Effect of Zinc Oxide nanoparticles on brightness of coated & calendered paper

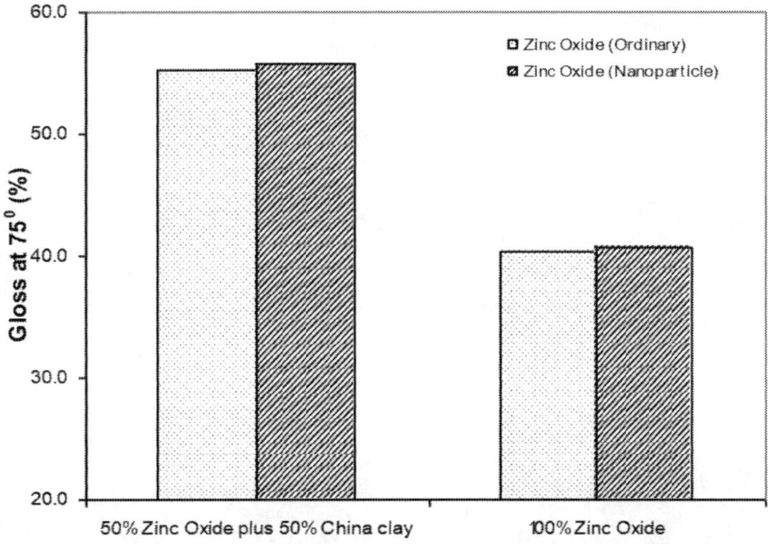

Fig. 2: Effect of Zinc Oxide nanoparticles on gloss of coated & calendered paper

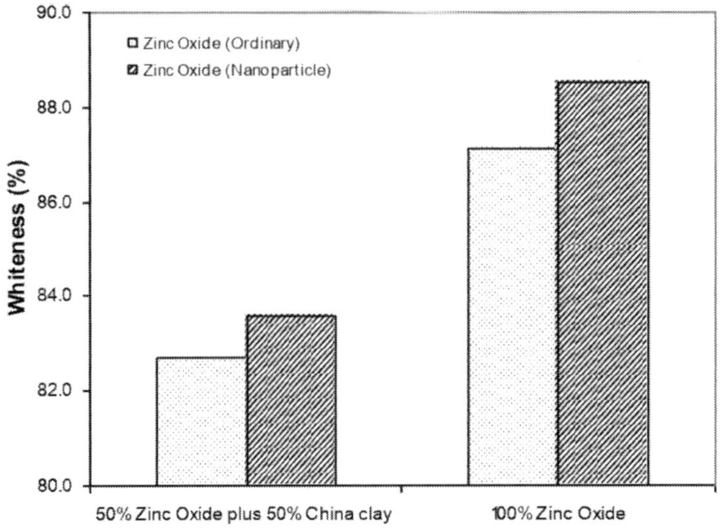

Fig. 3: Effect of Zinc Oxide nanoparticles on whiteness of coated & calendered paper

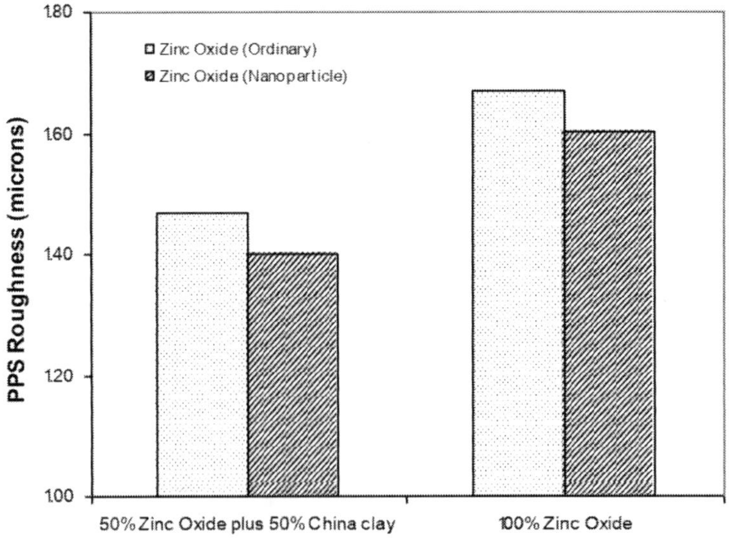

Fig. 4: Effect of Zinc Oxide nanoparticles on the roughness of coated & calendered paper

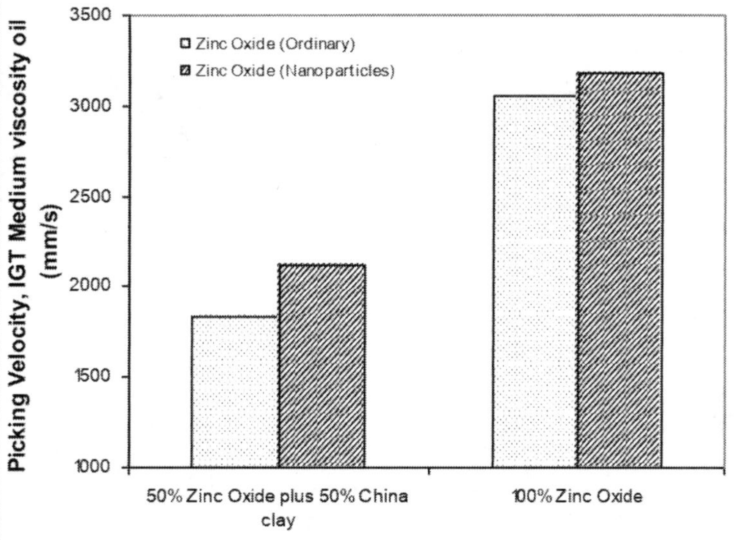

Fig. 5: Effect of Zinc Oxide nanoparticles on picking velocity of coated & calendered paper

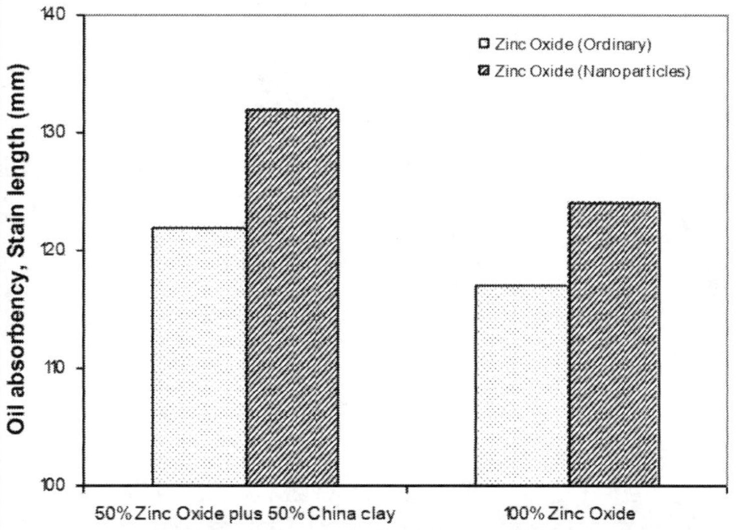

Fig. 6: Effect of Zinc Oxide nanoparticles on oil absorbency of coated & calendered paper

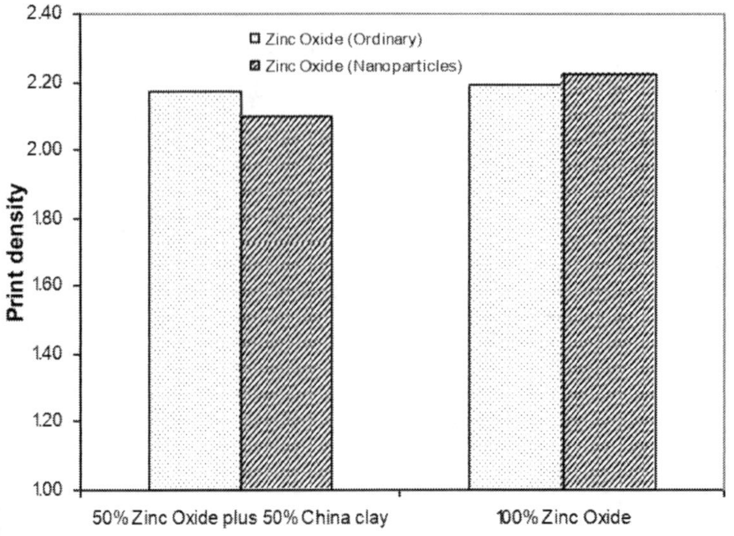

Fig. 7: Effect of Zinc Oxide nanoparticles on print density of coated & calendered paper

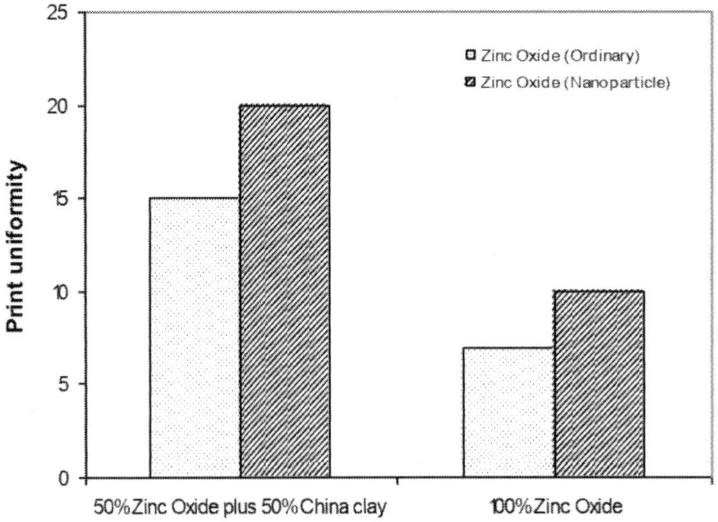

Fig. 8: Effect Zinc Oxide nanoparticles on print uniformity of coated & calendered paper

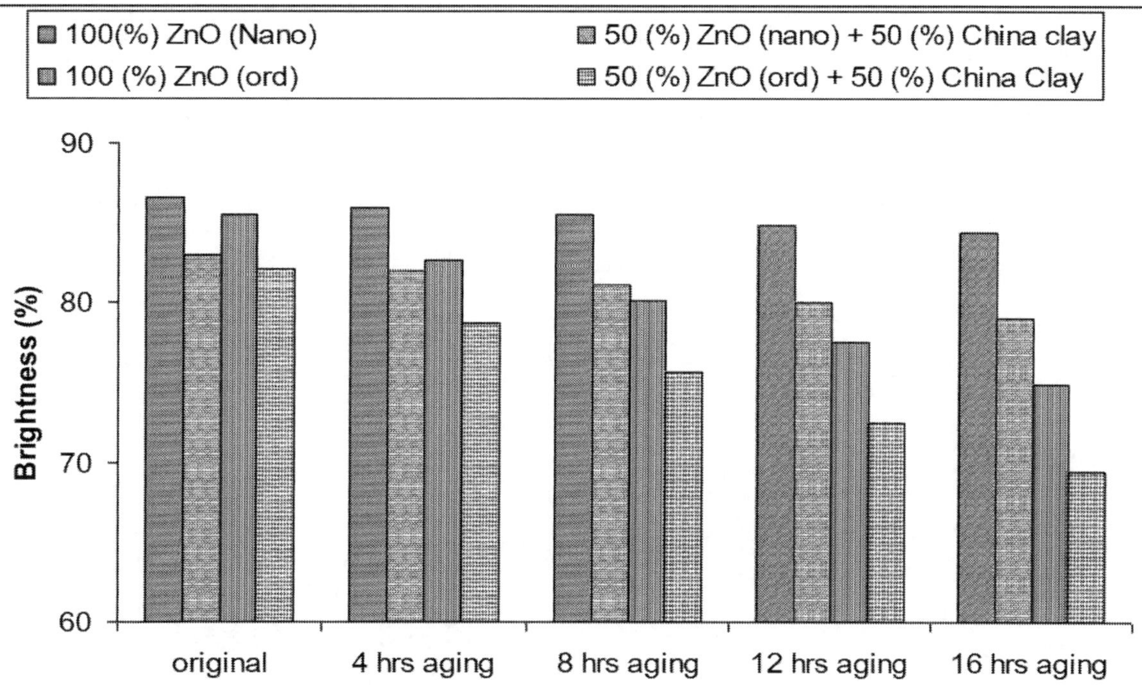

Fig. 9: Effect of zinc oxide nano particle on brightness reversion of coated and calendared paper

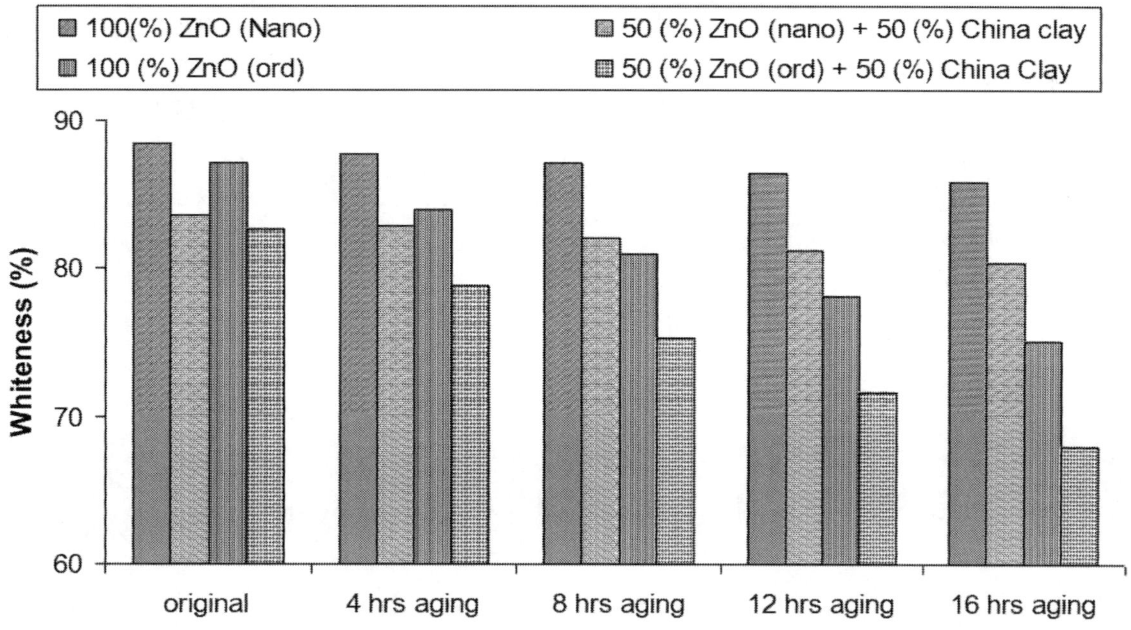

Fig. 10: Effect of zinc oxide nano particles on whiteness reversion of coated and calendared paper

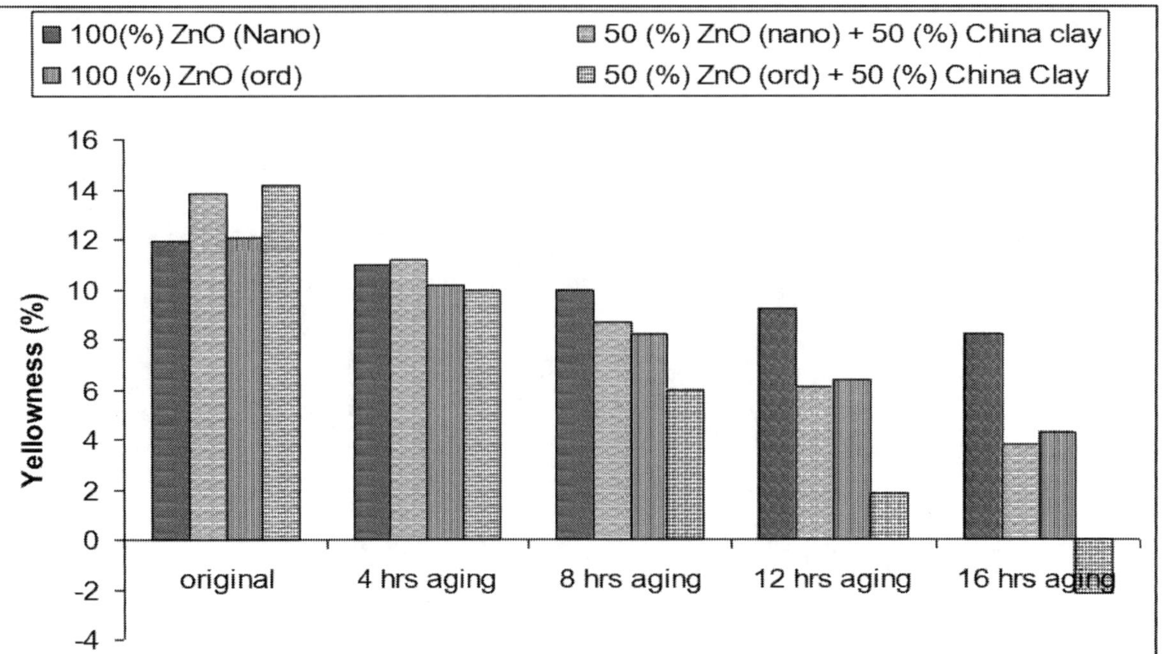

Fig.11: Effect of zinc oxide nano particle on yellowness reversion of coated and calendared paper

Table-I: Optical & printing characteristics of base paper used for coating

Property	Values obtained for the sample
Brightness (%)	88.5
Opacity (%)	80.0
Yellowness (%)	-5.13
Whiteness (%)	124.2
Gloss at 75^0 (%)	6.9
Porosity, Bendtsen (ml/min)	593

Table-II: Optical & printing characteristics of Zinc Oxide pigments used in coating

Property	Zinc Oxide (Ordinary)	Zinc Oxide (Nanoparticle)
Brightness (%)	97.8	96.2
Whiteness (%)	93.9	91.2
Yellowness (%)	1.64	1.70

Table-III: Different optical & printing characteristics of coated & calendared paper

Property	Pigment composition for coating			
	50% Zinc Oxide (Ord.) plus 50% China clay	50% Zinc Oxide (Nano) plus 50% China clay	100% Zinc Oxide (Ord.)	100% Zinc Oxide (Nano)
Coat weight (g/m^2)	12.7	12.6	12.8	12.4
Brightness (%)	82.1	83.0	85.5	86.6
Whiteness (%)	82.7	83.6	87.1	88.5
Gloss at 75°(%)	55.2	55.8	40.4	40.6
PPS Roughness (microns)	1.47	1.40	1.67	1.60
Oil absorbency, IGT Stain length (mm)	122	132	117	124

Picking velocity, IGT Medium viscosity oil (mm/s)	1830	2120	3060	3180
Picking Viscosity Velocity Product (VVP), IGT(Poise.m/s)	805	933	1346	1399
Print density	2.17	2.10	2.19	2.22
Print uniformity	15	20	7	10

Table-IV: Brightness reversion of coated and calendared paper after aging

Pigment Composition	Brightness (%)				
	Original	4 hrs aging	8 hrs aging	12 hrs aging	16 hrs aging
100(%) ZnO (nano)	86.6	86	85.5	84.9	84.4
50(%) ZnO (nano) + 50 (%) China Clay	83	82	81.1	80	79
100 (%) ZnO (ord)	85.5	82.7	80.2	77.5	74.9
50 (%) ZnO (ord) + 50 (%) China Clay	82.1	78.7	75.7	72.5	69.4

Table-V: Whiteness reversion of coated and calendared paper after aging

Pigment Composition	Whiteness (%)				
	Original	4 hrs aging	8 hrs aging	12 hrs aging	16 hrs aging
100 (%) ZnO (nano)	88.5	87.7	87.2	86.5	85.9
50 (%) ZnO (nano) + 50 (%) China Clay	83.6	82.9	82.1	81.2	80.4
100 (%) ZnO (ord)	87.1	84	81	78.1	75.1
50 (%) ZnO (ord) + 50 (%) China Clay	82.7	78.8	75.3	71.6	68

Table-VI: Yellowness reversion of coated and calendared paper after aging

Pigment Composition	Whiteness (%)				
	Original	4 hrs aging	8 hrs aging	12 hrs aging	16 hrs aging
100 (%) ZnO (nano)	11.9	11	10	9.2	8.2
50 (%) ZnO (nano) + 50 (%) China Clay	13.86	11.2	8.7	6.1	3.8
100 (%) ZnO (ord)	12.1	10.2	8.2	6.4	4.3
50 (%) ZnO (ord) + 50 (%) China Clay	14.2	10	6	1.8	-2.2

(394c) Investigation of Mass Transport Properties of Microfibrillated Cellulose (MFC) Films

Matteo Minelli[1], **Marco Giacinti Baschetti**[2], Ferruccio Doghieri[2], Mikael Ankerfors[3], Tom Lindstrm[3], David Plackett[4] and Istvn Sir[4], (1)Department of Chemical Engineering, Mining and Environment Technology, University of Bologna, Bologna, Italy, (2)Department of Chemical Engineering, Mining and Environmental technology, University of Bologna, Bologna, Italy, (3)Innventia AB, Stockholm, Sweden, (4)Solar Energy Programme, Ris National Laboratory for Sustainable Energy, Technical University of Denmark, Roskilde, Denmark

Investigation of masstransport properties of microfibrillated cellulose (MFC) films

M.Minelli, M. Giacinti Baschetti, F. Doghieri, M. Ankerfors, T. Lindstrm, I.Sir, D. Plackett

Cellulose-based nanocomposites are now emerging as promising newmaterials suitable for a variety of applications. One source for such newnanocomposites is microfibrillated cellulose (MFC), produced by delamination ofcellulosic fibers in high-pressure homogenizers. MFC is an organic andbiodegradable reinforcement in polymer nanocomposites because of its highaspect ratio, good mechanical properties and ability to form networks: promisingproperties that may make this material suitable for use in a number ofindustrial fields and products.

The structure and transport properties of a four different filmsbased on two different generations of microfibrillated cellulose (MFC), aloneor in combination with glycerol as plasticizer, were investigated throughFE-SEM analysis and sorption or permeation experiments. FE-SEM revealed the existenceof complex structures in the different samples. A porous, closely packed fibernetwork, more homogeneous in the samples containing glycerol, wascharacteristic of the surface of MFC films; while film cross-sections presenteda dense layered structure with no evidence of porosity. Water vapor sorptionexperiments confirmed the hydrophilic character of these cellulosic materialsand showed a dual effect of glycerol, which reduced the water uptake at lowwater activity while enhancing it at high relative humidity, as showed inFigure 1. The observed trends of water sorption can be described considering thatboth physical adsorption on fiber surfaces and absorption in the celluloseamorphous phase takes place and therefore a dual-model sorption model wasemployed.

The water diffusivity (in the panel in Figure 1) in dry samples wasremarkably low for a porous material (10^{-11}-10^{-12}cm^2/s), confirming the existence of complex structures below thefilm surface; however, when water is present in the system, *D* rapidlyincreases with an exponential trend. Diffusivity is also definitely affected byplasticization, being higher for glycerol-containing samples.

Similar behavior was observed in permeation experiments. Dry MFCfilms showed excellent oxygen barrier properties, comparable with those ofpolymers usually considered suitable for ultra-high barrier applications.Indeed, when dry, the cellulose network presents a particularly compact andstiff structure, causing the observed barrier effect. Furthermore, although MFCfilms are characterized by a certain porosity, FE-SEM images of sample surfacessuggests that the pores are not substantially interconnected

When the water content in the membrane is raised, however, adramatic decrease in these properties was observed (Figure 2).

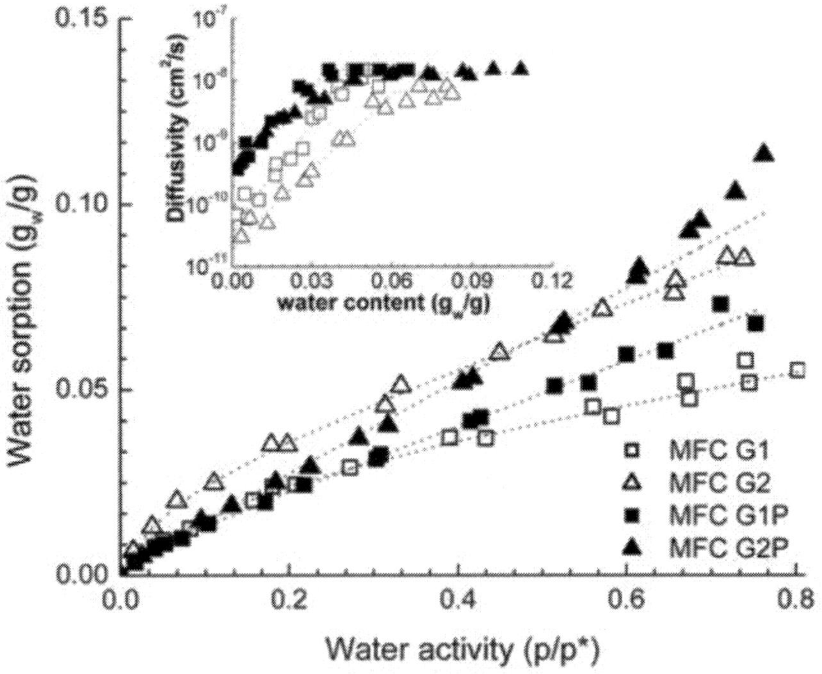

Figure 1. Water vapor sorption isothermsin MFC samples at 35°C, dotted lines are given by the best fitting of the data withthe dual-mode model equation. In the panel: water diffusivity at 35°C in the four different MFC samples as function of the average water concentration.

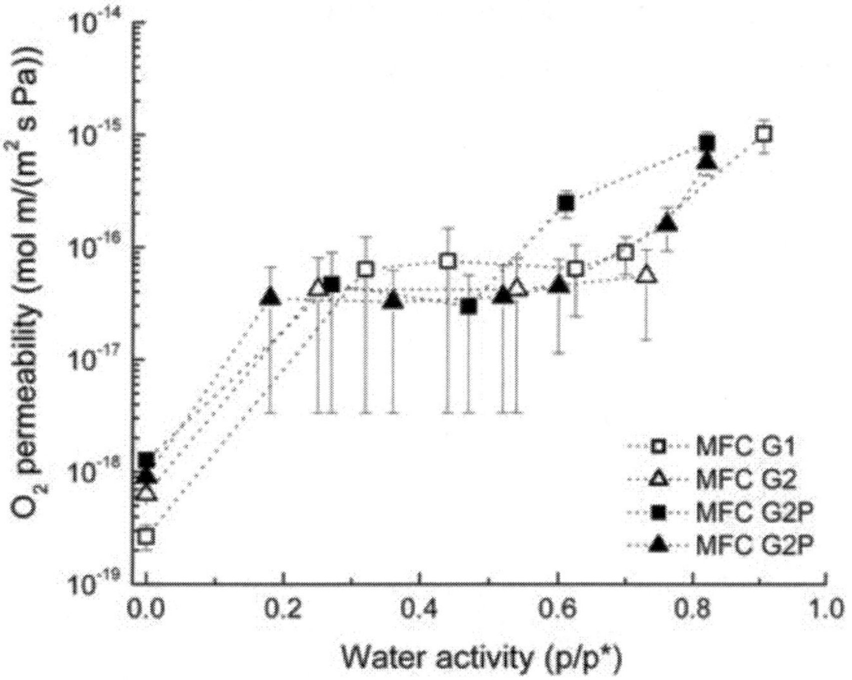

Figure 2. Oxygen permeability in thefour MFC samples at 35°C as a function of water activity.

See more of this Session: Nanocellulose-Based Materials and Composites
See more of this Group/Topical: Forest and Plant Bioproducts Division

(394d) Co-Products of Bioenergy System: Characterization of Saccharification Residuals

Han-Seung Yang, Shona Duncan, William T. Y. Tze and Jonathan Schilling, Department of Bioproducts and Biosystems Engineering, University of Minnesota, Saint Paul, MN

AIChE2011 Annual Meeting - Forest and PlantBioproducts Division

Co-Products of Bioenergy System: Characterizationof Saccharification Residuals

Han-Seung Yang[1], WilliamT. Y. Tze[1], Jonathan Schilling[1] and Shona Duncan[1]

[1]Departmentof Bioproducts and Biosystems Engineering

University of Minnesota

Saint Paul, MN 55108-6130 USA

Tel: 612-625-5200   Fax: 612-625-6286

e-mail: hsyang@umn.edu

The objective of this study is to investigate the effects ofdilute acid pretreatment and enzymatic saccharification on the reactivity,porosity, and crystallinity of solid residuals of biomass. Aspen wood flour was pre-treated with adilute acid treatment followed by enzymatic saccharification. The solid phase of the saccharificationprocess was analyzed using standard NREL protocols for amounts of lignin, pentosans,and hexosans. The functional groupsof solid residuals were analyzed using mid-infrared spectroscopy and hydroxylnumbers. The porosity andcrystallinity of the solids were also evaluated usingrespectively, nitrogen adsorption and X-ray diffraction. Cellulose nanocrystals were extractedfrom the solid residuals and verified using atomic force microscopy. The results will be discussed in thelight of reactivity, porosity, and crystallinity of saccharification residues topave the way for down-stream conversion into co-products of bioenergy (bio-alcohol) systems.

See more of this Session: Nanocellulose-Based Materials and Composites
See more of this Group/Topical: Forest and Plant Bioproducts Division

(394e) Bio-Nano Reinforcement of Polylactic Acid with Surface Modified Cellulose Nanocrystals

Jos Luis Orellana, Esteban E. Urea-Benavides and Christopher L. Kitchens, Department of Chemical and Biomolecular Engineering, Clemson University, Clemson, SC

Cellulose nanocrystals (CNC) have the potential to serve as a nanoscale reinforcement for polylactic acid (PLA), which has become a popular polymer but some of the deficiencies still need to be overcome. The main goal of this work is to improve the mechanical properties of the PLA films – especially toughness – by the addition of surface-modified CNC in the polymer matrix without sacrificing the thermal stability and processability.

CNC were isolated via acid hydrolysis from cotton and surface modified with an esterification process. Transmission Electron Microscopy (TEM) images demonstrated the isolation of the nanocrystals although aggregation can be evidently observed. CNC was further functionalized with different surfactants including decylamine (DA) and cetyltrimethylammonium bromide (CTAB). The dispersibility of the CNC was studied in different solvents such as THF, chloroform and ethyl acetate. Surface modified CNC were added in different concentrations to PLA. Tensile properties of the films were studied using an Instron universal testing machine.

It was found that the stability of the surface-modified CNC improved in organic solvents. DA greatly improved the stability of CNC in both THF and ethyl acetate; while CTAB did in chloroform. No settling was observed in these suspensions for at least three weeks. The toughness and elongation at break of PLA-CNC nanocomposites were improved by the addition of CNC by at least two-fold increase compared to plain PLA. Tensile strength and tensile modulus were not significantly affected by the addition of the crystals. The mechanical properties of the composites were compared and related to the CNC dispersion quality in the different solvents.

See more of this Session: Nanocellulose-Based Materials and Composites
See more of this Group/Topical: Forest and Plant Bioproducts Division

(396a) Review of Gasification Technologies

Ronald W. Breault, Computational Sciences Division, National Energy Technology Laboratory, Morgantown, WV

This work provides a single source that not only reviews the historical development of gasification but also compares the process to combustion. It also provides a short discussion on integrated gasification and combined cycle processes. The major focus of the paper is to describe the twelve major gasifiers being marketed today. Some of these are already fully developed while others are in various stages of development. The hydrodynamics and kinetics of each are reviewed along with the most likely gas composition from each of the technologies when using a variety of fuels under different conditions from air blown to oxygen blown and atmospheric pressure to several atmospheres.

See more of this Session: Panel Discussion on Emerging Areas In Fluidization and Fluid-Particle Systems
See more of this Group/Topical: Particle Technology Forum

(396b) NSF Programs

Ashok S. Sangani, Division of Chemical, Bioengineering, Environmental and Transport Systems, National Science Foundation, Arlington, VA

NSF Programs

See more of this Session: Panel Discussion on Emerging Areas In Fluidization and Fluid-Particle Systems
See more of this Group/Topical: Particle Technology Forum

Paper Unavailable

(396c) Advanced Equipments and Techniques

Ray A. Cocco, Particulate Solid Research, Inc., Chicago, IL
See more of this Session: Panel Discussion on Emerging Areas In Fluidization and Fluid-Particle Systems
See more of this Group/Topical: Particle Technology Forum

Paper Unavailable

(396d) Energy

Hamid Arastoopour, Department of Chemical and Biological Engineering, Wanger Institute for Sustainable Energy Research (WISER), Illinois Institute of Technology, Chicago, IL

Energy

See more of this Session: Panel Discussion on Emerging Areas In Fluidization and Fluid-Particle Systems
See more of this Group/Topical: Particle Technology Forum

Paper Unavailable

(396e) Chemical Looping Gasification - Challenges and Opportunities

Liang-Shih Fan, William G. Lowrie Department of Chemical and Biomolecular Engineering, The Ohio State University, Columbus, OH
See more of this Session: Panel Discussion on Emerging Areas In Fluidization and Fluid-Particle Systems
See more of this Group/Topical: Particle Technology Forum

Paper Unavailable

(396f) Computations

Madhava Syamlal, Office of Research and Development, National Energy Technology Laboratory, U.S. Department of Energy, Morgantown, WV
See more of this Session: Panel Discussion on Emerging Areas In Fluidization and Fluid-Particle Systems
See more of this Group/Topical: Particle Technology Forum

Paper Unavailable

(397a) Investigating Granular Milling In a Hammer Mill: Experiments and Simulation

Shivangi S. Naik and Bodhisattwa Chaudhuri, Pharmaceutics, University of Connecticut, Storrs, CT

Investigating granular milling in ahammer mill: experiments and simulation

Shivangi.Naik, Bodhisattwa. Chaudhuri,

Departmentof Pharmaceutical Sciences,

Universityof Connecticut, USA

Particle size reduction of dry granularmaterial by mechanical means, also known as milling or comminution, isundoubtedly a very important unit operation in pharmaceutical, agricultural,food, mineral and paper industries. A comprehensive understanding of thefragmentation mechanism that occurs during comminution paved the development ofa series of mechanistic and population balance based models. However, thesemodels completely ignore the inter-particle and particle boundary interactionwhich dictates granular flow behavior. We numerically model a pilot-plant scalehammer mill using a Discrete Element Method (DEM) to study the breakage andkinematics of the particle motion within the hammer mill.

Simulations were carried out to studythe effect of mill speed on kinetic energy of particles. In addition,parametric study was performed to understand the effect of hammer speed(rotational), feed rate and hammer-wall tolerance, on the final product sizedistribution. Below a critical hammer tip speed, a blending action rather thancomminuting is observed. Increase in hammer tip speed causes higher frequencyof impact of particles per unit time and higher specific energy of impactresulting in generation of much finer end product. With respect to feed rate, anarrow size distribution was obtained at lower feed rates. The feed rate determinesthe hold up of material in sizing chamber and hence energy required for sizereduction. At low hold-up, longer path lengths are achieved by particlesresulting in higher impact velocity and hence a finer size distribution. Athigher hold up the number of collisions is high, but the kinetic energy perparticle is low leading to poor breakage probability. Particle shape analysisrevealed fragmentation to be the dominant mechanisms of size reduction athigher speeds. We observe that both the specific kinetic and strain energy ofthe particles (colliding with hammer) increase as the impact point becomescloser to the hammer-tip. Further investigation will be carried out to estimatethe induced impact stress and specific energy of fragmentation (impact).

See more of this Session: Particle Breakage and Comminution Processes
See more of this Group/Topical: Particle Technology Forum

(397b) Impact of Processing and Formulation Parameters On Drug Nanoparticle Suspensions Produced Via Wet Stirred Media Milling

Ecevit Bilgili, Chemical, Biological and Pharmaceutical Engineering, New Jersey Institute of Technology, Newark, NJ and **Afolawemi Afolabi**, Chemical, Biological, and Pharmaceutical Engineering, New Jersey Institute of Technology, Newark, NJ

One way to improve the bioavailability of poorly water-soluble drugs is to reduce particle size of drug crystals down to nanoscale via wet stirred media milling. An increase in total surface area per mass loading of the drug and in specific surface area as well as reduced external mass transfer resistance allow a faster dissolution of the poorly-water soluble drug from nanocrystals. To prevent aggregation of nanoparticles, polymers and surfactants are dissolved in water acting as stabilizers via adsorption onto the drug crystals.

In the last two decades, ample experimental data were generated in the area of nanoparticle suspension production via media milling (see e.g. Bhakay et al., 2011; Bilgili et al., 2004, 2006; and references cited therein). Recirculation mode of milling operation, where the particle suspension in a holding tank continuously circulates through the stirred media mill, has been commonly used in lab, pilot, and commercial scales. To explain the impact of some of the processing parameters, stress intensity and stress number concepts have been widely used (Kwade, 1999), which do not account for the effect of suspension viscosity. We conducted a comprehensive study on the production of Griseofulvin nanosuspensions in a wet stirred media mill operating in the recirculation mode. Griseofulvin has been selected as a model poorly water-soluble BCS Class II drug. Impact of various processing parameters such as rotor speed, media loading, and media type (yttria stabilized Zirconia vs. crosslinked polystyrene) as well as of formulation parameters such as stabilizer type–loading and stabilizer addition method (initial vs. step-wise) was studied. A major novelty of the present contribution is that we have analyzed and interpreted the impact of processing and formulation parameters using a Microhydrodynamics analysis (Eskin et al., 2005). The key results suggest a minimum stabilizer-to-drug ratio, synergistic stabilizing effect of a hydrophilic polymer and surfactant, and a dampening effect of the polymer on the breakage kinetics above an optimum loading level. The paper also discusses applications of nanosuspensions and practical issues encountered during media milling.

References

A. Bhakay, M. Merwade, E. Bilgili, R. N. Dave, "Novel Aspects of Wet Milling for the Production of Microsuspensions and Nanosuspensions of Poorly Water-Soluble Drugs," Drug Dev. Ind. Pharm., doi:10.3109/03639045.2010.551775, (2011).

E. Bilgili, R. Hamey, B. Scarlett, "Nano-Milling of Pigment Agglomerates Using a Wet Stirred Media Mill: Elucidation of the Kinetics and Breakage Mechanisms," Chem. Eng. Sci., 61, 149–157 (2006).

E. Bilgili, R. Hamey, B. Scarlett, "Production of Pigment Nanoparticles Using a Wet Stirred Mill with Polymeric Media" China Particuology, 2, 93–100 (2004).

D. Eskin, O. Zhupanska, R. HameY, B. Scarlett, "Microhydrodynamics of Stirred Media Mills", Powder Technol., 156, 95–102 (2005).

A. Kwade, " Wet Comminution in Stirred Media Mills – Research and its Practical Application", Powder Technol., 105, 14–20 (1999).

See more of this Session: Particle Breakage and Comminution Processes
See more of this Group/Topical: Particle Technology Forum

(397c) Mitigating Electrostatic Chargeability of Ibuprofen by Wet Co-Milling with Polyvinylpyrrolidone

Wai Kiong Ng[1], Pei Liu[2], Prashant N. Balani[3], Sui Yung Chan[3] and Reginald Tan[4], (1)Crystallisation and Particle Sciences, Institute of Chemical and Engineering Sciences, Singapore, Singapore, (2)Department of Pharmacy, National University of Singapore, Singapore, (3)Department of Pharmacy, National University of Singapore, Singapore, Singapore, (4)Crystallization and Particle Sciences, Institute of Chemical and Engineering Sciences (ICES) Ltd, Singapore, Singapore

Milling, also known as comminution or grinding, is typically regarded as a particle size reduction process across many industries, including the manufacturing of pharmaceuticals and fine chemicals. In secondary pharmaceutical processing, milling is often used to increase the specific surface area of poorly water-soluble drugs, so as to improve their dissolution properties and bioavailability. However, besides reducing the particle size, the milling process is often accompanied by other unintended effects such as electrostatic charging. During the milling process, the contact between particles rubbing against each other and with the vessel wall causes undesirable triboelectrification of the milled material. This leads to generation of charge on the surfaces of active pharmaceutical ingredients and excipients. Charged materials tend to adhere firmly to metallic surface of containers and equipment reducing the process efficiency or even resulting in clogging and stoppages. Therefore, mitigating electrostatic charging is highly desirable for improving the pharmaceutical manufacturing process and ensuring consistent product quality.

In this study, wet co-milling is investigated as a potential method to reduce the electrostatic chargeability of ibuprofen (IBU). IBU is selected as the model compound as it possesses high cohesivity as well as poor powder flow during the manufacturing process. Sieved fraction (63-106 μm) of IBU prepared by a Retsch® sieve shaker was milled using a planetary ball mill in the presence of water and/or selected polymers for 15 min at ball-mass ratio of 50:1. After vacuum drying for 2 hours, physicochemical characterization of the milled material was conducted: electrostatic charge (Faraday's Cage and Electrometer), particle size (Laser Diffraction), moisture content (Karl Fischer Titration), crystallinity (X-Ray Powder Diffraction; XRPD), morphology (Scanning Electron

Microscopy; SEM), chemical interaction (Fourier Transformed Infrared Spectroscopy; FT-IR), surface adsorption behavior (Gravimetric Vapor Sorption) and thermal analysis (Differential Scanning Calorimetry; DSC).

Results showed that as compared to sieved IBU, (dry) milling increased the electrostatic charges by 10.3%, wet milling with de-ionised water alone (water content of 5 wt.%) showed a 60.3% charge reduction. Addition of polyvinylpyrrolidone (PVP) at 0.15 wt.% resulted in a further 96% charge reduction. The reduction in chargeability may be attributable to two effects: firstly, the increase in moisture content by wet-milling; and secondly, a possible reduction of residual solvent content by wet co-milling with PVP. The latter is still under investigation. This study demonstrated the potential of wet co-milling with a suitable polymer to significantly reduce the electrostatic chargeability of milled pharmaceutical materials.

See more of this Session: Particle Breakage and Comminution Processes
See more of this Group/Topical: Particle Technology Forum

(397d) Investigation of Scaling Parameters On the Particle Size Reduction Profile In Rotor Stator Wet Mills

Joshua D. Engstrom[1], Deniz Erdemir[1], Chenchi Wang[1], Aghogho Pedro[2], Chiajen Lai[1] and Jason Sweeney[2], (1)Drug Product Science and Technology, Bristol-Myers Squibb Company, New Brunswick, NJ, (2)Late Phase Chemical Development, Bristol-Myers Squibb Company, New Brunswick, NJ

A well defined formulation process requires understanding and controlling the powder properties of key inputs, most notably, the active pharmaceutical ingredient (API) particle size. In many instances, API crystallization control is necessary to achieve the desired particle size. However, it is sometimes necessary to reduce the particle size further. In these instances wet milling offers an opportunity to reduce particle size below what can be achieved by crystallization alone. The current literature offers limited insight into the fundamentals of particle size reduction during wet milling as well as translation of lab scale proof of concept to the pilot and manufacturing scale. In this investigation, comparisons of lab and pilot plant scale data for a given API are explored in reference to mill type, flow rate, and shear frequency to demonstrate the impact of scale on the feasibility of meeting the target API particle size. For lab and plant experiments, the particle size was measured real-time using focused beam reflectance measurement (FBRM) and correlated to spot slurry samples analyzed by light scattering as well as microscope images. The benefit of a fundamental understanding of this technology will aid in building a platform to translate lab scale development to the pilot plant and, subsequently, the manufacturing scale.

See more of this Session: Particle Breakage and Comminution Processes
See more of this Group/Topical: Particle Technology Forum

FACTORS INFLUENCING PARTICLE BREAKAGE CHARACTERISTICS IN THE PRESENCE OF ULTRASOUND

1. INTRODUCTION

Acoustic cavitation induced through ultrasound (US), sound of frequency greater than 20 kHz, has several multi-disciplinary applications. While the focus has been on intensifying rates and altering chemical reaction pathways through ultrasound, relative less focus has been given to its physical effects. While ultrasound has been routinely used in industries for cleaning purposes, opportunities exist for extending its application to several applications involving surface area creation from particle breakage and transport through the interface. Ultrasound, though a feasible option, induces complex phenomena, which collectively contribute to faster transport rates. These involve microjets, shock waves, interparticle collisions and localized turbulence at the vicinity of the liquid-solid interface (microstreaming).

2. SCOPE AND OBJECTIVES OF THIS WORK

Studies involving the effect of power ultrasound on simultaneous size reduction and dissolution of particles are scarce in open literature. Effect of ultrasound on various aspects of mass transfer involving both dispersed particle systems and whole solid objects such as cylinder and disk has been reported by Kannan and Pathan (2004), Sandilya and Kannan (2010, 2011). The scope of this presentation is to understand the process of ultrasonic particle size-reduction in solid–liquid systems where the solid is sparingly soluble and dispersed as particles in the liquid. The effect of power ultrasound (20 kHz) on the particle breakage dynamics of sparingly soluble benzoic acid was studied by dispersing it as particles (variable surface area) in aqueous solvents viz. distilled water and aqueous glycerol. The objectives of this work are to

a. Determine the particle size distribution of benzoic acid particles in different solvents at different power levels of ultrasound
b. Quantify the particle size distributions in form of statistical distributions
c. Understand the mechanism of particle breakage through close-up photographs of the benzoic acid particles subject to ultrasound
d. Illustrate the particle motion and breakage in the ultrasonic field through high speed photography

3. EXPERIMENTAL METHODS

Experiments pertaining particle breakage were carried out in a jacketed cylindrical vessel. A cryostat (Ultra Cryostat Circulator, Scientific Instruments, Chennai) was used to maintain constant temperature of the process vessel contents. An ultrasonic probe (VCX-500, Sonics and

Materials Inc., USA), rated at 500 W with a resonating frequency of 20 kHz and a tip diameter of 13 mm was used. Also, the particle size distributions are measured using a laser particle size analyzer (S3500, Microtrac Inc., USA), which has a measuring range from 0.025 to 1408 □m. An optical microscope (Axioskop MAT2, Carl Zeiss Inc., Germany) was used to acquire close-up images of the benzoic acid particles treated under ultrasonic irradiation.

Particle breakage runs

Ultrasound–assisted particle breakage experiments were carried out both in the presence and absence of solute transfer. In the runs with mass transfer the particle sizes were influenced by simultaneous breakage and dissolution. To delineate the effect of ultrasound on the particle size reduction through breakage alone, a separate set of experiments was carried out with solutions of distilled water or 24% (w/w) aqueous glycerol pre–saturated with benzoic acid so that only particle breakage is allowed to take place. These were termed as non–mass transfer runs. The particle size distributions at regular time intervals were obtained under various experimental conditions using the laser particle size analyzer.

4. RESULTS AND DISCUSSION

4.1.Ultrasound assisted particle dissolution and particle breakage characteristics

The cumulative particle size distribution data were analyzed for the lognormal distribution trends using the procedure outlined by Allen (2003). Data pertaining to every experimental condition was plotted on a log-probability graph and the modality of the distribution (i.e. unimodal or bimodal or trimodal) were identified. The lognormal distribution was fitted to the experimental data and the distribution parameters were obtained by using the *nlinsq* toolbox of MATLAB® (The Mathworks Inc., 2007). The multi-modal distribution was fitted using a linear combination of lognormal distributions (Meier et al., 2009):

$$\frac{\mathrm{d}\phi}{\mathrm{d}\left(\ln d_p\right)} = \sum_{i=1}^{n} \frac{A_i}{B_i\sqrt{2\pi}} \exp\left(-\frac{\left[\ln(d_p) - C_i\right]^2}{2B_i^2}\right) \tag{3}$$

where ϕ represents the cumulative volume percentage of particles added at each interval up to the particle size d_p of interest. The parameters of the model given by Eq. (3) viz. blend factor (A_i), standard deviation (B_i) and mean (C_i) were thus estimated for each experiment. They were then used to predict the overall distribution of the particles. The lognormal distributions were found to fit the experimental data satisfactorily and R^2 value better than 0.97 in few cases and better than 0.98 in the remaining were obtained.

The particle size distributions and their statistical parameters were considerably influenced by the presence or absence of solute transfer. The mean particle size, the spread and number of peaks were affected by the presence of solute transfer. Even though the mass transfer effects only occurred until saturation of the solute in the solvent, it was sufficient to create a considerable difference in the particle size distributions and their associated parameters.

4.2. Effect of ultrasound on the solid morphological characteristics

A qualitative treatment of the effect of sonication on the solid particles has been carried out using an optical microscope. The pictures indicates several indentations on the particle surface confirming the pitting action created by microjects impinging on the solid surface. The size of the indentation on the particle surface varied between 59 μm and 167 μm. The nature of the indentations reveal that multiple impingements on the same particle by microjets may have occurred. The orientation of the particle at the time of impingement may also influence the diameter and shape of the pits. Some indentations show a spread along the major axis of the particle which may have been due to the merging of the adjacent pits. It was clear from the images that the ultrasound microjects were active along the frontal surface of the particles and were not that effective along the lateral edges. This may be attributed to the frontal surface having higher surface area than the lateral surface. In addition to pits, cracks were also visible along the particle surface.

In addition to particles having pits and cracks, the images revealed that a few particles had ridges and valleys on the frontal surface which may be attributed to the shear effects caused by shockwaves. The valleys created on the surface of the particles may reduce their strength and hence become sites for particle breakage. A microjet impingement or collision with another particle may lead to its fragmentation. The curved edges observed on the particle images may be attributed to microstreaming phenomena.

4.3 Effect of ultrasound on particle motion and breakage

Video footages were obtained using a high speed camera (X-PRI, AOS Technologies AG, Switzerland) at a frame rate of 500 frames per second. Usually the particles were found to be fragmented as they came under the ultrasound probe.

5. CONCLUSIONS

Microstreaming, microjets and inter–particle collisions hold a complex interplay in the size reduction process of the particles in both the mass transfer and non mass transfer solid–liquid systems investigated. Linear combination of lognormal distributions fitted the experimental particle size distribution data satisfactorily. These ultrasound induced effects also influenced the shape of the particles. Video footages showed that the particles were primarily broken when they came under the ultrasonic jet created by ultrasound probe tip.

REFERENCES

Allen, T., "Powder Sampling and Particle Size Distribution", Elsevier, Amsterdam (2003).

Kannan, A. and S. K. Pathan, "Enhancement of solid dissolution process", *Chemical Engineering Journal* 102, 45-49 (2004).

Meier, M., E. John, D. Wieckhusen, W. Wirth and W. Peukert, "Generally applicable breakage functions derived from single particle comminution data", *Powder Technology* 194, 33-41 (2009).
Krishna Sandilya, D., A. Kannan, "Effect of ultrasound on the solubility limit of a sparingly soluble solid", Ultrasonics Sonochemistry, Vol 17, pp. 427-434, 2010.
Krishna Sandilya, D., A. Kannan, "Intensification of the dissolution of a sparingly soluble solid from a spinning disk in the presence of power ultrasound", Industrial & Engineering Chemistry Research (accepted, 2011)

(397f) Effects of Mill Designs and Process Parameters In Milling of Alumina-Magnesia Extrudates

Fernando J. Muzzio and Aditya U. Vanarase, Chemical and Biochemical Engineering, Rutgers, The State University of New Jersey, Piscataway, NJ

Milling process is one of the most common processes involved in solids processing industries. Generally selection a mill for a given application is based on its capacity, and its ability to produce the desired particle size. Choice mill and the necessary operating parameters is often made based on heuristic information. In solids processing plants, the capacity of manufacturing is subjected to change which leads to changes in the throughput and also the material properties of incoming material to the mill, and it is necessary to adjust the operating parameters (e.g. Rotor Speed and Flow rate) to maintain the intermediate material/ product specifications under desired limits. In such cases, it is beneficial to have a predictive understanding of the particle size distribution (PSD) as a function of operating parameters. The work presented in this paper focuses on optimization of a milling process involved in manufacture of ceramic (Alumina-Magnesia) supports. The main objectives of this work were to increase the capacity of the milling process while maintaining a tighter particle size distribution in an industrial application.

A set of mills (Jaw crusher, co-mill, hammer-mill, knife-mill and reciprocating mill) exhibiting different milling mechanisms under a range operating conditions (different screens and rotor speeds) were examined. Two types of initial materials in the form extrudate pallets of alumina-magnesia, differing in the content of dry matter were used for the milling experiments. Material response to the milling conditions over the entire PSD range was examined. The relationships between various statistical parameters such as (d10 vs. d50) and (d90 vs. d50) were assessed for all possible experimental conditions to understand capability of a given mill produce more/less fines (d10) or coarse particles (d90) at a desired particle size (d50). The scaling behavior of the PSD was also analyzed by computing its poly-dispersity index (d90-d10)/d50 and assessing its behavior over the entire operating space.

This work enhances the predictive understand of the milling process of ceramic materials, it also offers quantitative methodologies to compare the milling performance of industrial mills.

See more of this Session: Particle Breakage and Comminution Processes
See more of this Group/Topical: Particle Technology Forum

(397g) Aspect Ratios and Modeling In Fragmentation and Attrition

Priscilla Hill, Chemical Engineering, Mississippi State University, Mississippi State, MS

It is often desired to be able to model the evolution in particle size during particle breakage. Population Balance Models (PBMs) are commonly used to model this evolution in particle size. However, they can also be used to model changes in particle shape using shape parameters such as aspect ratios [1]. The PBMs use breakage distribution functions that account for the distribution of particle sizes formed when a particle breaks, and they can account for changes in aspect ratio.

Previous work [2] presented comparisons between experimental data and PBMs using two breakage distribution functions – one for attrition and one for fragmentation. However, changes in particle shape were not presented. This paper presents breakage distribution functions for both fragmentation and attrition, where the breakage distribution functions include particle shape as expressed as aspect ratios. Simulation results are compared with experimental data.

1. P. J. Hill, "Statistics of Multiple Particle Breakage Accounting for Particle Shape", *AIChE J.*, **50**, 937-952 (2004).
2. P. J. Hill and S. M. Reeves, "Breakage Distribution Functions in Fragmentation and Attrition", presented at the AIChE National Meeting in Salt Lake City, November 2010.

See more of this Session: Particle Breakage and Comminution Processes
See more of this Group/Topical: Particle Technology Forum

(411a) Control Emulsions, Suspensions, and Solids Separation with Inline Particle Size Measurements In Dark Crude Oil

Anjan P. Pandey and Des O'Grady, METTLER TOLEDO AutoChem, Columbia, MD

Hydrocarbon pipeline flow is typically a complex multi-phase system comprised of bubbles, droplets, and particles including asphaltenes, waxes, gas hydrates, emulsions, and inorganic particles. To control separations, flow, and process efficiency, analytical measurements are often applied by petroleum engineers and chemists. Yet extreme temperature and pressure conditions make offline sampling and analysis difficult and time consuming. Offline measurements can also alter or destroy multiphase components and cannot be applied to make real-time process optimization and control decisions. With the implementation of inline particle characterization technology, particle or droplet phase behavior can be measured quickly, in real time, without pulling samples.

This paper focuses on the application of real time, in-process particle and droplet measurement tools to avoid processing problems in the following R&D and full scale production applications:

- Controlling asphaltene and wax precipitation in dark crude oil
- Improving oil-water and water-oil separation through in-process droplet measurements
- Visualizing and inhibit gas hydrate formation to ensure consistent flow rates
- Improving oil/sand tailings settling rates and optimizing flocculent and chemical surfactant effectiveness

See more of this Session: Solids-Liquid Separation Processes for the Petroleum Industry
See more of this Group/Topical: Separations Division

(411b) Alternatives to Tailings Ponds for Oil Sand Tailings Disposal

Yuming Xu and Tadeusz Dabros, CanmetENERGY, Natural Resources Canada, Devon, AB, Canada

ABSTRACT

Oil sand resources are playing an increasingly important role in maintaining energy supplies for North America. Bitumen is produced from oil sand, mainly by two processes: in situ recovery by SAGD (steam assisted gravity drainage) for deep reservoirs and open-pit mining operations for shallow deposits. In open-pit mining operations, bitumen is separated from sand grains by a hot-water extraction and flotation process that produces large volumes of tailings. In the existing commercial plants, tailings from the extraction operation are pumped into large settling ponds. The coarse solids settle out quickly as a sand beach while the fine particles settle very slowly and finally form mature fine tailings (MFT), a stable suspension containing about 30% solids. Over the years, large volumes of MFT have accumulated in the ponds and the continuing accumulation is a cause of growing economic and environmental concern.

In order to address these concerns we are actively investigating alternative methods of dry and stackable tailings disposal such as flocculation and thickening, filtration, and centrifugation. Coarse oil sand tailings are flocculated using polymeric flocculants and the sediments are either disposed in the mined-out pit or further filtered. Various factors affecting flocculation and filtration performance were investigated in this work. The results demonstrate that the tailings have much better filterability after appropriate flocculation, and filtration becomes a viable method of producing stackable tailings. Another method for treating fine tailings and MFT is centrifugation. Centrifugal force enhances the separation of fine clays from water and speeds up further consolidation. Pilot-scale test results show that centrifugation of the fine tailings produces consolidated cake that can be disposed of without the need for pond containment. These methods provide potential alternatives for eliminating or reducing the use of tailings ponds and for the recovery of water and the associated thermal energy.

See more of this Session: Solids-Liquid Separation Processes for the Petroleum Industry

(411c) Gas Flow Behavior In FCC Disengagers Vessel with Different Outlet Tube Length of Primary Cyclone Separator

Jiangyun Wang, Yu Mao, Juan Wang and Meili Liu, College of Chemical Engineering, China University of Petroleum,Beijing, Beijing, China

To deeply understand the effect of primary cyclone separator's outlet tube length on gas oil flow behavior in fluid catalytic cracking disengager space, the gas oil flow in a full-size disengager is numerically studied on the platform of a commercial computational fluid dynamics (CFD) code Fluent 6.1.22. The Reynolds stress model (RSM) is applied to simulate the gas oil flow field. As no geometry simplification is made in generating the structured mesh for the space both inside and outside the two-stage separator system, the whole flow behavior of the gas oil in the disengager is obtained. Coupled with the RSM, the stochastic tracking mode is used to quantify the gas oil into the secondary cyclone separator. It is shown that the growth of the outlet tube length of primary cyclone separator contributed to oil gas flowing into the inlet of secondary cyclone separator, effectively reducing the gas oil quantity into the disengager and the gas oil residence time, which avoids unwanted overreaction of hydrocarbon.

See more of this Session: Solids-Liquid Separation Processes for the Petroleum Industry
See more of this Group/Topical: Separations Division

(411d) Reclamation of Spent Drilling Fluids by Cross-Flow Microfiltration

Yanling Wu, Environmental Solutions Applied Research, M-I SWACO, A Schlumberger Company, Houston, TX

Oil- and synthetic-based drilling fluids can become contaminated with significant quantities of low-gravity solids, which cannot be removed by conventional solids control equipment and methods. The accumulation of low-gravity solids degrades the performance of costly oil- and synthetic-based drilling fluids in several ways and often generates large quantities of spent drilling fluids not suitable for drilling activities. As a result of rigorous environmental regulations moving toward zero discharge, drilling wastes are the focus of attention in the oil and gas exploration industry.

In this research, pressure driven liquid-phase microfiltration (MF) was investigated for the first time as a potential solution to the removal of low-gravity solids from spent oil- and synthetic-based drilling fluids, where fine solids are retained by the membrane and clean oil passes through the membrane and is recovered as permeate. The results demonstrated that cross-flow MF is capable of breaking the strong emulsions of drilling fluids and recovering clean base oil in the permeate, independent of drilling fluid formulation and composition while minimizing chemical handling and the variability in chemical-based processes. Understanding the impact of each operating parameter on permeate flux will allow us to realize the full potential of cross-flow MF as a spent drilling fluids treatment technology. In this study several operating parameters such as trans-membrane pressure, back-pulse, cross flow velocity and temperature were investigated to optimize permeate flux, minimize membrane fouling and maximize oil recovery using a membrane with a nominal pore size of 0.1 micron. A membrane-cleaning protocol for the fouled membranes was also developed.

See more of this Session: Solids-Liquid Separation Processes for the Petroleum Industry
See more of this Group/Topical: Separations Division

(411e) Technology Development and Demonstration of a Low-Temperature Fluidized-Bed Biomass Dryer

Charles J. Coronella, Mike P Matheus and Victor R. Vasquez, Chemical Engineering, University of Nevada, Reno, Reno, NV

At the University of Nevada, Reno, we initiated a research project to study mechanisms of drying wastewater solids (sludge) in 2007. Preliminary studies led to the development of a new low-temperature fluidized-bed sludge dryer. The sludge dryer has been tested in batch and in continuous studies, and operates at very low temperatures. We report results of a parametric study based on air flow rate and dryer temperature. The dryer produced dried solids at unusually low temperatures, between 125 ° F and 175 ° F, with heat provided at 210 ° F. Collected sludge has been characterized in terms of its moisture content, particle size, and ash content. The moisture of the collected product ranges between 4% and 15%, and can be controlled by proper selection of operating conditions.

The dryer is a novel configuration, and consists of a bed of inert solids operating in the bubbling bed regime. Heat is provided to the bed by internal heat transfer tubes. Wet sludge falls into the warm bed, which is bubbling vigorously. The rapid particle movement causes the wet sludge to break apart, increasing surface area, and thereby enhancing heat transfer. As the sludge dries and fragments, it becomes relatively light dust particles. The small dry particles have specific density 1.2, and are entrained in the fluidizing air. Solid particles are removed from the air in a filter outside the bed, and collected and stored.

The capacity of the dryer is limited by the rate of air flow. Air leaving the bed can remove a finite amount of moisture, up to its saturation of humidity. Increasing too high has a deleterious effect on the performance of the fluidized bed.

We have successfully completed a demonstration of our dryer with a continuous feed of up to 20 lb/hour of wet (85% moisture) sludge onsite at the Truckee Meadows Water Reclamation Facility (TMWRF). We are currently seeking partners for a 200 lb/hour demonstration. Trials and lessons from commercialization with the support of the UNR Tech Transfer Office will be discussed.

(418a) Controllable Synthesis of ZnO Nanocrystals

Xin Liu, Xianliang Wang and Mark T. Swihart, Chemical and Biological Engineering, University at Buffalo (SUNY), Buffalo, NY

In this report, we present a one-pot chemical solution method based upon gradual heating of premixed precursors, in contrast to more common hot-injection methods, for controllable synthesis of ZnO nanocrystals. In general, this type of method is much more amenable to scale-up than hot-injection methods that can be very sensitive to mixing rates, which in turn are difficult to maintain at larger scales. By tuning the mixture of solvent and surfactant (surface capping) agents, we have prepared ZnO nanocrystals with diverse morphology, size and photoluminescence properties. These include needle-like, branched, conical, and porous nanostructures as well as quantum dots with controllable size from 2 to 10 nm. Moreover, by systematically varying factors such concentration of monomer, identity of surfactants, temperature and reaction time, qualitative understanding of the nucleation and growth process has been developed. Ostwald ripening induced shape evolution is also observed and discussed.

See more of this Session: Particle Technology Forum Poster Session
See more of this Group/Topical: Particle Technology Forum

(418b) Controlling Aggregation In Non-Polar Environments: Applications for Energy and Enhanced Oil Production

Sara M. Hashmi and Abbas Firoozabadi, Chemical & Environmental Engineering, Yale University, New Haven, CT

The problem of asphaltene precipitation falls at the intersection of two important chemical engineering topics: nanomaterials for enhanced oil production applications and the fundamentals of nanoparticle and colloidal aggregation. Furthermore, investigations of asphaltenes reveal unique electrostatic characteristics which may prove to be important in understanding and controlling asphaltene aggregation and precipitation.

Asphaltenes, the heaviest and most aromatic component of crude oils, comprise a broad chemical class defined only by their solubility: asphaltenes are soluble in aromatic solvents such as toluene and xylene, but insoluble in light alkanes such as hexane and heptane. The precipitation process takes asphaltenes from the molecular scale in solution to a fully separated phase which can coat rock walls and metal equipment. Asphaltene formation in reservoirs, on bulkheads and wellbores, and in pipelines can greatly impede oil production operations. While the application of bulk aromatic solvents can facilitation the dissolution of asphaltenes to the molecular level, their use is neither economically viable nor environmentally friendly. Mechanisms which can stabilize asphaltenes at the nanoparticle or colloidal scale would not impart full thermodynamic stability to systems containing asphaltenes, but the metastability provided could greatly enhance oil production and allow for asphaltene removal. For instance, several polymeric dispersants have been shown to stabilize asphaltenes at the colloidal scale, through surface adsorption onto the particles.

The phenomenon of nanoparticle and colloidal aggregation provides the key to understanding the phase separation of asphaltenes. Unstable asphaltene systems undergo a cascading aggregation process: molecular association is quickly followed by aggregation to the nano- and micron scale, followed by aggregation to even larger length scales, sedimentation, and full separation. Truncating this aggregation process could arrest the precipitation process entirely. At the same time, asphaltene colloids and nanoparticles occur naturally in non-polar systems: hydrocarbon reservoirs. Colloidal aggregation typically occurs through long-range electrostatic attraction and short-range van der Waals forces. The control of aggregation in aqueous environments can be accomplished through surface charge functionalization of colloids to induce repulsion, or the addition of counterions in solution to screen electrostactic attractions. However, non-polar suspension are more energetically difficult to stabilize than aqueous suspensions due to the low dielectric constant of non-polar solvents and the lack of counterions in solution to screen electrostatics interactions. Studies in non-polar suspensions of carbon black, silica, and polytstyrene colloids with surfactant additives suggest that charging can occur even in non-polar suspensions. While some polymeric dispersant can stabilize asphaltenes at the colloidal scale, the involvement of electrostatic interactions remains an open question.

In this study, we investigate the aggregation of colloids to the micron scale and beyond, as well as its arrest through the use of polymeric dispersants. While some dispersants can actually dissolve asphaltenes at sufficient concentrations (~1% by weight), our measurements demonstrate that other dispersants can truncate asphaltene aggregation and stabilize colloidal asphaltenes on the sub-micron scale. Furthermore, the dispersants which arrest aggregation do so at much smaller concentrations (~10-100 ppm by weight), showing great promise for industrial applications. The dynamics of the truncated aggregation lend some clues as to the nature of the stabilization provided. We measure both colloidal asphaltene aggregation rates as well as the electrophoretic mobility of the asphaltene colloids to investigate the possibility that electrostatic repulsion accounts for the stabilization of colloidal asphaltenes.

See more of this Session: Particle Technology Forum Poster Session
See more of this Group/Topical: Particle Technology Forum

(418c) Locally Triggered Nanoparticle Aggregation by Interfacial Colloidal Destabilization

Serdar Ozturk[1], Yassin A. Hassan[2] and Victor M. Ugaz[1], (1)Artie McFerrin Department of Chemical Engineering, Texas A&M University, College Station, TX, (2)Departments of Nuclear and Mechanical Engineering, Texas A&M University, College Station, TX

A classic indispensible technique to study transport and flow involves injecting small-molecule tracers into a fluid stream. But unexpected challenges emerge when this method is adapted to investigate colloidal nanomaterials (i.e., nanofluids) because the delicate balance among physiochemical interactions governing suspension stability is easily disrupted by these additives. Failure to consider these effects can lead to conflicting and erroneous interpretations about the observed phenomena. Here we introduce a microfluidic probe that enables these interactions to be examined by locally imposing steep chemical gradients in a laminar flow, mimicking the environment created during tracer injection. Unexpectedly, we find that localized aggregation can be readily triggered in low concentration (< 1 vol%) Al_2O_3 nanoparticle suspensions that would otherwise appear to be highly stable based on conventional bulk characterization. In addition to providing a new straightforward method to assess suspension stability, interfacial colloidal destabilization can also be exploited to deposit uniform lines of dense nanoparticle aggregates along the interface between co-flowing streams, potentially laying the foundation for a robust non-lithographic method to precisely pattern particulate-based nanomaterials on surfaces.

See more of this Session: Particle Technology Forum Poster Session
See more of this Group/Topical: Particle Technology Forum

(418d) The Impact of Shear Thickening Fluid Volume Fraction On the Rheological Response of a Suspension Emulsion

Jeremy N. Fowler, Anthony Pallanta, John E. Kirkwood and Norman J. Wagner, Chemical Engineering, University of Delaware, Newark, DE

Shear thickening fluid (STF) suspension emulsions are immiscible polymer blends where a shear thickening fluid (Φ=0.42-0.55 silica in 200 g/mol polyethylene glycol) is dispersed in a Newtonian Homopolymer, in this case PDMS. These so called suspoemulsions exhibit unique rheological behavior that makes them well suited for impact resistant material applications. Suspoemulsions containing volume fraction of STF as low as 10-20% show significant shear thickening, but the mechanism for this behavior is not currently well understood. The magnitude of the shear thickening response is sensitive to the composition of the shear thickening fluid and is therefore tunable. The bulk suspoemulsion experiences an onset of shear thickening at approximately the same shear stress as the neat STF. As volume fraction of shear thickening fluid increases, shear induced phase inversion becomes an important consideration in approaching a mechanism for bulk suspoemulsion shear thickening. Rheo-microscopy is coupled with standard rheology experiments to show that there are fundamental changes in droplet deformability and distribution at the onset of shear thickening. The Utracki model gives a reasonable approximation for the point of phase co-continuity for suspoemulsions under shear flow and is verified experimentally. The composition of the STF and the bulk suspoemulsion can be used in conjunction with the processing conditions to completely tune the response of the final material across a broad range of shear stresses.

See more of this Session: Particle Technology Forum Poster Session
See more of this Group/Topical: Particle Technology Forum

(418e) Compressed Dense Pharmaceutical Excipient Blends: Size Dispersity and Void Structure

Meenakshi Dutt, Chemical and Petroleum Engineering, University of Pittsburgh, Pittsburgh, PA and Bruno C. Hancock, Pfizer Inc., Groton, CT
See more of this Session: Particle Technology Forum Poster Session
See more of this Group/Topical: Particle Technology Forum

Paper Unavailable

(418f) Compressed Dense Powder Blends: Size Dispersity and Void Structure

Meenakshi Dutt, Chemical and Petroleum Engineering, University of Pittsburgh, Pittsburgh, PA

Using Discrete Element Method (DEM) simulatons (Dutt et al.,Comp. Phys. Comm.), we explore the effect of particle sizedistribution on the microstructure evolution of compressed densepowder packings. We consider four powder samples with discrete sizedistributions representative of pharmaceutical blends: 200 microns,195-225 microns , 170-260 microns, 150-295 microns. We generate thepowder packings by allowing the particles to settle under gravity fora fixed interval of time, or until a cut-off packing fraction isattained. The gravitationally settled powder is then uni-axiallycompressed at a constant strain rate for a fixed interval of time. Weobserve that higher compressive strains are required to attain a givenvalue of packing fraction, for increasing number of components in thepowder blends (Dutt et al., submitted). We also compute the voidstructure evolution via a dynamically tesselating algorithm whichcalculates the pore network in dense particulate systems. We find therange of pore volumes to increase with the number of components in theblend (Dutt et al., Physica A). Our approach can be used to designparticulate systems with a desired microstructural response tocompressive strains.

See more of this Session: Particle Technology Forum Poster Session
See more of this Group/Topical: Particle Technology Forum

(418g) Size Control of Copper Oxide Nanoparticles In Flame Spray Pyrolysis

Kosi Aroh, Chia-Ying Chiang and Sheryl Ehrman, Chemical and Biomolecular Engineering, University of Maryland, College Park, MD

As technology grows, more interest has been placed on semiconductors, specifically copper oxide (CuO) semiconductors, due to their electronic and optoelectronic applications. For these applications the size of CuO is very important. In this study, flame spray pyrolysis (FSP) was used to create nearly spherical CuO nanoparticles with an aspect ratio 1.2-1.3. To control the size of these particles flame temperature, residence time and precursor concentration were varied. As the precursor concentration increased from 0.5% to 35% w/w, primary particle size increased from 7±2 to 20±11nm in diameter. Larger primary particle size was observed in the low gas flow system (set B) due to the long residence time in high temperature zone. For the dependence of temperature on particle size, particles grew to similar size, i.e. ~11 nm, in both flame conditions within the hot temperature zone (80% of melting point of CuO) but particle having longer residence time, i.e. 550 ms in set B, the standard deviation of particle size is 45% larger than particle with 66 ms as residence time in set A. Liquid nitrogen was added as a quench to the post temperature zone of the system. Particle size decreased down to 35% of original size.

See more of this Session: Particle Technology Forum Poster Session
See more of this Group/Topical: Particle Technology Forum

(418i) Color Tunable Nanophosphors by Codoping Y2O3 with Tb and Eu

Georgios A. Sotiriou, Melanie Schneider and Sotiris E. Pratsinis, Department of Mechanical and Process Engineering, Particle Technology Laboratory, ETH Zurich, Zurich, Switzerland

Biological labels are used in bio-imaging and the ability to tag cells and biomolecules facilitates diagnosis and therapy. Even though organic dyes are used for such applications, their photobleaching inhibits their further employment. Quantum dots do not exhibit photobleaching and their emission color can be tuned by their size, however, their optical blinking and the high toxicity of their components restrict their use in bio-applications. An alternative strategy for such biolabeling applications are inorganic nanoparticles consisting of a host crystal matrix doped with lanthanide ions, the so-called nanophosphors. The color of these particles depends on the doping ion and not on their size, while their optical properties are superior to organic dyes and quantum dots. The target of this project is the flame synthesis of Y_2O_3 nanoparticles doped with Tb^{3+} and Eu^{3+}. The co-doping of these ions in the Y_2O_3 matrix allows for the fine color tuning of the emission bandwidth [1].

[1] Sotiriou, G. A., Schneider, M. & Pratsinis, S. E. Color-Tunable Nanophosphors by Codoping Flame-Made Y_2O_3 with Tb and Eu. *The Journal of Physical Chemistry C* **115**, 1084-1089 (2011).

See more of this Session: Particle Technology Forum Poster Session
See more of this Group/Topical: Particle Technology Forum

(418j) Anoxic Regions In Gas-Liquid-Solid Fluidized Bed Bioreactor

Kal Renganathan Sharma, Department of Chemical Engineering, Prairie View A & M University, Prairie View, TX

Fluidized bed bioreactors can be used for wastewater treatment. Fan and Wen (1982) patented a process with the region near the distributor plate fluidized and a region near the top of the bed that is fixed bed. A restraining sieve plate is used to achieve the fixed bed at the top. Tand and Fan (1987) discuss use of fludizied bed bioreactors for fermentation and wastewater treatment processes. Fluidized bed biofilm reactor has been demostrated to outperform other reactor configurations used in wastewater treatment such as activated sludge system and trickling filter bioreactor. Causative factors identified for better performance are; (i) better capacity to handle higher biomass concentration due to immobilization of cells onto the solid particles; (ii) relaxation of limit on operational liquid flow rates on account on CSTR limiations on operational flow rates; (iii) intimate contact achievable between the liquid phase and solid phase achievable and; (iv) higher microbial activity achievable by use of support particles. Mathematical mode is developed in order to describe the simultaneous diffusion and reaction of oxygen. Michaelis and Meten kinetics is assumed to be obeyed. The anoxic regions are identified in the asymptotic limit of zeroth order and maximum reaction rate at steady state. The turbulence in the fluidized bed has been noted as a causative factor in poorer solids capture. More solids can be captured when the fluidization conditions such as particle size and flow regimes are identified. Transient concentration profile of oxygen, organics, microbes and nutirents are modeled using: (i) Fick's law of diffusion; (ii) Damped Wave Diffusion and Relaxation (Sharma, 2005). Penetration distances and inertial lag times are calculated. When to use Monod kinetics and when to use Michaelis and Menten kinetics are shown clearly (Levenspiel, 1999). Efforts are underway to find the muliplicity even at steady state when Haldane equation that is nonlinear need be used.

See more of this Session: Particle Technology Forum Poster Session
See more of this Group/Topical: Particle Technology Forum

(418k) Multiscale Design of Aerosol Coating Reactors

Beat Buesser and Sotiris E. Pratsinis, Department of Mechanical and Process Engineering, ETH Zurich, Zurich, Switzerland

Multiscale Design of Aerosol Coating Reactors

B. Buesser and S.E. Pratsinis

Particle Technology Laboratory, Institute of Process Engineering,

Department of Mechanical and Process Engineering, ETHZrich, 8092 Zrich, Switzerland

Core-shell particles facilitateincorporation of functional particles into host (e.g. liquid or polymer)matrices like silica-coated TiO_2 pigments[1], carbon coated Cu forsensors or superparamagnetic[2] Fe_2O_3.Typically core-shell particles are made in the liquid phase[3] but there is keeninterest to develop gas-phase or aerosol coating processes that do not generateliquid by-products, offer fewer process steps, easier particle collection andhermetic[2] shells. Coating ofparticles in the gas phase, however, is challenging, as particle motion andgrowth are much faster than in liquids. As a result, it is difficult to controland develop a scalable gas phase coating process. So, even commerciallyproduced particles made by aerosol routes (e.g. pigmentary TiO_2 madeby the °Chloride-process) are coated by wet processes[3].

Here[4], gas-phase (aerosol)coating is elucidated in considerable detail, for the first time to ourknowledge, by computational fluid and particle dynamics for core particles (TiO_2)and coating shells (SiO_2). Emphasis is placed on understanding theinfluence of process variables (coating weight fraction and mixing intensity(Figure 1) and geometry of core aerosol & shell precursor vapor) oncore-shell product characteristics by a trimodal aerosol particle dynamics model[5] accounting for SiO_2 monomergeneration, coagulation and sintering. The predicted extent of complete (orhermetic) coating shells is compared to the measured photocatalytic oxidationof isopropanol by such particles[6,7] and release of acetone.As hermetic SiO_2 shells prevent the photocatalytic activity of TiO_2,the performance of coated particles is explained by the spatial distribution ofshell thickness on core particles with detailed reactor flow field analysis.

Financial support from the Swiss National ScienceFoundation (SNF) grant # 200021-119946/1 and European Research Council isgratefully acknowledged.

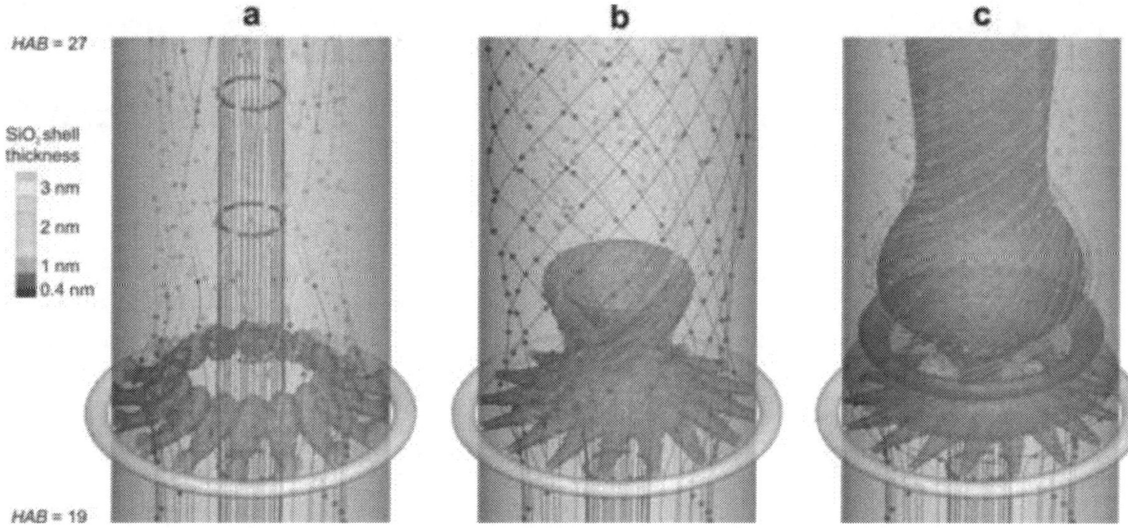

Figure 1 Influence of nitrogen flow rate (mixing intensity) a) 5.8 l/min, b)15.8 l/min and c) 30.8 l/min on the coating precursor and coating shellthickness distribution inside the aerosol coating reactor.

1. SubramanianNS, Diemer RB, Gai PL; E. I. du Pont de Nemours and Company (Wilmington, DE,US); Process for making durable rutile titanium dioxide pigment by vapor phasedeposition of surface treatment. US patent 200627303(A1). 2006.

2. Teleki A, Suter M, Kidambi PR,Ergeneman O, Krumeich F, Nelson BJ, Pratsinis SE. Hermetically coatedsuperparamagnetic Fe_2O_3 particles with SiO_2nanofilms. *Chem. Mater.* 2009; 21, (10), 2094-2100.

3. Egerton TA. The modification offine powders by inorganic coatings. *KONA.* 1998; 16, 46-59.

4. Buesser B, Pratsinis SE. Design ofgas-phase synthesis of core-shell particles by computational fluid aerosolcynamics. *AIChE J.* 2011; DOI: 10.1002/aic.12512,

5. Buesser B, Pratsinis SE. Design ofAerosol Particle Coating: Thickness, Texture and Efficiency. *Chem. Eng. Sci.*2010; in Press, doi: 10.1016/j.ces.2010.07.011,

6. Teleki A, Heine MC, Krumeich F,Akhtar MK, Pratsinis SE. *In-situ* coating of flame-made TiO_2particles by nanothin SiO_2 films. *Langmuir.* 2008; 24, (21),12553-12558.

7. Teleki A, Buesser B, Heine MC,Krumeich F, Akhtar MK, Pratsinis SE. Role of gas-aerosol mixing during in situcoating of flame-made titania particles. *Ind. Eng. Chem. Res.* 2009; 48,(1), 85-92.

(418l) Number and Size of Primary Particles In Agglomerates From Mass and Mobility Measurements

Max L. Eggersdorfer[1], Dirk Kadau[2], Hans J. Herrmann[2] and Sotiris E. Pratsinis[1], (1)Particle Technology Laboratory, Institute of Process Engineering, Department of Mechanical and Process Engineering, ETH Zurich, Zurich, Switzerland, (2)Department of Civil, Environmental and Geomatic Engineering, ETH Zurich, Zrich, Switzerland

Numberand Size of Primary Particles in Agglomerates from Mass and MobilityMeasurements

Real-time characterization of nanoparticlesis necessary for continuous monitoring of aerosol manufacturing and airbornepollutant particles, but is still challenging as these particles restructure[1]and sinter or coalesce[2]. Gas-borne nanoparticles tend to cluster andform irregular structures which influence their transport properties, effectivesurface and density or scattering behavior, to name a few. Here a method isdeveloped to determine the constituent (primary) particle number, n_p,and diameter, d_p, in such structures from their mass andmobility diameter in the free molecular and transition regime. Such data aretypically obtained by differential mobility analyzer (DMA) and aerosol particlemass analyzer (APM) measurements of high temperature aerosols encountered inmaterials processing and engine emissions. Emphasis is placed in determining n_p and$d_p$in fractal-like agglomerate structures. The method is applied and compared tosuch measurements, microscopic images, and correlations in the literature.Primary particle size estimations from prior models show a significantoverestimation if the agglomerate size is much larger than that of the primaryparticle. Reasonable agreement between the present method and primary particlediameters from counting electron micrographs is found[3]. The proposedmethod allows characterizing nanoparticle agglomerates with respect to primaryparticle size and number without any fitting together with their structure fromDMA-APM measurements. Figure 1 shows such agglomerateconsisting of 512 primary particles generated by diffusion-limitedcluster-cluster agglomeration (DLCA) with its radius of gyration and mobilityradius in the free molecular and transition regime.

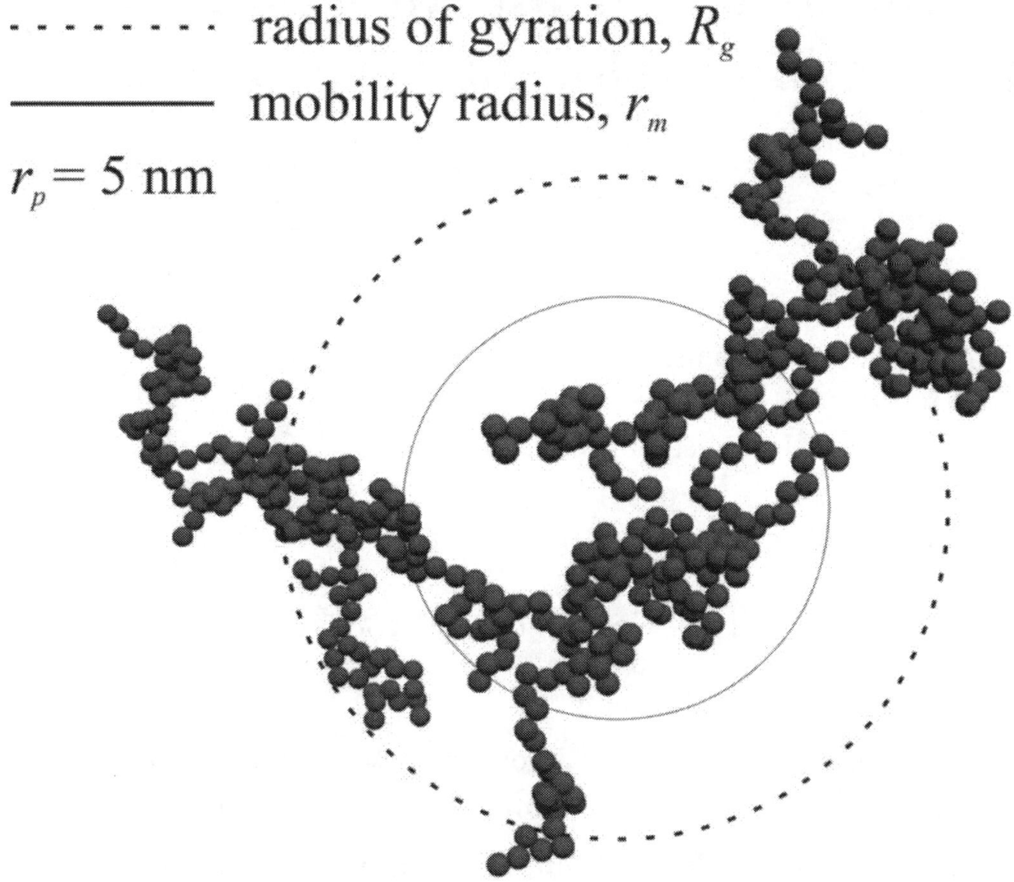

Figure 1:The radius of gyration $R_g$and mobility radius r_m(free molecular and transition regime) of an agglomerate consisting of 512primary particles generated by DLCA.

1. Eggersdorfer, M.L., Kadau, D.,Herrmann, H.J. and Pratsinis, S.E., Fragmentation andrestructuring of soft-agglomerates under shear. *J. Colloid Interface Sci.* 342 (2010) 261-268.

2. Eggersdorfer, M.L., Kadau, D.,Herrmann, H.J. and Pratsinis, S.E., Multi-ParticleSintering Dynamics: from Fractal-like Aggregates to Compact Structures. Langmuir in press (2011).

3. Scheckman, J.H., McMurry P.H. andPratsinis, S.E., Rapid characterization ofagglomerate aerosols by in-situ mass-mobility measurements. *Langmuir* 25 (2009) 8248-8254.

See more of this Session: Particle Technology Forum Poster Session
See more of this Group/Topical: Particle Technology Forum

(418m) Silica Sintering Rate and Mechanism by Molecular Dynamics

Sebastian Conrady, Max L. Eggersdorfer, Beat Buesser and Sotiris E. Pratsinis, Department of Mechanical and Process Engineering, ETH Zurich, Zurich, Switzerland

Silica SinteringRate and Mechanism by Molecular Dynamics

S.Conrady, M.L. Eggersdorfer, B. Buesser and S.E. Pratsinis

ParticleTechnology Laboratory, Institute of Process Engineering,

Departmentof Mechanical and Process Engineering, ETH Zrich, 8092 Zrich, Switzerland

Fumedsilica is the third largest industrial aerosol commodity by value (after carbonblack and TiO_2 and forth by volume after ZnO[1]). Its mainapplication is as a flowing aid including paints, microelectronics,pharmaceutics, cosmetics etc. It is typically manufactured by combustion of$SiCl_4$ resulting in mostly amorphous SiO_2 particles due tothe high temperature cooling rates in the process. Silica (SiO_2) particles grow bysimultaneous coagulation and sintering where their relative rates determine ifaggregates (primary particles connected by sinternecks) or agglomerates (particles connected by van der Waals forces) areformed. The particle morphology has a major influence on the productperformance, e.g. mobility, scattering, mechanical strength etc. and evenpossible health effects.

In order to control the structure it isimportant to control the dynamics of particle growth, most importantly thecoagulation and sintering rate as a function of particle diameter andtemperature. The sintering model of Koch and Friedlander [2] describesthe surface area evolution using characteristic times for sintering, which is thetime needed to reduce the excess surface area of an aggregate over that of anequal mass sphere by 63% [3].Amorphous silica is usually sintering by viscous flow. For this characteristicsintering times have been proposed in literature accounting for a nanoeffect onsintering for the smallest particles by modifying the viscosity to account forthe reduced melting temperature of silica[4].

Here a molecular dynamics simulation ongraphic processing units (GPU) is used to investigate amorphous silicananoparticles undergoing sintering. The Si and O ions interact by a pairwisepotential presented in the literature that has been originally developed forcrystalline SiO_2.

After equilibrating each particleseparately, the influence of particle size and temperature on the evolution ofthe surface area is discussed. Moreover, the differences and similaritiesbetween the sinter mechanism of amorphous silica and crystalline $TiO_2$are investigated[5] and a newsintering rate is proposed and compared to literature.[3,4]

1. Wegner K, Pratsinis SE. Scale-upof nanoparticle synthesis in diffusion flame reactors. *Chem. Eng. Sci.* 2003;58(20):4581-4589.

2. Koch W, Friedlander SK. The effectof particle coalescence on the surface area of a coagulating aerosol. *Journalof Colloid and Interface Science.* 1990;140 (2):419-427.

3. Xiong Y, Pratsinis SE. Formationof agglomerate particles by coagulation and sintering--Part I. Atwo-dimensional solution of the population balance equation. *Journal ofAerosol Science.* 1993;24(3):283-300.

4. Tsantilis S, Briesen H, PratsinisSE. Sintering time for silica particle growth. *Aerosol Sci. Technol.* 2001;34(3):237-246.

5. Buesser B, Grhn AJ, Pratsinis SE.Sintering Rate and Mechanismof TiO_2 Nanoparticles by Molecular Dynamics. *J. Phys. Chem. C.* 2011(accepted).

See more of this Session: Particle Technology Forum Poster Session
See more of this Group/Topical: Particle Technology Forum

(418n) Skeletal Alumina Supports Produced Via Atomic Layer Deposition and Applications In Renewable Fuels

Paul Lichty, Department of Chemical and Biological Engineering, University of Colorado, Boulder, CO, Xinhua Liang, Chemical and Biological Engineering, University of Colorado at Boulder, Boulder, CO, Carl Bingham, National Renewable Energy Laboratory, Golden, CO and Alan W. Weimer, Department of Chemical and Biological Engineering, University of Colorado at Boulder, Boulder, CO

Fuels production through metal oxide redox cycling can be completely renewable if reduction temperatures are achieved using concentrated solar energy. Many combinations of metal oxides have been investigated over the years and cobalt ferrites show much promise as they have good thermodynamic properties and high melting points. However, bulk ferrites degrade when cycled due to sintering of starting materials and resulting diffusion limitations. Recent work has found that coating the ferrites on rigid supports helps to reduce sintering. When using alumina as a support several Co/Fe/Al spinel species are formed which are capable of being cycled at lower temperatures. Cobalt ferrite has been coated on alumina nano powder and high surface area alumina shells. Large, high surface area polymer particles (~600μm, 43.5 m^2/g) were coated via atomic layer deposition (ALD) with several layers of alumina. The alumina layers were applied using TMA/water ALD chemistry. The particles were then heated to 800°C in air to burn out the polymer substrate. The alumina substrates were then coated with alternating layers of Fe_2O_3 and Co-O using ferrocene and cobaltocene ALD chemistries. ICP results also show metal oxide mass loadings up to 27%. Thermo-chemical cycling was performed in a Netzch DIL 402 thermogravimetric analyzer with CO production measured using an in-situ Netzch QMS 403C mass spectrometer. Particles with alumina supports maintain cycling efficiency as compared to unsupported particles. Testing has been conducted on sun in a multi-tube reflective cavity reactor at the National Renewable Energy Laboratory in Golden, Colorado. Cycling results help to estimate reactor efficiencies.

See more of this Session: Particle Technology Forum Poster Session
See more of this Group/Topical: Particle Technology Forum

(418o) Continuous Synthesis of Freestanding ZnO Nanorods In a Flame Reactor

Vishwanath Gandikota, Department of Materials Science and Engineering, Missouri University of Science and Technology, Rolla, MO and Yangchuan Xing, Dept. of Chemical and Biological Engineering, Missouri University of Science and Technology, Rolla, MO

ZnO is an II-VI semiconductor material with a large band gap (3.37 eV), which can be excited by UV radiation and hence can be used in making optoelectronic devices. ZnO belongs to the non-centrosymmetric point group 6mm. This property inherently present in ZnO can be exploited for its use in piezoelectric applications and pyroelectric applications. ZnO exhibits a wide variety of growth morphologies such as nanoparticles, nanorods, nanobelts, nanocages etc., Literature shows that these wide variety of morphologies of ZnO can be synthesized by using different liquid phase and gas phase routes. The current work discusses the synthesis of ZnO nanorods by using a Counterflow Diffusion Flame (CDF) Reactor.

Unlike other conventional methods used for production of ZnO nanorods, we propose a technique that can be used to synthesize ZnO nanorods in the absence of a substrate. The substrate free growth process makes the process continuous and hence can be used for large-scale production of ZnO nanorods. We show that ZnO nanorods of various aspect ratios can be synthesized using our process by just varying process parameters like gas flow rate and precursor flow rate. The effect of adding dopants like SnO_2 and Al_2O_3 on the growth of ZnO nanorods is also being studied. Synthesized ZnO nanorods have been characterized by X-ray diffraction (XRD), Transmission Electron Microscopy (TEM), and Selected Area Electron Diffraction (SAED). The surface growth direction and normal growth directions of the ZnO nanorods have been determined by using the SAED patterns. Also, a probable growth mechanism for the synthesized ZnO nanorods is proposed in the current work. The orientation and the homogeneity of the ZnO nanorods is an important factor for its use in optoelectronic devices.

See more of this Session: Particle Technology Forum Poster Session
See more of this Group/Topical: Particle Technology Forum

(418p) Experiments and Numerical Modeling of An Aqueous Film Coating Process In a Pan Coater

Ekneet Sahni and Bodhisattwa Chaudhuri, Department of Pharmaceutical Sciences, University of Connecticut, Storrs, CT

To better understand the influence of processing parameters on coating variability by experimentally validated first principles model. Lactose non-pareils (3.3-3.5 mm) are used as the model compound to perform coating experiments at previously determined optimal mixing conditions. The non-pareils are coated upto 6% (~250 μm) theoretical weight gain using Black Opadry II. Discrete element based numerical model is developed to study the effect of granular mixing and coating in a pan coater where the post processing data from simulations is used to calculate the residence time distribution which in turn predicts the effect of process variables on coating variability. Moreover, numerical modeling of spray dynamics is performed for the same operational parameter set and similar spray characteristics (center and the radius of the spray zone) used in the experiments. The coating uniformity is estimated for three levels of pan (tilt, rotational speed and fill load) and spray (flow rate and coating fluid concentration) variables by calculating the change in diameter of the non-pareils using a vernier caliper. A balance between the particle motion and the rate of spray application is maintained to achieve a minimum variability in coating. The mass distribution of the coated particles is quantified in the numerical model by the total number of particles passing through the spray zone and the frequency distribution of the residence time of the particles. The coating variability decreased with the increase in pan tilt and an optimum pan speed. Pan fill level does not seem to have much effect under the conditions observed. Further, increase in the spray concentration and flow rate at high tilt and optimum speed decreased the variability. Simulations show good agreement with the experiments regarding the effect of orientation (tilt) of the pan coater on coating variability. However, the simulations over predicted the effect of speed as compared to the experiments to obtain the minimum coating variability. The coating variability is calculated for various processing parameters. A priori performance predictive model is developed which provides a better understanding of the particle dynamics and hence the factors affecting coating uniformity.

See more of this Session: Particle Technology Forum Poster Session
See more of this Group/Topical: Particle Technology Forum

(418q) Protein Trapping of Silica Nanoparticles

Peter N. Yaron[1], Joo Chuan Ang[2], Jhih-Min Lin[2] and John W. White[2], (1) Department of Chemical Engineering, Carnegie Mellon University, Pittsburgh, PA, (2)Research School of Chemistry, Australian National University, Acton, Australia

The ways in which nanoparticles may enter living cells and the toxic responses of those cells are currently of interest. The commercial interest in nanoparticle technology for optoelectronics, composites and energy capture and storage as well as medical uses suggests an eventual wide dispersion of nanoparticulate material in the environment. It has been proposed that cell entry is facilitated by body proteins forming a ''corona'' around an exogeneous nanoparticle upon entry to the lungs or blood stream. The corona is supposed to facilitate cell entry by compatiblizing the nanoparticle to living tissue. The phenomena may be more general, however, since it has been demonstrated that entry of foreign bodies into membranes and whole cells can be promoted by a variety of peptide and macromolecular ''opsonins'', or hindered by ''dysopsonins''. Previous work by others have identified a wide range of biomolecules in human plasma that bind to polystyrene nanoparticles forming a nanoparticle–protein corona. The identity of the corona was found to be significantly affected by the size and surface chemistry of the nanoparticles. We report in this paper the difference in organization and structure of the nanoparticle–protein corona depending on the method of sample preparation (free protein in solution and oriented protein at the air–water interface). Nanotoxicology may be a problem in the future unless the causes and prevention of toxic responses can be identified. The exploration of these types of response is the focus of our current work.

Here we show the extent to which nanoparticle–protein binding can be studied quantitatively on the nanoscale by X-ray and neutron reflectometry (using isotopic contrast variation) at the air–water interface. The method allows structural changes in ''wet'' conditions to be followed in real time. The structures produced by the interaction of hydrophilic silica nanoparticles of diameter about 8 nm with β-casein are described for the proteins held at the air–water interface, as isolated monolayers and at this interface after binding to the silica in solution.

See more of this Session: Particle Technology Forum Poster Session

(418r) Synthesis and Assembly of Anisotropic Particles

Sijia Wang, **Fuduo Ma** and Ning Wu, Chemical Engineering, Colorado School of Mines, Golden, CO

Due to the presence of anisotropy in the physical properties, chemical compositions, surface functionalities, or geometric shapes, anisotropic particles can assemble via directional interactions, forming large numbers of superstructures with novel collective properties that are not manifested in single particle or in bulk material. As such, anisotropic particles are excellent candidates for more than molecular models and have broad ranges of industrial-level applications.

Here, we present our synthetic strategies to make one particular kind of anisotropic particles: colloidal dimers. By combination of phase separation, elastic contraction/expansion, and surface coating, we have made a variety of colloidal dimers with different characteristics, such as (1) homogeneous dimers with identical chemical composition but different sizes on each lobe; (2) heterogeneous dimers with different surface functionalities on each lobe; (3) heterogeneous dimers with different chemical compositions on each lobe; and (4) heterogeneous dimers with opposite signs of charges on each lobe. The precise control of (physical, chemical, geometrical) anisotropy on colloidal dimers allows us to study the self- and guided assembly of colloidal dimers systematically. We will also demonstrate intriguing results of assembling those anisotropic particles under different external fields, such as the electric field and convective flow field. Rich phase behaviors and transient states will be presented.

See more of this Session: Particle Technology Forum Poster Session
See more of this Group/Topical: Particle Technology Forum

(418s) Investigating Contact Drying In a Filter Dryer: Experiments and Simulation

Ekneet Sahni[1], Bodhisattwa Chaudhuri[1] and John C. Strong[2], (1)Department of Pharmaceutical Sciences, University of Connecticut, Storrs, CT, (2)Abbott Laboratories, Abbott Park, IL

Pharmaceutical manufacturing processes often consist of a series of unit operations, each intended to alter certain properties of the materials being processed leading to the embodiment of certain desired properties of the final product. To ensure acceptable variations, significant importance should be given to the mechanistic and parametric attributes of each of the unit processes. Our study focuses on quantitative investigation of mixing and heat transfer in a filter dryer in the quest to determine the optimum drying conditions. Fundamental process understanding and efficient process modeling schemes needed for the development of robust process are lacking. We use discrete element method (DEM) based numerical model, to improve the understanding of agitated vacuum drying process, and hence develop quantitative approach for optimization. DEM explicitly considers inter-particle and particle-boundary interactions, providing an effective tool to solve the transient heat transfer equations.

Once the dryer vessel is charged with the slurry (1L), nitrogen is purged through the drying chamber. The granular bed is then dried by applying heat to the inner wall of the vessel. During each batch run, at repeated intervals, the impeller is stopped and a thermocouple is placed through one of the ports in the dryer to measure the temperature of the bed as a function of time. At the same time, the samples of wet glass beads/Lactose are withdrawn using a cup sampler and stored in tared vials to measure the loss in weight. This leads to calculation of drying rate as a function of material properties and operational conditions of the filter dryer. Simulations are performed to check the effect of various system variables: impeller speed (5rpm, 15rpm, and 25rpm), fill ratio (25%, 45%, and 65%) and the wall temperature (313 K, 323 K, and 333 K) on the drying performance of the two granular systems (glassbeads or lactose with ethanol). Typical system with glass beads and lactose powder are numerically simulated using appropriate material properties.

Higher wall temperature showed an increase in the drying rate for both glass beads and lactose systems, thereby decreasing the total time for drying operation. Also at given wall temperature, the increase in fill volume (bed depth) resulted in a decline in the drying rate. Furthermore, as the rotational speed of the impeller decreases, the heat transfer in the granular bed increases. The heat transfer coefficient are also calculated for different parameters under study. Hence the optimal drying condition is determined. Current study can subsequently be used to direct future design of experiments based on a rationalized approach rather than the simple and usual trial and error method.

See more of this Session: Particle Technology Forum Poster Session
See more of this Group/Topical: Particle Technology Forum

(418t) Rate-Based Separation of Cohesive Particles

Diana M. Livano, Department of Chemical Engineering, University of Pittsburgh, Pittsburgh, PA and J. J. McCarthy, Department of Chemical and Petroleum Engineering, University of Pittsburgh, Pittsburgh, PA

Achieving an efficient separation of individual components in a granular mixture is as much of a challenge as achieving a complete mixture, particularly, when dealing with non-invasive methods. This study focuses on separation of cohesive particle flow via vibration. Separation is therefore the outcome of the combined effect of capillary and viscous forces on particle-particle as well as particle-wall interactions due to the fluid presence, and the kinetic energy due to the vibration. We will analyze the importance of fluid selection, based on its properties, and how it relates to the vibrational velocity of the bed in order to attain effective separation. Particle Dynamics (PD) simulations will be used as a means of demonstrating the validity of the theory and determining the optimal separation parameters, in order to ultimately corroborate results experimentally.

See more of this Session: Particle Technology Forum Poster Session
See more of this Group/Topical: Particle Technology Forum

(431a) Development of a Continuous Filter-Drier for Lab to Kilo-Lab Scale

David J. am Ende[1], David Pfisterer[1], Kevin Girard[1], Daniel Blackwood[2] and Ed Plocharczyk[3], (1)Chemical R&D, Pfizer, Inc, Groton, CT, (2)Pharmaceutical Development, Pfizer, Inc, Groton, CT, (3)Flow technology, D&M Continuous Solutions LLC, Greenwood, IN

Continuous isolation and drying is relatively rare in the pharmaceutical industry. There are several advantages to continuous filtration and drying provided the challenges of conveying and moving wet and dry solids can be overcome. These advantages include: 1) consistency of particle properties (eliminate batch to batch variability), 2) higher efficiency and reduced energy utilization, 3) smaller equipment size and portability. Highly consistent particle properties of API enable lower-cost drug product options such as direct compression operations. Direct compression and tableting operations require consistent API particle size and flow properties and high accuracy of drug-excipient blend compositions. Other motivating factors include the design of fit for purpose portable continuous equipment. The prototype filter drier system hardware design and system performance will be discussed.

See more of this Session: Applications of Continuous Processing In Manufacture of Drug Substance
See more of this Group/Topical: Food, Pharmaceutical & Bioengineering Division

(431b) Development of a Continuous Process for Tasisulam Sodium

Jennifer M. Groh, Timothy D. White, K. Derek Berglund and Martin D. Johnson, Chemical Product Research and Development, Eli Lilly and Company, Indianapolis, IN

The development of a synthetic route to tasisulam sodium (5-bromo-thiophene-2-sulfonic acid 2, 4-dichlorobenzoylamide sodium salt, hereafter referred to as tasisulam) utilizing continuous Schotten-Baumann reaction conditions is presented. This process was envisioned with the goal of creating a process with fewer unit operations utilizing the direct formation of the final sodium salt from a sulfonamide and acid chloride without isolation of the free acyl sulfonamide. Very tight release specifications were in place for the tasisulam API batch process and the challenges of meeting these requirements are detailed. Another processing goal was to minimize worker exposure to the cytotoxic API. This was accomplished by eliminating a solid isolation and reducing the total material in the system at any one time. Operating the Schotten-Baumann conditions in a continuous fashion gave two main advantages. It allows for inexpensive, dedicated lab hood equipment for API production, and it enables higher yield and better impurity rejection in the extraction, solvent exchange, and crystallization for this route. Batch development, continuous proof of concept, 50 g/day lab scale demonstration and 5 kg/day commercial scale runs will be presented.

See more of this Session: Applications of Continuous Processing In Manufacture of Drug Substance
See more of this Group/Topical: Food, Pharmaceutical & Bioengineering Division

(431c) API Attribute Control by Continuous Fast Crystallization/ Precipitation

Dimitrios Zarkadas[1], Christopher Pridgen[2], Mourad Hamedi[1], Plamen Grigorov[3] and Michael Whittington[4], (1)Chemical Process Development & Commercialization, Merck and Co. Inc., Rahway, NJ, (2)Pharmaceutical Commercialization Technology, Merck and Co. Inc., Summit, NJ, (3)Chemical Process Development & Commercialization, Merck and Co. Inc., Union, NJ, (4) Chemical Process development & Commersialization, Merck and Co. Inc., Union, NJ

Increased bioavailabiltyand/or good API particle size and attribute control is needed to enhance orrender robust the pharmacokinetic profile of an increased number of currentdrugs and/or new molecular entities. Continuous fast crystallization/precipitationis a more effective method compared to batch crystallization/precipitation tocontrol API attributes and produce small particles for a number of reasons.Continuous fast crystallization offers higher mixing intensity and lower mixingtimes as well as a wider range of these two parameters. This results in increasedoperational flexibility. In addition, supersaturation is constant during acontinuous process thus producing a narrower PSD compared to batchcrystallization. Finally, the scale up factor from laboratory to commercialscale is two to three orders of magnitude lower for continuous crystallization.This paper describes a systematic approach for the implementation of continuousfast crystallization processes. Individual case studies showcase the advantagesoffered by continuous processing.

We identifiedthree main drivers for the implementation of continuous fast crystallizations.The first is manufacturing API with small particle size to enhancebioavailability. The second is particle size and/or other API attribute controlwithin a tight range. The third is the increased throughput and associated costreduction offered by continuous processes. We also identified a systematicapproach for the implementation of continuous fast crystallization and theirscale up from laboratory to pilot plant and commercial scale equipment. Thisapproach includes the determination of the fundamental kinetic constants of thesystem (mixing and induction time) and the subsequent selection of an appropriatemixer design for scale up.

The first case studyinvolves the isolation and attribute control of an amorphous API. Itsprecipitation ^{2}/g for 10 batches. Theequivalent average particle sizes are 220 and 170 nm, values indicating thegood PSD control offered by the continuous precipitation process.

The secondcase study involves a dissolution limited drug substance, which requiredmicronization to provide adequate exposure. However, we encountered severeclogging of the jet mill chamber during micronization, which severely limitedprocess throughput (3 days to micronize 12 kgs of drug substance). Theseproblems were overcome by altering the input to the micronization process. Wefound that producing API with a co-axial injection continuous crystallizationresulted in high API surface areas. The latter was proven considerably moreamenable to jet milling; the process throughput increased more than tenfold,from 200 g/hr to about 1.5-2.0 kg/kr.

See more of this Session: Applications of Continuous Processing In Manufacture of Drug Substance
See more of this Group/Topical: Food, Pharmaceutical & Bioengineering Division

(431d) Efficient and Cost Effective Dehydration of Pharmaceutical Streams by Pervaporation

Dimitrios Zarkadas[1], Ronnie Feng[1], Michael Wu[1], Plamen Grigorov[2], Azzeddine Lekhal[1] and Kevin Molter[1], (1)Chemical Process Development & Commercialization, Merck and Co. Inc., Rahway, NJ, (2)Chemical Process Development & Commercialization, Merck and Co. Inc., Union, NJ

Pervaporationis a continuous process that has been used successfully the last two decadesfor the dehydration of organic solvents at large scales and for improvingdistillation efficiency of azeotropic systems like ethanol-water. Similarly toother membrane processes, pervaporation offers several advantages. It oftenresults in lower costs, it is modular and therefore easy to scale up, offersmuch higher mass transfer rates per unit volume compared to conventional separationequipment and can effect separations difficult to achieve otherwise. Despitethe growing importance of pervaporation in the chemical industry, there are noreported systematic efforts to apply the technology in the dehydration ofpharmaceutical streams. The latter is important in several cases: reactionengineering (condensation reactions, moisture sensitive reactions, dehydrationof catalysts), crystallization (improve yield by water removal, generation ofsupersaturation) and solvent recovery. Dehydration of pharmaceutical streams iscurrently performed by atmospheric or vacuum distillation, which is inherentlyan inefficient process; large solvent volumes are required to drive off a smallamount of water and the separation often takes a long time which results inhigh operating costs and decreased throughput. In this paper we report ourongoing efforts to implement pervaporation as an alternative yet powerful toolfor dehydration applications in the pharmaceutical industry.

A systematicanalysis of frequently encountered dehydration problems in the pharmaceuticalindustry has revealed that pervaporation would be a better choice compared todistillation when the stream contains less than about 10% wt. water, forms anazeotrope with water or needs to be dehydrated to water contents less than0.5-1.0 % wt. In all other cases the selection between pervaporation anddistillation can be made on economic grounds. Two major operating costs havebeen identified for distillation: solvent (usage and disposal) and labor andoverhead costs associated with increased operating time. Pervaporation obviatesthe need to use solvent to remove water and therefore does not incur solventusage and disposal costs. If the membrane configuration is appropriatelydesigned, pervaporation can result in shorter operating times compared todistillation and therefore reduced operating costs. We have developed detailedperformance and economic comparison tools, which enable the design of processand cost efficient pervaporation processes.

The attractiveness ofpervaporation for dehydration applications is illustrated with severalexamples. These examples include the removal of water from a 2-MeTHF and anIpAc stream to improve yield during the subsequent isolation steps as well asthe improvements to a THF recovery process including a dehydration step. Theabove analysis and examples clearly illustrate that pervaporation can deliverthe required separation goal in a robust and more efficient manner compared todistillation, reduce solvent usage resulting in a greener process, reduceoperating time and overall result in a more cost efficient solution fordehydration applications in the pharmaceutical industry..

See more of this Session: Applications of Continuous Processing In Manufacture of Drug Substance
See more of this Group/Topical: Food, Pharmaceutical & Bioengineering Division

(431e) Failure Mode Analysis of Continuous Process with Process Models

Sze Wing (Candice) Wong, Chemical Process R&D, Eli Lilly and Company, Indianapolis, IN, Prashant B. Kokitkar, Engineering Technology Center, Eli Lilly and Company, Indianapolis, IN, K. Derek Berglund, Chemical Product Research and Development, Eli Lilly and Company, Indianapolis, IN and Martin Johnson, Chemical Process Research & Development, Eli Lilly &Co., Inc., Indianapolis, IN

Failure Mode Analysis of Continuous Process with Process Models

Sze-Wing Wong[1], Prashant Kokitkar, K. Derek Berglund, Martin D Johnson, and Shekhar K. Viswanath

Chemical Product R&D, Lilly Research Laboratories, Eli Lilly and Company, Indianapolis, IN 46285

Abstract: This presentation described the use of process models to drive process development of a continuous process at Eli Lilly and Company. In this case study, the continuous process is represented with two separate models: a reaction kinetics model and a counter-current extraction model. The reaction kinetics parameters were put into a three-in-series CSTR (Continuous Stirred Tank Reactor) model in which it can simulate different failure modes such as pump flow rate off-set or total pump failure. The resulting impurity profile of the reaction mixture becomes the input to the counter-current extraction model where the equilibrium data of all the components at each stages is known. With the combination of both process models, one can estimate how much process disturbances the system can handle. The simulation results could then be used to identify potential critical process parameters and also be used to set the process control strategy. This case study shows that the models can positively impact development of a continuous process even at early development phase, which is consistent with the quality by design principles.

[1] Corresponding author

E-mail address: wongsz@lilly.com

See more of this Session: Applications of Continuous Processing In Manufacture of Drug Substance
See more of this Group/Topical: Food, Pharmaceutical & Bioengineering Division

(431f) Novel Concepts for Continuous Purification of APIs Using Electro-Chromatography

Marie-G. Braunbruck[1], Heidrun Gruber-Woelfler[1], Peter Feenstra[1], Rene Laskowski[2], Hans-Joerg Bart[2] and Johannes G. Khinast[1], (1)Institute for Process and Particle Engineering, Graz University of Technology, Graz, Austria, (2) Lehrstuhl fr Thermische Verfahrenstechnik, Technische Universitt Kaiserslautern, Kaiserslautern, Germany

In pharmaceutical industry batch processing still dominates the manufacturing of the active
pharmaceutical ingredients (APIs). One example is liquid chromatography, which is one of
the basic separation methods for biosynthetic products but is still operated in batch mode. As
continuous processes offer many advantages, approaches to develop continuous chromatographic processes have become increasingly important.
An effort in this area is the CAEC project, which deals with the development of a continuous
annular electro-chromatography (CAEC) prototype [1]. This prototype combines the
advantages of electrophoresis and preparative liquid chromatography in a continuous
operation mode, thus leading to both, a high throughput and a high separation efficiency.
Furthermore, the eluting analytes are monitored with an online detection system to ensure a
continuous quality control.
We present the design, characterization and implementation of the stationary phase in this
novel prototype. The stationary phase is based on a silica monolith which can be adjusted to
the separation problem with functional groups, such as amino groups for ion exchange
chromatography or alkyl chains for reversed phase chromatography [2]. In this way, not
only the liquid phase but also the stationary phase can be adapted to specific

problems
which makes the CAEC universally applicable. The monolith is supported with silica particles
that prevent the stationary phase from shrinking. This combination of particles and the
monolithic structure increases the mechanical and chemical stability of the stationary phase
while preserving a high porosity to achieve a high throughput.

The macroporous one-piece material was implemented in several moulds, such as a planar
test cell that represents a fraction of the annular prototype and already allows a continuous
separation via electro-chromatography. In this planar test cell hydrodynamics and heat
transfer [3] were investigated and standard test systems could be successfully separated
achieving theoretical plate numbers of over 20000 per meter and a throughput of up to
50 ml/h.

In addition, we present a simulation program that is implemented in MATLAB to support the
design of the chromatographic experiments. The program is based on a stochastic (Monte
Carlo) approach and algorithms for convection, diffusion and retention are implemented [4].

With this program chromatograms can be simulated based on the population distribution of
the analytes. Therefore, these simulations can help to determine the suitable stationary as
well as mobile phase for a given separation problem. Moreover, the molecular interactions of
the analytes with the stationary phase were studied using molecular dynamics (MD)
simulations.

Overall, the outcomes of our approaches are powerful tools for future separation tasks in
pharmaceutical industry and can be individually assigned to a broad spectrum of problems.

This work is supported by the 7th Research Framework Programme of the European
Commission (Grant agreement number: NMP2-SL-2008-206707)

[1] https://caec.bci.tu-dortmund.de/

[2] Gruber-Woelfler, H.; Braunbruck, M.-G.; Feenstra, P.W.; Laskowski, R.; Bart, Khinast, J.G., Oral
Presentation, 9th International Congress of Chemical and Process Engineering CHISA 2010,
Prague, August, 2010
[3] Laskowski, R.; Bart, H.-J.; Gruber-Wölfler, H.; Feenstra, P.; Braunbruck, M.-G.; Khinast, J.;
Hofmann, C.; Menges, G.; Werner, B.; Löb, P.; Hessel, V.:, 17th International Symposium on
Electro- and Liquid Phase-separation Techniques (ITP), 2010, Baltimore, USA.
[4] McGuffin, V.L.;Wu, P.R.; Journal of Chromatography A, 722 (1996) 3

See more of this Session: Applications of Continuous Processing In Manufacture of Drug Substance
See more of this Group/Topical: Food, Pharmaceutical & Bioengineering Division

(457a) Wet Flows In a Bladed Mixer: Experiments and Simulations of Monodisperse Spheres

Brenda Remy, Process Research and Development, Bristol-Myers Squibb Co., New Brunswick, NJ, Johannes G. Khinast, Institute for Process and Particle Engineering, Graz University of Technology, Graz, Austria and Benjamin J. Glasser, Chemical and Biochemical Engineering, Rutgers, the State University of New Jersey, Piscataway, NJ

Bladed mixers are commonly used in a variety of pharmaceutical processes such as wet granulation, agitated drying, and mixing and blending. Cohesion due to the presence of a liquid phase is often present in many industrial scenarios where bladed mixers are used (i.e. wet granulation, agitated drying). For bladed mixers, the majority of past studies have focused on dry flows. Therefore bladed mixer processes involving wet flows are not fully understood. The lack of process understanding can lead to operational problems and can cause complications during scale-up.

The flow and agglomeration of wet particles in a bladed mixer was studied experimentally using Particle Image Velocimetry and computationally using the discrete element method. The experimental and computational work showed that particle beds at low moisture contents are characterized by enhanced convective and diffusive particle motion as well as enhanced mixing kinetics when compared to dry particle beds. This behavior is attributed to the development of small particle agglomerates which behave like rough, non-spherical particles and enable the transfer of energy from the blades to the particle bed. At higher moisture contents, a different behavior was observed. Particle convective and diffusive motion was hindered by the presence of moisture at higher levels leading to a decrease in mixing performance. This occurs as large agglomerates are formed and are not broken apart by shear leading to poor mixing. The extent of agglomeration at different moisture contents was quantified via the discrete element simulations. Agglomerate size distributions and morphology were shown to be strong functions of moisture content.

See more of this Session: Industrial Application of Computational and Numerical Approaches to Particle Flow I
See more of this Group/Topical: Particle Technology Forum

CPFD Flow Pattern Simulation in Downer Reactors

Alireza Abbasi[1,2], Mohammad Ashraful Islam[1], Paul E. Ege[2], Hugo I. de Lasa[1*]

(1) Department of Chemical and Biochemical Engineering, The University of Western Ontario, London, ON, Canada
(2) Reactech Process Development Inc., Markham, ON, Canada
*Corresponding author (hdelasa@eng.uwo.ca.)

1. Abstarct

Computational Particle Fluid Dynamics (CPFD) simulation can assist considerably towards flow pattern characterization. The present study contributes with a fluid dynamic simulation based on the numerical solution of continuity and momentum balance equations in a 3D framework using Barracuda's software. The proposed model includes both a reactor configuration and specific C_D drag coefficients recommended for downflow reactors. In this study, the CPFD simulation conditions are selected to represent actual flow conditions likely to be used in a gas-solid downflow reactor unit with FCC particles. Model results are established in the context of stabilized solid and fluid flow patterns. They involve numerical calculations in excess of 1.5 seconds. The CPFD model predicts local densification of solids and asymmetric "wavy flows". In addition, this model forecasts higher particle velocity than gas velocity, once the flow reaches an axial position larger than 1 meter ($z > 1$m) from the gas injector. These expected findings are accompanied with valuable original observations about the intrinsic rotational character of the flow pattern in downer reactors. Numerical 3D calculations show that both gas and particle velocities involve the following: a) an axial velocity component, b) a radial velocity component (about 5% of axial velocity component), c) an angular velocity component. The relevance of these calculated velocity components for determining the rotational flow pattern are established in a significant range of operational conditions and solid flux/gas flux ratios.

2. Introduction

The characterization of flow patterns in downflow reactors has progressed significantly in recent years.[1-2] A diversity of techniques have been applied to characterize both particle and gas flow in downer reactors.[1-4] In this respect, there have been significant advances in the development of fiber optic sensors and in the characterization of particle clusters.[3,4]

These reactors can be considered as displaying a relatively uniform flow structure and an expected plug-flow behavior for both gas and solid phases.[5] Due to low back-mixing and short residence times, a downer reactor is suitable for the enhancement of a diversity of chemical processes such as fluid catalytic cracking (FCC) process, residual oil cracking, and biomass and coal pyrolysis.[6,7] In this respect, it is essential to understand the macro as well as the micro-structural behavior of gas-solid flow in a downer reactor for reactor scale-up.[8]

The fluid dynamics in downflow reactors have been considered using a number of approaches. However, and due to the complex bed down flow hydrodynamics, proposed approaches still show limitations.[9] There are in fact, two types of theoretical approaches which one can use in computational fluid dynamics (CFD) to describe gas and solid flows. These are the Eulerian–Eulerian model and the Eulerian–Lagrangian model. The Eulerian–Eulerian model considers the solid phase as a continuous medium interacting with the fluid phase. Conservation equations for each phase are derived to obtain a set of equations, which have a similar structure for all phases. Conservation equations are related via constitutive relationships obtained from empirical information and/or kinetic theory.[10] Such a fluid dynamic modeling method has been partially successful in simulating multiphase flow phenomena in risers in which gas and solids move concurrently upwards. This Eulerian–Eulerian model presents major limitations in describing the physical model of the gas–particle, particle–particle and particle–wall interactions in the downer units.

In downflow reactors, the gas-solid flow can be predicted using the most rigorous computation particle fluid dynamic (CPFD) numerical method. This numerical method belongs to the direct element method (DEM) class of solutions. The Eulerian reference frame allows the modeling of forces using spatial gradients of properties.[11] On the other hand, the dispersed phase is treated by tracking a large number of particles through the calculated flow field. The complexity of calculating dense particle–particle interactions makes traditional DEM methods limited to the number of the particles to be included in the calculation. It also makes these methods to be usually restricted to two-dimensional solutions.[11] However; The CPFD method overcomes these limitations.

This work aims to describe downer hydrodynamics in 3-D (three dimensions) using the advanced CPFD method. This approach allows describing particle motions and particle distributions providing snap shots along the downer unit. With this aim, the upcoming sections of this paper address various issues of modeling and simulation as follows: a) Section 3 considers specifically the proposed mathematical model as well as the selected boundary conditions, b) Section 4 describes the configuration of the experimental set up as well as the selected dimensions and operating conditions, c) Section 5 reports the CPFD simulations and includes discussions of various computational results obtained.

3. Experimental Setup, Simulation Parameters and Verification of Numerical Procedures

A detailed description of the experimental unit is provided in Nova et al. [1] This unit is equipped with a feeder section where air flow is fed via inclined 45 degrees openings forming high velocity jets. These high velocity jets intimately mix gas and solid particles. This allows supporting the idealized assumption of uniform solid and gas distribution at the downer entry, as adopted in the present model. The unit ends in a cyclone where gas and particles are separated. The downer unit as implemented in the CREC laboratories is provided with CREC Optiprobes allowing experimental characterization of gas-solid flow and measurements of particle clusters and solid densification. To proceed with the simulations in the downer reactor, computational domains were established in the downer using the following mesh: $7\times7\times557$. Air as well as 84 micron FCC particles were considered to be fed from the top of the column via the feeding section.

Preliminary work of the present study, involved the testing of the adequacy of numerical calculations. It is confirmed that at least 1.27 seconds of computational time is required for reaching what can be considered as a representative description of solid flows. From that point and on, the particle solid fluxes remain close to stabilized average values as follows: 9.41 kg/m^2s

for Case 2, 43.66 kg/m^2s for Case 3, 62.51 kg/m^2s for Case 4 and 80.03 kg/m^2s for Case 5 within a standard deviation of +/-10%. Thus, the necessary computational time needed for a stable numerical calculation at a set location was established.

4. ***Results and Discussions***

Once the fluid dynamic model for the downer reactor was established as adequate, given that it passed the various numerical tests as described in the previous section, the model was utilized extensively to provide macroscopic descriptions of the flow field as reported in Figure 1.

For instance, Figure 1 describes the average axial velocities for both gas and particles of 2.25 kg/m^2s air flux and of 43.7 kg/m^2s particle flux. The selected feeding temperature was 300K. One can see that as anticipated, the cross-section average fluid gas velocity remained fairly constant along the downer. On the other hand, the cross-section average axial particle velocity goes from being smaller than the gas velocity at the injection section and evolves towards superseding the gas velocity. This is expected given the influence of gravity exerted on the particles of the evolving gas-solid suspension. Towards the end of the downer, this velocity difference between phases or so-called slip velocity stabilizes and becomes particle or particle cluster terminal velocity.

These results are consistent with the ones observed by other researchers in the technical literature and provide in this respect, additional value to the model proposed.[12-17] Thus, on the basis of the above, one can conclude that the fluid dynamic cross-sectional average simulation results of the present model are consistent with expected fluid dynamic results in downer units as already documented in the technical literature.

Furthermore, Results confirm when gas velocity increases particle velocity is augmented and when gas velocity decreases particle velocity is reduced. In summary, a CFPD model using both 2D and 3D plots, becomes an essential and valuable tool to identify local and time dependant fluid dynamic property changes, which otherwise are missed using cross-sectional average property based downer simulation.

Figure 2 reports another interesting description of downer fluid dynamics, under the scope of the 3D CFPD plots. This figure reports an axial cross section snap shot of the downer column at various axial levels. It can thus, be observed that there is a uniform down flow suspension flow close to the gas injection section. This flow becomes however progressive "wavy" with solid densification leading to regions of higher solid fraction followed by regions of lower solid fraction. This "wavy" flow becomes more apparent towards the end of the downer where the particle fraction variations are observed across the radial position. One should note that these 3D simulations are supported by downer suspension flow property variations as recorded by CREC Optiprobe sensors.[1,4] These authors measured "particle clusters" or solid densification evolving in the axial direction of the downer. Furthermore, these observations are in agreement with 2D fluid dynamic downer simulations as reported by Zhao et al.[17] Zhao et al.[17] simulations were able to describe densifications along the axial position in the downer under the constraints of the 2D model.

The CPFD simulation can also be used to provide snap shots of particle and gas velocities at various axial and radial downer locations. Figures 3 and 4 display these local properties for 2.25 kg/m^2s air and 43.7 kg/m^2s particle fluxes at a 300K feeding temperature. One can see that there are correspondences of gas and particle velocity components. Both gas and particles, display radial velocity vectors as described in Figure 3 and 4. These radial velocity components force the particle suspension not only to move downwards but also circumferentially.

A rationalization of this downward and circumferential flow can be provided given the asymmetric "wavy flow" described in the previous sections of this paper. Particle clusters move at a higher velocity than the gas. This tends to compress and redirect the gas fluid towards regions of lower particle concentrations. This flow redirection generates in the context of "wavy flows", radial fluid velocity components. Radial gas velocity also leads given the associated fluid drug on particles, to impart on particles radial velocity components. In this respect, one can see that local particle and gas velocities can display, as described in Figure 3 and 4, radial components of about 5-15% magnitude of the axial velocity component.

Furthermore, and in order to have confirmation of the gas-solid circumferential flow in the 3D downer, the following additional simulations were developed. The results focus on the downer outlet section located at 0.2-0 m axial position and on the following conditions: a) 2.25 kg/m^2s air flux and b) 0 kg/m^2s, 9.2 kg/m^2s, 43.7 kg/m^2s, 62.5 kg/m^2s, 80.0 kg/m^2s particle flux.

One can note that for 0 kg/m^2s or a down flowing of gas free of solid particles, there is no circumferential flow with this being consistent with the expected classical single gas-phase fluid dynamics. However, as soon as the particle flow increases from 9 to 43.7 kg/m^2s, the circumferential flow gains importance. Finally, as soon as the solid flux surpasses the 43.7 kg/m^2s level and moves up to 62.5 kg/m^2s and 80.0 kg/m^2s, the circumferential flow is reduced in importance. Particle and gas radial velocities are then reduced to 4% and 2% of the axial velocity values. Thus, it can be hypothesized, that when at lower/moderate particle fluxes, the presence of solids and variation of suspension density promotes circumferential flow. It is also expected that this effect is reduced at higher solid fluxes. This dampening effect at higher solid fluxes can be assigned to the increasing influence of particle-particle interactions.

In summary, it is shown in the present study, that the simulation of the 3D time dependant gas-solid suspension in a downer becomes a most valuable tool to identify essential features of the downer fluid dynamics such as "wavy" suspension and circumferential flow. It is anticipated that better understanding of 3D downer fluid dynamics will facilitate downer scale up and extensive application for catalytic gas-solid reactions.

5. Conclusions

a) The proposed gas-solid downer based CFPD model complies with a number of numerical simulation stability and overall mass balance tests.

b) The 3D downer simulations of this study are very valuable to describe the local and time variations of fluid dynamic properties in downers including particle/particle cluster velocities, gas velocity and pressure.

c) The 3D simulations lead to the identification of particle clusters and an asymmetric "wavy flow" . This asymmetric "wavy flow" promotes circumferential gas flow with radial velocity components both for the gas and the particles.

d) The identified circumferential flow across solid fluxes leads to radial velocity components. These radial velocity components are dampened when solid particle concentrations reach high particle concentrations.

6. References

1. Nova SR, Krol S, deLasa H. Particle velocity and particle clustering in down-flow reactors. Powder Technol. 2004;148:172–185.

2. Luo B, Yan D, Ma YL, Barghi S, Zhu J. Characteristics of gas-solid mass transfer in a cocurrent downflow circulating fluidized bed reactor. Chem Eng J. 2007;132:9-15.

3. Qi XB, Zhang H, Zhu J. Solids concentration in the fully developed region of circulating fluidized bed reactor. Powder Technol. 2008:183:417-425.

4. Vaishali S, Roy S, Mills PL. Hydrodynamic simulation of gas-solids down flow reactors. Chem Eng Sci. 2008;63:5107-5119.

5. Zhang H, Zhu J-X, Bergougnou MA. Hydrodynamics in downflow fluidized beds (1): solids concentration profiles and pressure gradient distributions. Chem Eng Sci. 1999;54:5461-5470.

6. Cheng, Y., Wu, C., Zhu, J., Wei, F., Jin, Y., 2008. Downer reactor: From fundamental study to industrial application. Powder Technol. 183, 364-384.

7. Zhu J-X, Yu Z-Q, Jin Y, Grace JR, Issangya A. Cocurrent downflow circulating fluidized bed (downer) reactors -A state of the art review. Can. J. Chem. Eng. 1995;73:662–677.

8. Lu X, Li S, Du L, Yao J, Lin W, Li H. Flow structures in the downer circulating fluidized bed. Chem Eng J. 2005;112:23-31.

9. Sunun L, Nattha T, Terdthai V, Parinya K., DEM modeling and simulation of a down-flow circulating fluidized bed. Chem Eng Comm. 2008;17:1328-1344.

10. Huilin L, Yurong H, Gidaspow D. Hydrodynamic modeling of binary mixture in a gas bubbling fluidized bed using the kinetic theory of granular flow. Chem Eng Sci. 2003;58:1197- 1205.

11. Snider DM, Three fundamental granular flow experiments and CPFD predictions. Powder Technol. 2007;176:36–46.

12. Wang Z, Bai D, Jin Y. Hydrodynamics of cocurrent downflow circulating fluidized bed (CDCFB). Powder Technol. 1992; 70: 271–275.

13. Tuzla K, Sharma AK, Chen JC, Schiewe T, Wirth KE, Molerus O. Transient dynamics of solid concentration in downer fluidized bed. Powder Technol. 1998;100:166-172.

14. Johnston PM, deLasa HI, Zhu J–X. Axial flow structure in the entrance region of a downer fluidized bed: Effects of the distributor design, Chem Eng Sci. 1999;54:2161- 2173.

15. Cao C, Weinstein H. Characterization of downflowing high velocity fluidized bed, AIChE J 2000;46(3):515-522.

16. Bolkan Y, Berruti F, Zhu J, Milne B. Modeling circulating fluidized bed downers, Powder Technol. 2003;132:85–100.

17. Zhao Y, Ding Y, Wu C, Cheng Y. Numerical simulation of hydrodynamics in downers using a CFD-DEM coupled approach. Powder Technol. 2010;199:2-12.

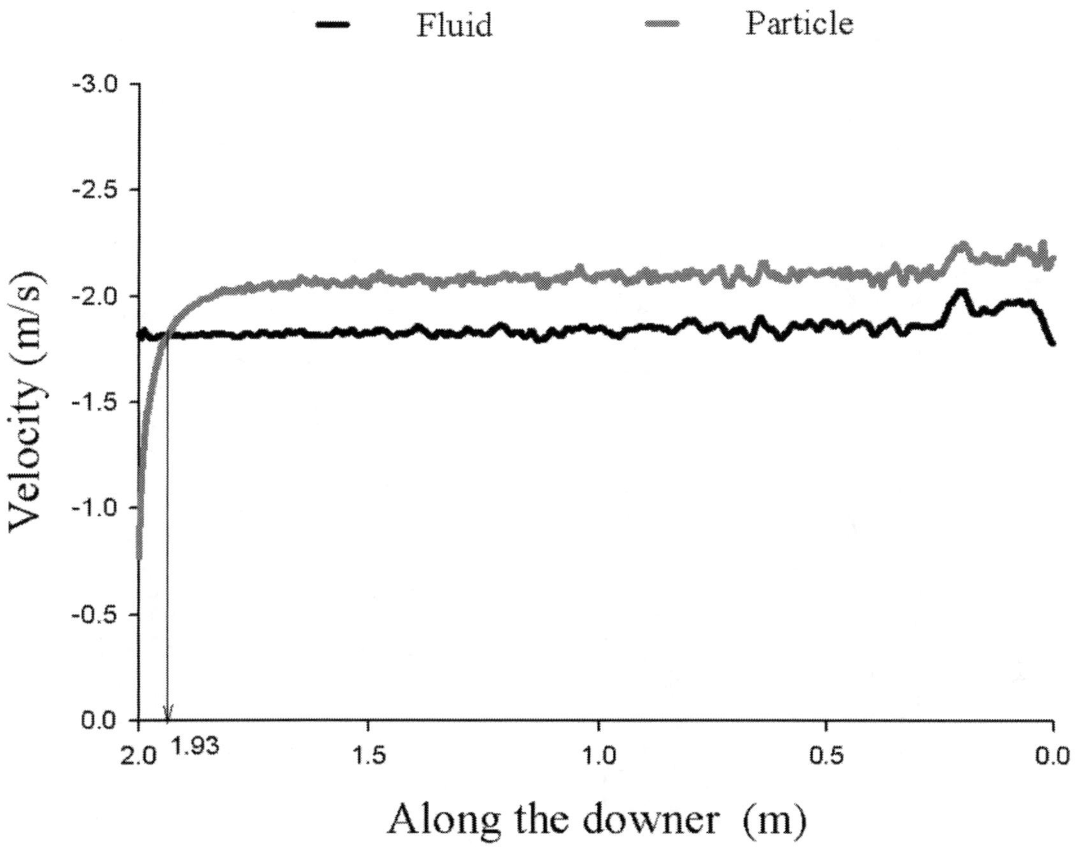

Figure 1: Average axial velocity profile along the downer. Operating Conditions: a) 2.24 kg/m^2s air flux and b) 43.7 kg/m^2s particle flux. The feeding temperature: 300K.

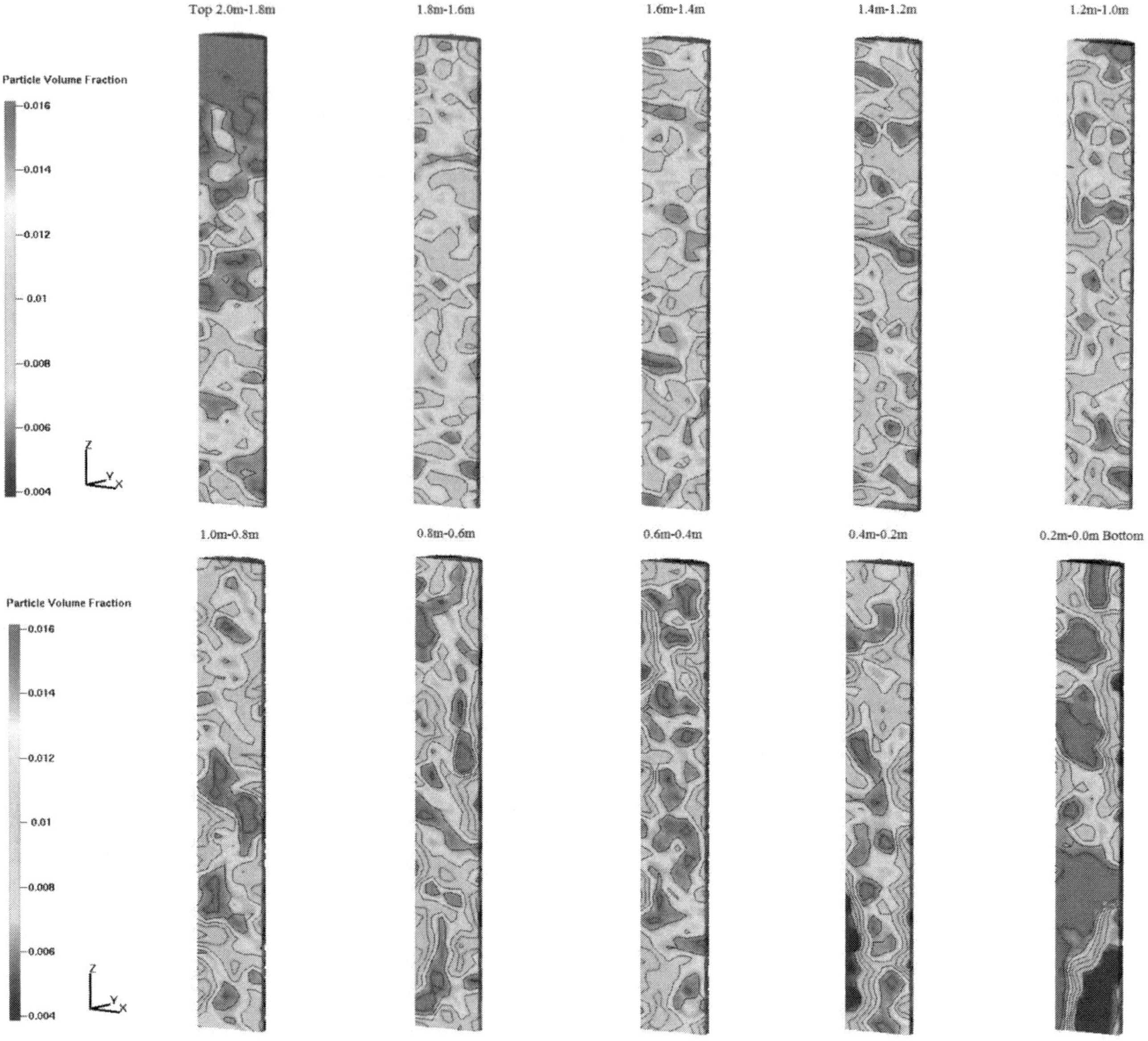

Figure 2: Contour of Particle Volume Fraction. Operating Conditions: a) 2.25 kg/m^{2}s air flux and b) 43.75 kg/m^{2}s particle flux. The feeding temperature: 300K

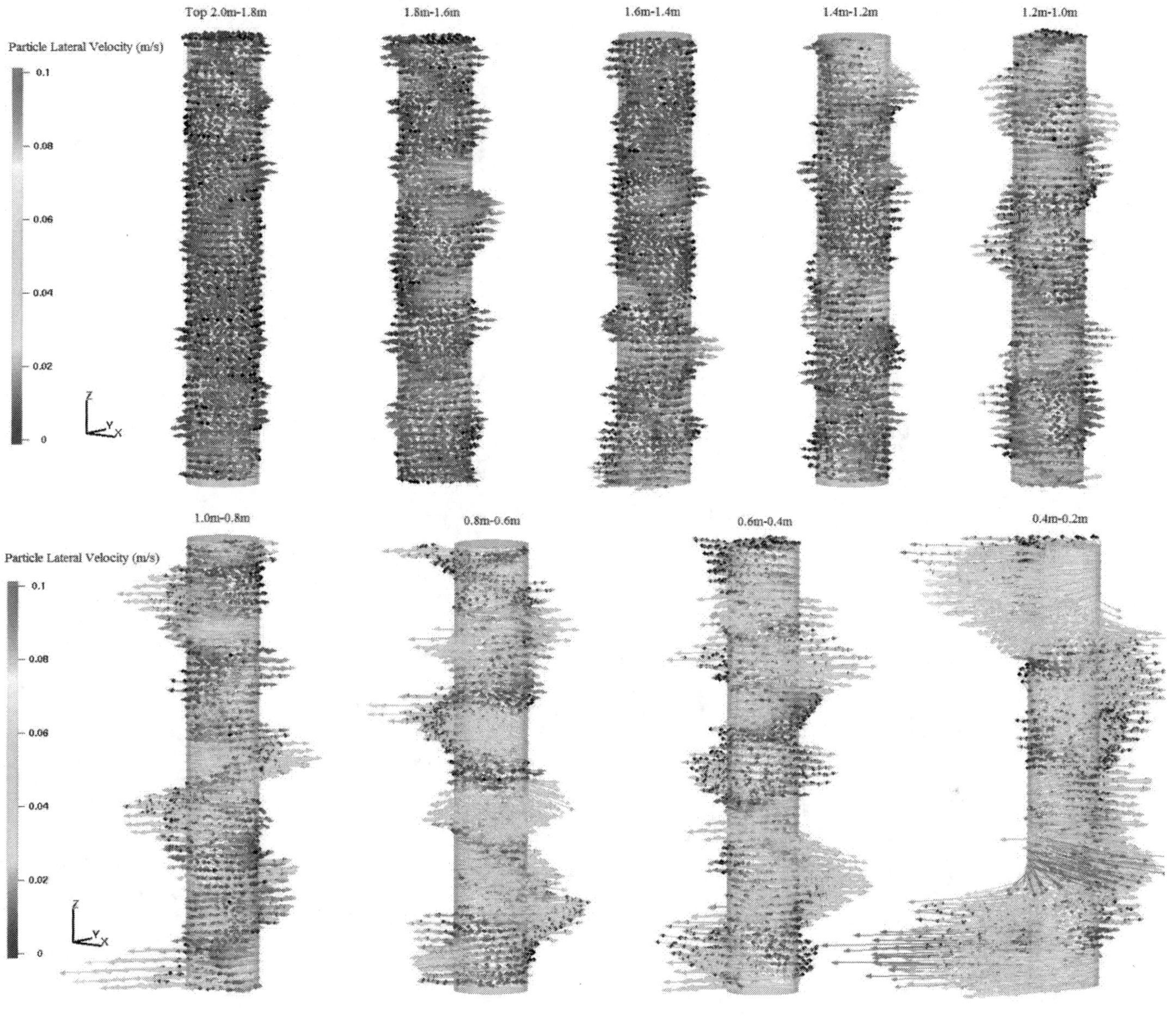

Figure 3: Particle Lateral Velocity. Operating Conditions: a) 2.25 kg/m^2s air flux and b) 43.75 kg/m^2s particle flux. The feeding temperature: 300K

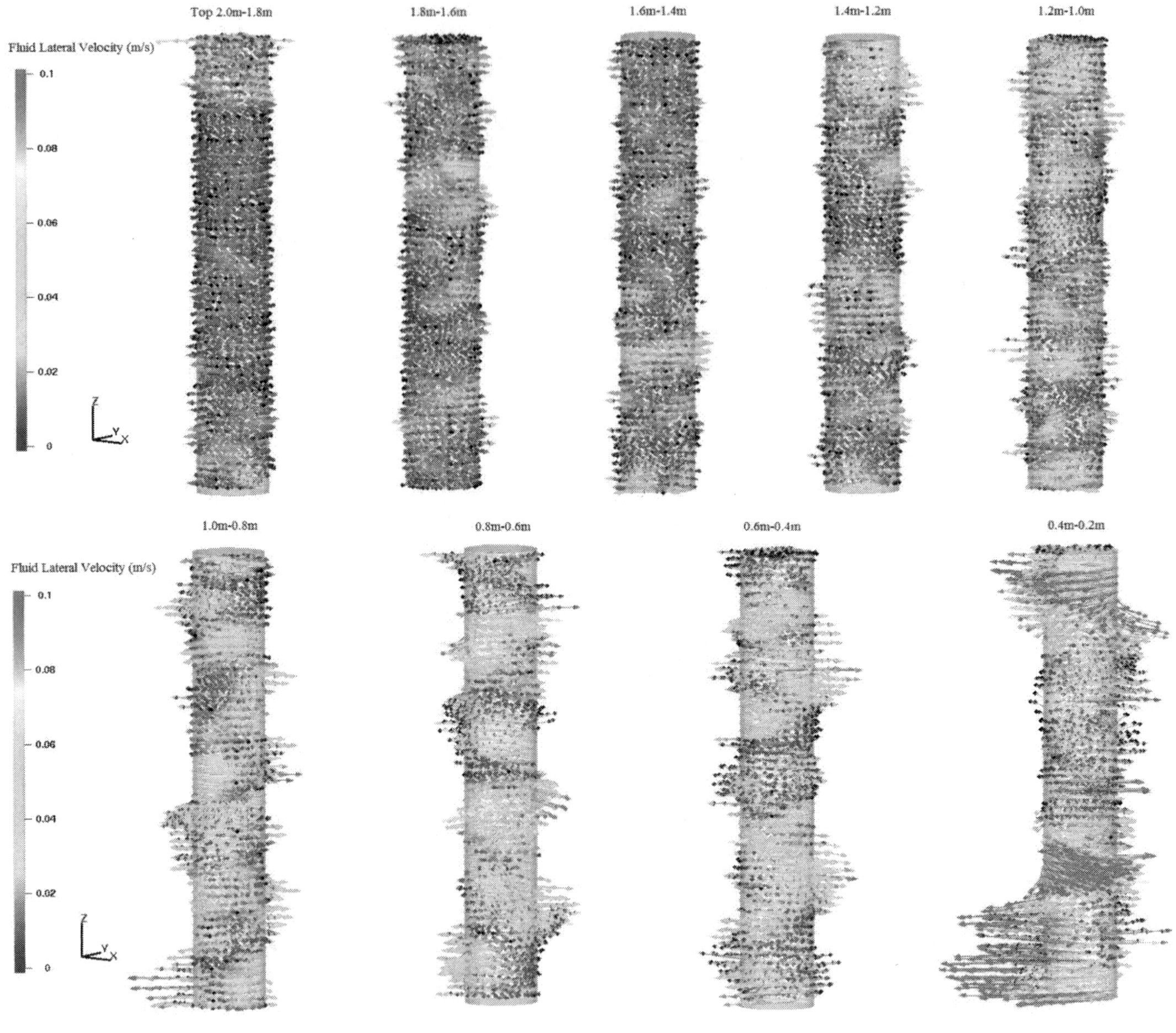

Figure 4: Fluid Lateral Velocity. Operating Conditions: a) 2.25 kg/m^2s air flux and b) 43.7 kg/m^2s particle flux. The feeding temperature: 300K

(457c) Direct Simulations of Flocculation In Sedimenting Solid-Liquid Suspensions

Jos Derksen, Department of Chemical & Materials Engineering, University of Alberta, Edmonton, AB, Canada

Direct simulations of flocculation in sedimentingsolid-liquid suspensions

Gravity-based separation (settling) is common practice inmany processes involving solids handling, specifically in mining and related(mineral processing) applications. A widely used way to increase settlingrates, and to allow for gravity based separation of fine particles is topromote aggregation so that flocs are formed thatfall faster through liquid than the primary particles. We study flocculation insedimenting solid-liquid suspensions numerically byperforming direct simulations with full resolution of the solid-liquidinterfaces and with full (two-way) coupling between liquid flow and solidparticle motion. In the simulations uniformly sized solid spheres are released.They fall under the action of (net) gravity, they collide with one anotheraccording to a hard-sphere model, they interact through the interstitialliquid, and they tend to aggregate as a result of a square-well potential surroundingeach sphere. By means of the simulations we study how the settling velocity dependson the strength of the square-well potential and the other key parameters inthe process: the solids volume fraction and the particle size. The observationsregarding settling velocities are interpreted in terms of the micro structureof the suspension (radial and angular distribution functions), and in terms ofthe floc size distribution and the fractal dimensionof the flocs.

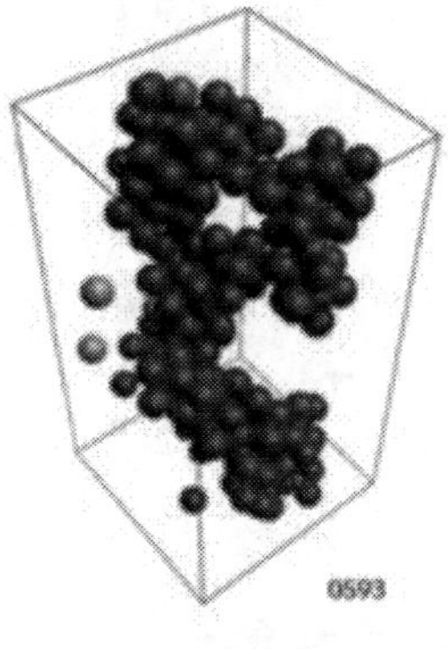

See more of this Session: Industrial Application of Computational and Numerical Approaches to Particle Flow I
See more of this Group/Topical: Particle Technology Forum

(457d) DEM Simulation On the Effect of Particle Size Distribution On the Bubble Size

Pradeep Gopalakrishnan, Mechanical Engineering, Virginia Tech, Blacksburgh, VA

Fluidized bed has been widely used for many applications such as coal combustion, fluidized catalytic cracking and biomass gasification. The advantages of fluidized beds are its simplicity, high heat and mass transfer and continuous operation. Most of these applications involve particles with different size distribution (PSD) and its effects on the hydrodynamics are critical. Previous studies showed that the particles with wide PSD or particles with fines added offer better heat transfer and mass transfer characteristics. Beetstra et al.[1] analyzed the effect of PSD for Geldart A particles and showed that with increase in PSD width, bubble size decreases at higher velocities and increases at lower velocities. Chew et al.[2] observed a monotonic increase in bubble frequency, velocity and chord length with PSD width for Geldart B particles.

The mechanism behind the enhanced performance due to addition of fines is least understood. One popular surmise is that the fines added produces lubrication effect which leads to low granular viscosity and better fluidization. Alternate suggestions proposed are increased porosity in the dense phase resulting in high flow through dense phase and higher solids holdup in voids which alters the bubble characteristics. The current study aims at analyzing the effect of the PSD using Discrete Element Method (DEM). DEM resolves particle-particle interaction and particle-wall collisions and provides detailed view of the bubble dynamics.

The current study analyzes two-dimensional bubbling fluidized bed with cross section 20cm by 15cm height. The simulation is carried for Geldart B particles with mean diameter 500 mm. The analysis is carried out for monodisperse, bidisperse and a continuous PSD with standard deviation to the mean diameter ratio equal to 15%. The results will be analyzed to estimate effect of particle size on the bubble frequency, speed and size. The granular temperature will be computed which will identify the effect of the fines over the granular viscosity.

1. Beetstra, R., Nijenhuis, J., Ellis, N., and van Ommen, J. R. "The influence of

the particle size distribution on fluidized bed hydrodynamics using high-throughput experimentation," *AIChE Journal* Vol. 55, No. 8, 2009, pp. 2013-2023.

2. Chew, J. W., and Hrenya, C. M. "Link between bubbling and segregation patterns in gas-fluidized beds with continuous size distributions," *AIChE Journal*, 2011.

See more of this Session: Industrial Application of Computational and Numerical Approaches to Particle Flow I
See more of this Group/Topical: Particle Technology Forum

(457e) Numerical Study of the Segregation of Binary Mixture In Dense Gas-Solid Fluidized Bed

N'dri Arthur Konan[1], David E. Huckaby[2] and Thomas J. O'Brien[2], (1) Computational Science Division, National Energy Technology Laboratory, Morgantown, WV, (2)Computational Science Division, National Energy technology Laboratory, Morgantown, WV

Numerical study of the segregation of binary mixture in dense gas-solid fluidized bed

N'dri A. Konan, E. David Huckaby and Thomas J. O'Brien

National Energy Technology Laboratory, Morgantown, WV 26505, USA

n'dri.konan@or.netl.doe.gov, E.David.Huckaby@netl.doe.gov

Keywords: Polydispersion, kinetic theory, segregation, drag law, restitution coefficient

Powders involved in the engineering flow systems consist of particle mixtures of various densities and especially of a wide distribution of the size of the particles. Depending on the flow conditions, the behavior of the larger/heavier particles assimilates to a migration towards the lower region of the system, while the upper region populates of the smaller/lighter particles. That typically results in local (or spatial) concentration of the species or type of the particles significantly higher than the other particle phases within the different regions of the system, in the situation of the severe segregation.

Successful operation of chemical looping using solid fuels (e.g. coal) is dependent on designing a fuel reactor in which the oxygen carrier is returned to the air reactor with minimal unburned fuel and ash. One proposed approach (e.g. Berguerand & Lyngfelt, 2008 and Shen et al., 2009) is to design the oxygen carrier, fuel reactor and fuel grind such that the denser and larger oxygen carrier settles to bottom of the bed, the solid fuel moves toward the upper portion of the bed and eventual is entrained in the exhaust after the fuel is consumed. Accurate simulation segregation is critical for predicting the operation of a reactor and associated chemical looping system based on this design.

The origin of such a segregation phenomenon, although extensively investigated numerically as well as in the experiments in the frame of the dense gas-particle fluidized beds, is not yet completely understood. Numerically, the pertinence of the predictions is mainly attributed to the accurate description of:

- · the exchange of the momentum between the gas and the particles, i.e. the drag force, which acts differently on the inclusions of a cloud of polydisperse or monodisperse particles (in size), at both the same porosity and particle Reynolds number (see e.g. Beetstra et al., 2007; Leboreiro et al., 2008, Sarkar et al., 2009,...);
- · the energy dissipation rate due to the inelastic collisions between the particles via the restitution coefficient. Indeed, the local decrease of the particle kinetic energy that can be expected because of the higher dissipation will cause the accumulation of the particles at those locations. That consistently gives rise to denser regions within the bed and thus the emergence of the regions where the volume fractions of the particles are rather small, i.e. of the bubbles, whose dynamics dominates the hydrodynamics of the bed (see e.g. Goldschmidt et al., 2001, Dahl & Hrenya, 2005, ...);
- · the momentum transfer between the particle phases through contact. Following Gera et al. (2004), the solid-solid momentum transfer term consists of a collisional term derived from kinetic theory and an additional term to account for the *hindrance effect*. This term is to account for the fact that the different particle phases will begin to move as a single phase while approaching the packing limit. Although the model successfully predicted the segregation rate (Gera et al., 2004; Fan & Fox, 2008, Azizi et. al. 2010), the modeling of the hindrance effect requires of the case to simulate for setting of the coefficient relevant to that effect. However, other authors have had similar success using kinetic theory models without an explicit model for non-collisional solid-solid momentum transfer (see e.g. Gourdel et al., 1999; Lathouwers & Bellan, 2001; Iddir et al., 2005, Garzo et al., 2007, ...). It should be noted that these studies employ radial distribution functions which diverge at the packing limit in contrast to the studies using the hinderance model which use plastic flow models and non-divergent radial distribution functions.

The present paper investigates the segregation occurring within the dense laboratory-scale fluidized bed of binary mixture of Geldart D-type particles studied by Goldschmidt et al. (2003). The hydrodynamics of the bed (considered as 2D) is modeled using the n-fluid Eulerian approach (Simonin, 1996). The polydisperse model relies upon the relative velocity between the particle phases and their kinetic energies as the main mechanisms that drive the collisions between phases (Gourdel et al., 1999). Two polydisperse gas-particle drag models are assessed: another combination of the drag models of Ergun and Wen & Yu based on their minimum for porosity smaller than 70% (Gobin et al., 2003) and the model of Beestrat et al. corrected by an effect of size polydispersity (Beestrat et al., 2007). Different fluidization conditions are simulated (ranging from smaller, slightly above and sufficiently larger than the minimum fluidization velocity of the largest particles) to investigate the influence of drag law on the segregation in terms of the fluidization velocity. Various compositions of the mixture and the influence of the restitution coefficients are also considered. The predictions of the segregation rates are compared against the experiments and discussed.

References

Berguerand N., Lyngfelt A., *Design and operation of 10kWth chemical-looping combustor for solid fuels – Testing with South African coal, Fuel*, N°87, 12, pp.2713-2726, 2008.

Shen L., Wu J, Gao Z and Xiao J, *Experiments on chemical looping combustion of coal with a NiO based oxygen carrier*, Combustion and Flame, Vol. 156, issue 3, pp. 721-728, 2009

Beestra R., van der Hoef M. A., Kuipers, J. A. M.: *Numerical study of segregation using a new drag force correlation for polydisperse systems derived from lattice-Boltzmann simulations*. Chem. Eng. Science, N°62, p. 246-255, 2007

Leboreiro J., Joseph Gustavo G., Hrenya C. M., Snider D. M., Banerjee S. S., Galvin J. E.: *The influence of binary drag laws on simulations of species segregation in gas-fluidized beds*. Powder Tech., N°184, p. 275-290, 2008

Sarkar S., van der Hoef M. A., Kuipers, J. A. M.: *Fluid-particle interaction from lattice-Boltzmann simulations for flow through polydisperse random arrays of spheres*. Chem. Eng. Science, N°64, p. 2683-2691, 2009

Goldschmidt M. J. V., Kuiper J. A. M., van Swaaij W. P. M.: *Hydrodynamic modelling of dense gas-fluidised beds using the kinetic theory of granular flow:*

effect of coefficient of restitution on bed dynamics. Chem. Eng. Science, N°56, p. 571-578, 2001

Dahl S. R., Hrenya C. M.: *Size segregation in gas-solid fluidized beds with continuous size distributions*. Chem. Eng. Science, N°60, p. 6658-6673, 2005

Gera D., Syamlal M., O'Brien T. J.: *Hydrodynamics of particle segregation in fluidized beds*. Int. J. Multiphase Flow, N°30, p. 419-428, 2004

Fan R., Fox R. O.: *Segregation in polydisperse fluidized beds: Validation of a multi-fluid model*. Chem. Eng. Science, N°63, p. 272-285, 2008

Azizi S., Hosseini S., Ahmadi G., Moraveji N., *Numerical Simulation of Particle Segregation in Bubbling Gas-Fluidized Beds,* Chemical Engineering Technology, Vol. 33, N°3, pp. 421-432 2010

Gourdel C., Simonin O., Brunier E.: *Two-Maxwellian equilibrium distribution function for the modelling of a binary mixture of particles*. In Proc. of the 6th Int. Conference of Circulating Fluidized Beds, J. Werther (Editor), DECHEMA, Frankfurt am Main - Germany, p. 205-210, 1999

Lathouwers D., Bellan J.: *Modeling of dense gas-solid reactive mixtures applied to biomass pyrolysis in fluidized bed*. Int. J. Multiphase Flow, N°27, p. 2155-2187, 2001

Iddir H., Arastoopour H.: *Modeling of multitype particle flow using the kinetic theory approach*. AIChE J., Vol. 51, N°6, p. 1620-1632, 2005

Iddir H., Arastoopour H., Hrenya C. M.: *Analysis of binary and ternary granular mixtures behavior using the kinetic theory approach*. Powder Tech., N° 151, p. 117-125, 2005

Garzo V., Dufty J.W., Hrenya C. M.: *Enskog theory for polydisperse granular mixtures. I. Navier-Stokes order transport*. Phys. Rev. E, N°76, 031303:1-27, 2007

Garzo V., Dufty J.W., Hrenya C. M.: *Enskog theory for polydisperse granular mixtures. II. Sonine polynomial approximation*. Phys. Rev. E, N°76, 031304:1-20, 2007

Goldschmidt M. J. V., Link J. M., Mellema S., Kuiper J. A. M.: *Digital image analysis measurements of bed expansion and segregation dynamics in dense gas-fluidised beds*. Powder Tech., N°138, p. 135-159, 2003

Simonin O.: *Statistical and continuum modelling of turbulent reactive particulate flows*. Von Karman Institute Lecture Series 1996-02, 1996

Gobin A., Neau H., Simonin O., Llinas J. R., Reiling V., Selo J. L.: *Fluid dynamic numerical simulation of a gas phase polymerization reactor*. International Journal for Numerical Methods in Fluids, 43, pp. 1199-1220, 2003

See more of this Session: Industrial Application of Computational and Numerical Approaches to Particle Flow I
See more of this Group/Topical: Particle Technology Forum

(457f) Analysis of Phase Holdup Characteristic of a Gas-Liquid-Solid Fluidized Bed by CFD Simulation and Experiment

H. M. Jena, Chemical Engineering, National Institute of Technology, Rourkela, Rourkela, India

Three-phase (gas-liquid-solid) reactors are used extensively in chemical, petrochemical, refining, pharmaceutical, biotechnology, food and environmental industries. Among various types of gas-liquid-solid reactors, the co-current gas-liquid-solid fluidized bed reactor using coarser solid particles is the vital and most widely used three-phase contactor finds application in numerous industrial applications. The successful design and operation of such a system depends on the prediction of the fundamental characteristics. The gas holdup is one of the most important characteristics for analyzing the performance of a three phase fluidized bed. For chemical processes where gas-liquid mass transfer is the rate limiting step, it is important to be able to estimate the gas holdup as this relates directly to the mass transfer. The complex hydrodynamics of these reactors are not well understood due to complicated phenomena such as particle–particle, liquid–particle and particle–bubble interactions. For this reason, computational fluid dynamics (CFD) has been promoted as a useful tool for understanding multiphase reactors for precise design and scale up.

In the present investigation an attempt has been made to study the gas holdup characteristics of a three phase fluidized bed in a broader range of operation. Experiments have been conducted to examine the gas holdup of a co-current gas-liquid-solid three-phase fluidized bed with a modified air sparger using liquid as the continuous phase and gas as the discontinuous phase. Spherical glass beads have been used as the solid phase. An antenna type air sparger has been used for the generation of fine bubbles. By the use of such air sparger the pressure drop in the distributor section can be avoided. A correlation for overall gas holdup has been developed from dimensional analysis. A two dimensional transient model has been developed to simulate the phase holdup behaviour of the mentioned gas–liquid–solid three-phase fluidized bed using the computational fluid dynamics (CFD) method. An Eulerian Granular Multiphase model has been used in the present study and simulations are carried out using the commercial CFD package

Fluent 6.2.16. Comparisons have been made between the overall gas holdup values calculated from the developed and the existing correlations and the experimental ones and with those obtained from CFD simulation.

***Keywords*: three-phase fluidization, cfd, phase holdup, multiphase flow.**

See more of this Session: Industrial Application of Computational and Numerical Approaches to Particle Flow I
See more of this Group/Topical: Particle Technology Forum

(457g) Cartesian Grid Simulations of Gas-Solids Flow Systems with Complex Geometry

Jean-Francois Dietiker[1], Tingwen Li[2], Rahul Garg[2] and Mehrdad Shahnam[3], (1) West Virginia University Research Corporation, Morgantown, WV, (2)National Energy Technology Laboratory, Morgantown, WV, (3)Computational Science Division, National Energy technology Laboratory, Morgantown, WV

Complex geometry encountered in many applications of gas-solids flow need special treatment in most legacy multiphase flow solvers with Cartesian numerical grid. This paper briefly outlines the implementation of a cut cell technique in the open-source multiphase flow solver---MFIX for accurate representation of complex geometries. Specifically, applications of the Cartesian cut cell method to different gas-solids fluidization systems including a small-scale bubbling fluidized bed with submerged tube bundle [1] and a complete pilot-scale circulating fluidized bed [2] will be presented. In addition to qualitative predictions on the general flow behaviors inside each system, quantitative comparison with the available experimental data will be presented. Furthermore, some results on extending the current cut-cell technique to Lagrangian-Eulerian simulations will be presented.

1. Kim, S. W., J. Y. Ahn, S. D. Kim and D. Hyun Lee (2003). Heat transfer and bubble characteristics in a fluidized bed with immersed horizontal tube bundle. International Journal of Heat and Mass Transfer 46(3): 399-409.

2. Shadle, L., R. Cocco, A. Issangya, and R. Panday, (2010). NETL/PSRI Challenge Problem 3, https://mfix.netl.doe.gov/challenge/index.php.

See more of this Session: Industrial Application of Computational and Numerical Approaches to Particle Flow I

See more of this Group/Topical: Particle Technology Forum

(463a) Magnetic Plasmonic Core-Shell Nanoparticles for Fast Condensation and Detection of Bacteria Using SERS

Lei Zhang, Jiajie Xu, Shaoyi Jiang and Qiuming Yu, Chemical Engineering, University of Washington, Seattle, WA

Noble metal nanoparticles (NPs) such as gold and silver have been widely used to enhance the Raman scattering of molecules adsorbed on their surfaces to the order of 10^6 - 10^{14} folds, known as surface-enhanced Raman scattering (SERS). These colloid NPs have also been used as SERS-active substrates for the detection and identification of microorganisms such as bacteria and spores. In the real applications, however, samples containing microorganisms are always in low concentration. Therefore, the ability to condense microorganisms is highly desired. In addition, the uniformly and densely covering microorganisms with colloid NPs could greatly increase the detection sensitivity and reproducibility. We developed a novel method that is capable of condensation and detection bacteria simultaneously by using the Fe_3O_4-Au magnetic plasmonic core-shell NPs. The superparamagnetic Fe_3O_4 NP clusters with the average size of 70 nm were synthesized and then a gold shell was attached by reducing chloroauric acid. The size and morphology of the final multi-functional NPs were examined by SEM and TEM, showing the NPs are about 200 nm with spiky gold shells. After plasma cleaning of the NPs, they were mixed with 10^4-10^6 cfu/mL bacteria. A 10 μL drop of the solution was added on a clean silicon substrate. By putting a corner of the magnet underneath the silicon substrate, the NPs were concentrated to a ~500 μm spot and thereby condensing the bacteria. SEM images showed that the number of bacteria in the condensed spot area is about 120 folds higher than those outside the NP spot. Moreover, all bacteria were surrounded by uniformly and densely packed NPs. SERS spectra were taken from the spot area by using a 785 nm laser focused to a 2 μm x 15 μm rectangle spot. Multiple spectra were taken for each sample and highly reproducible spectra of each bacterium were obtained. To further use this method to identify bacterial species, *Escherichia coli K12, Pseudomonas aeruginosa, and Acinetobacter calcoaceticus* at the concentration of 10^5 cfu/mL were tested. The results showed that they could be clearly discriminated by the Raman spectra due to the high sensitivity of this method. The bacteria can be further differentiated by applying the principle component analysis (PCA). The success of the preliminary tests shows that this novel platform holds great potential to be applied in environment monitoring, food safety, and homeland security.

See more of this Session: Magnetic Particle Synthesis and Properties
See more of this Group/Topical: Particle Technology Forum

Assembly of Magnetic Microcomposites From Low pH Precursors Using a Novel Micro-Fluidic-Jet-Spray-Dryer

Ria Amelia[1], Winston Wu[1], Xiao Dong Chen[1,2], and **Cordelia Selomulya[1]**
[1] Department of Chemical Engineering, Monash University, Clayton VIC 3800 Australia
[2] Department of Chemical and Biochemical Engineering, College of Chemistry and Chemical Engineering, Xiamen University, 361005 Fujian Province, P.R. China

Extended Abstract
Spray drying is a convenient method to produce particles in high yields within a short period of time. The benefits include efficient energy and solvent usages, allowing for scale-up. However, a major challenge in most conventional spray drying is the polydispersity of the produced particles. A novel spray dryer called micro-fluidic-jet-spray-drier (MFJSD) coupled with micro-fluidic-aerosol-nozzle (MFAN) could generate droplets in a single trajectory pattern, to produce monodisperse particles. The drying temperature ranges from 90-300°C, so that it can be used to accommodate heat sensitive materials if necessary. Here we investigated the use of MFJSD to generate uniform magnetic microcomposites, specifically focusing on the effects of precursor composition, droplet size, and secondary heat treatment on the final properties of the particles. The presence of silica nanoparticles in the precursor was demonstrated to directly affect the morphology of the particles. Precursor containing silica nanoparticles generated particles with bowl-like shapes due to slower redistribution of solutes to support the particle skeleton during drying. In the absence of silica nanoparticles, the particles were almost perfectly spherical albeit with dimpled surfaces (**Figure 1**). After being subjected to calcination after drying, iron oxide crystals were found on the particle surfaces accounting for the overall magnetic property of the microcomposites (**Figure 2**), with lower magnetisation observed for particles containing non-magnetic silica. Calcination period was shown to affect the magnetisation of the microcomposites, with increasing strength observed at longer periods. As the initial solutions were prepared in acidic condition, the formation of β-FeOOH in the precursors that was later converted into 6-line ferrihydrite would be the starting point in the reactions to form the iron oxide crystals on the particle shells.

For more information, please refer to:

Amelia, R., Wu, W. D., Chen, X. D., and Selomulya, C. (2011) Assembly of magnetic microcomposites from low pH precursors using a novel micro-fluidic-jet-spray-dryer Chemical Engineering Research and Design, doi:10.1016/j.cherd.2011.06.014 (*in press*)

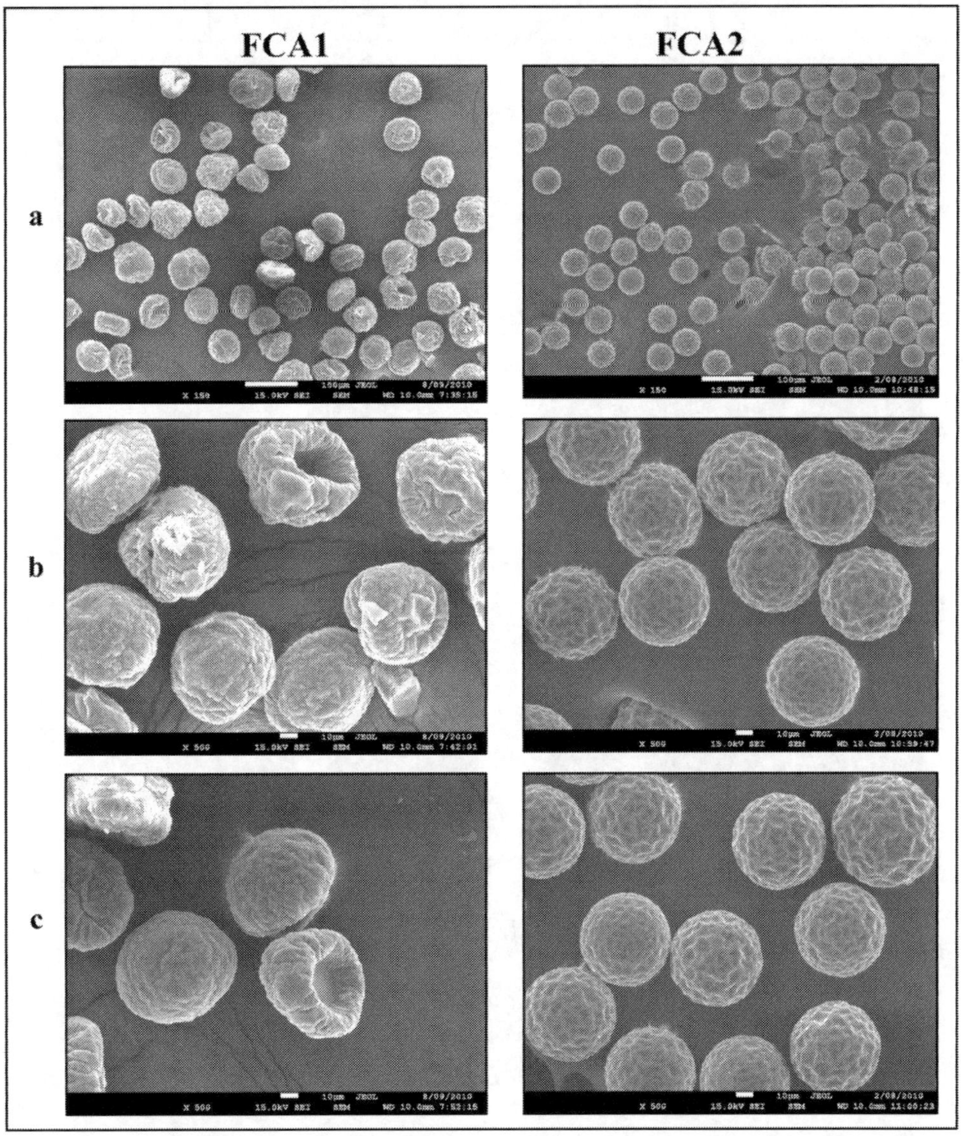

Figure 1 Uniform microparticles generated via microfluidic spray drying from precursors containing: (a) 0.5 wt % $FeCl_3$, 1.00 wt %, SiO_2, 2.50 wt % lactose (FCA1); (b) 0.5 wt % $FeCl_3$, 3.50 wt % lactose (FCA2)

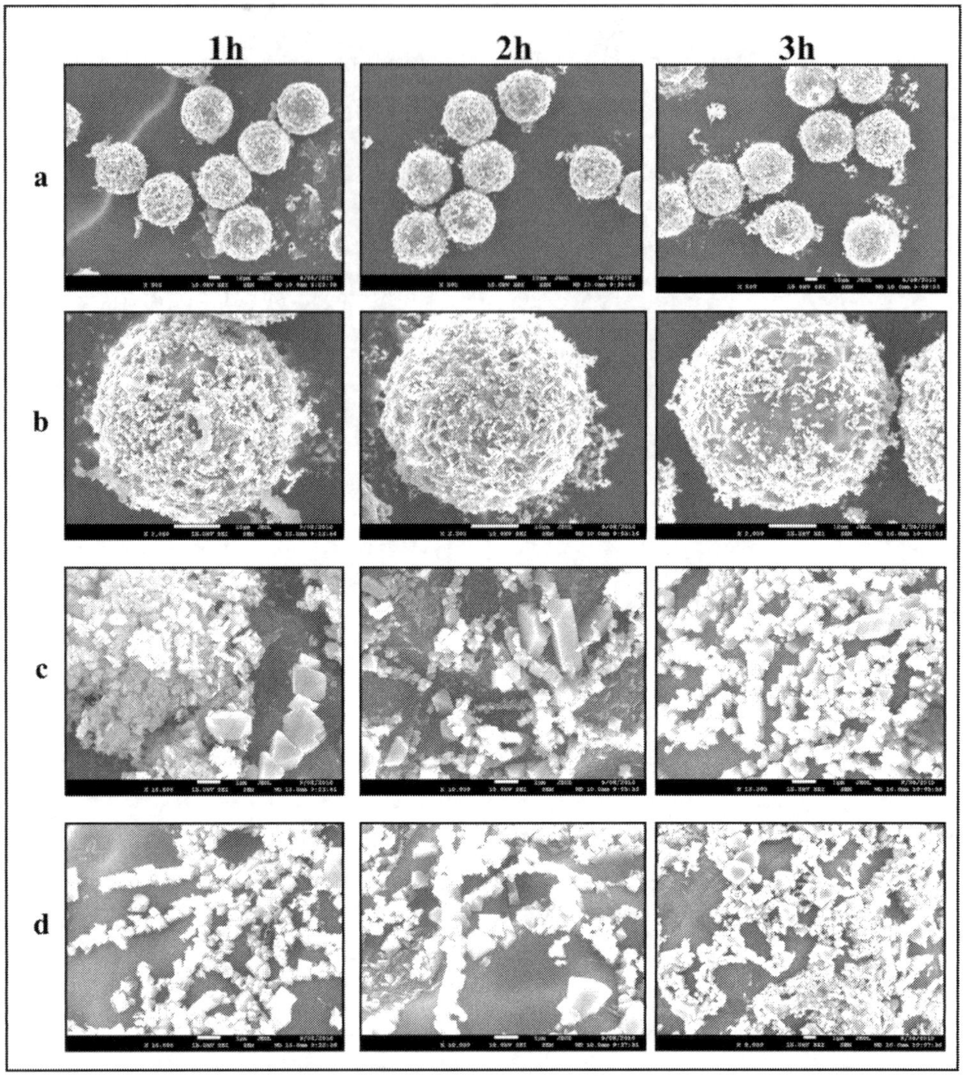

Figure 2 SEM images of FCA2 samples with different calcinations time. Scale bars of 10μm (a-b) and 100nm (c-d).

(463c) Development of Adequate Analytical Methods for Reducing Risks of Magnetic Nanoparticles In Therapeutic Applications

Evagelos K. Athanassiou, Department of Chemistry and Applied Biosciences, ETH Zurich, Zurich, Switzerland, Christoph M. Schumacher, Institute for Chemical- and Bioengineering, ETH Zurich, Zurich, Switzerland, Inge Herrmann, Institute of Anesthesiology, University Hospital Zurich, Zurich, Switzerland, Robert N. Grass, Institute for Chemical- and Bioengineering, ETH Zurich, 8093 Zurich, Switzerland and Wendelin J. Stark, Institute for Chemical and Bioengineering, ETH Zurich, 8093 Zurich, Switzerland

The colorful variety of nanoparticles offers numerous attractive opportunities to industry and enthusiastically written technical papers provide the motivation for a rapid implementation of such novel materials into industrial products and therapeutic strategies. The rapidly growing applications of nanotechnology, however, require a detailed understanding of both benefits and risks [1-3]. Particularly magnetic nanoparticles have attracted increasing interest as they might revolutionize current drug delivery and make cancer remission therapy more effective [3, 4]. Successful application strongly relies on a safe implementation that goes along with detailed knowledge of interactions and effects that nanomagnets might impart once entering the body as well as effective strategies for recycling and removing them from the body once the therapeutic action is over. As the application of such nanomagnets is usually in the ng-mg scale, adequate analytical methods need to be developed that exhibit such high resolution and can detect these ultra-low concentrations.

Since most bio-compatible nanomagnets (metallic or oxidic) are based on iron and the background level of iron in most biological fluids (e.g. human blood) is very high, straight forward elemental analytics (degradation of the nanomagnets followed by AAS, ICP-OS or ICP-MS) are impeded. In this study we demonstrate strategies to detect iron based nanomagnets at very low levels. We will discuss strategies based on precious metal spiking, fluorescence, enzymatic analytics and magnetometry in terms of ease of applicability, biocompatibility and detection limits.

Reference:

[1] L.K. Limbach, Y. Li, R.N. Grass, T.J. Brunner, M.A. Hintermann, M. Muller, D. Gunther, W.J. Stark, **Environ. Sci. Technol.** 39(23), 9370-76 (2005).
[2] L.K. Limbach, P. Wick, P. Manser, R.N. Grass, A. Bruinink, W.J. Stark, **Environ. Sci. Technol.** 41, 4084-9 (2007).

[3] I.K. Herrmann, R.N. Grass, W.J. Stark, **Nanomedicine** 4(7), 787-98 (2009).
[4] I.K. Herrmann, M. Urner, F.M. Koehler, M. Hasler, B. Roth-Z´Graggen, R.N. Grass, U. Ziegler, B. Beck-Schimmer, W.J. Stark, **Small**, 6(13), 1388-92 (2010) and **Nature Nanotechnology** 5, 628 (2010).

See more of this Session: Magnetic Particle Synthesis and Properties
See more of this Group/Topical: Particle Technology Forum

(463d) Single Magnet Directed Synchronization of Magnetic Janus Particles

Sangyeul Hwang[1], Srijanani Bhaskar[2], Jaewon Yoon[2] and Joerg Lahann[3], (1) Chemical Engineering Department, University of Michigan, Ann Arbor, MI, (2) Macromolecular Science and Engineering, University of Michigan, Ann Arbor, MI, (3)Department of Chemical Engineering, Macromolecular Science and Engineering, Biomedical Engineering, and Materials Science and Engineering, University of Michigan, Ann Arbor, MI

We, herein, introduce preparations of various types of magnetic Janus particles (MJPs) with different sizes and chemical components by means of the electrohydrodynamic (EHD) co-jetting, which has been recently known for producing multi-compartmentalized particles. Two distinctive polymers, a water soluble poly(acrylamide-*co*-acrylic acid) (p(AAM-*co*-AA) and an organic solvent soluble poly(lactic-*co*-glycolic acid) (PLGA), were respectively used to prepare MJPs. The controlled diameters of MJPs were obtained from submicron to tens of micron in this study. Along with the magnetite compartment having the superparamagnetic property in the MJPs, two different pigments/fluorescence dyes including carbon black (CB) and titanium dioxide (TiO2), and various fluorescence polymers were also compartmentalized by designed manners in the process of EHD co-jetting, which can be used as secondary functional materials such as demonstrating different optical/fluorescent colors.

Magnetic responses of MJPs toward various orientations and rotations of a given magnetic field are clearly described in this work. Magnetic anisotropy in MJPs that is originated from the shape anisotropy plays a key role in those magnetic responses. Based on those studies, we found a new methodology to ensemble colloidal behaviors of MJPs, which can effectively express the secondary functions embedded in the anisotropic particles. Detailed experimental conditions for such synchronization of MJPs' dynamics via a single magnet are also presented.

See more of this Session: Magnetic Particle Synthesis and Properties
See more of this Group/Topical: Particle Technology Forum

(463e) Precision Magnetophoresis Separation of Superparamagnetic Colloids

Jordi S. Andreu[1], Jordi Faraudo[1] and Juan Camacho[2], (1)Materials Simulation and Theory, Institut de Cincia de Materials de Barcelona (ICMAB-CSIC), Bellaterra (Barcelona), Spain, (2)Physics Department, Universitat Autnoma de Barcelona, Bellaterra (Barcelona), Spain

PRECISION MAGNETOPHORESIS SEPARATION OF SUPERPARAMAGNETIC COLLOIDS.

The manipulation of magnetic particles by the use of inhomogeneous magnetic fields
(magnetophoresis) has emerged as a topic of great interest in a wide range of research and
technological areas [1,2]. The idea behind magnetic separation is to take advantage of the
distinctive magnetic response of the particles in solution to remove them from complex
mixtures by the use of applied inhomogeneous magnetic fields [3]. In the different applications,
magnetic particles are typically functionalized with proper chemical groups, designed to bind to
specific non-magnetic components, thus enabling the separation of non-magnetic materials by
combining the use of magnetic particles and magnetic fields.
Current standard magnetophoretic techniques (such as High Gradient Magnetic Separation)
suffer from different difficulties, including lack of reproducibility and scalability and the loose
of control over the magnetic conditions under which the magnetic particles are removed.
Basically, the external magnetic field applied induces highly inhomogeneous gradients (as large
as 10^4 T/m) in the separator [4]. The magnetic fields generated in this way are not predictable or
reproducible and the magnetic force experienced by the colloids is not uniform

across the
system. These inhomogeneous conditions common to the HGMS approach makes difficult to
develop numerical and/or analytical solutions to the problem, which would help in a better
understanding of the magnetophoretic mechanisms and a higher performance; for instance, by
means of a better design of separators or a better choice of the magnetic particles used in
specific applications.
In order to overcome these limitations and facilitate the use of magnetic colloids in biotechnological applications we have made use of a new concept of magnetic separation (the so
called Precision Magnetophoresis) to effectively remove different types of superparamagnetic
nanoparticles from solution. The process is based on the use of a uniform magnetic gradient,
allowing better quality control and scalability, together with a better control over the
experimental conditions, providing a proper framework for the development of theoretical
models (analytical or numerical).
Within this approach, we have studied the magnetic separation of superparamagnetic colloids
by theory and experiments. Under sufficiently large magnetic fields, a fast
separation process occurs in which the superparamagnetic colloids form chain-like structures
(reversible aggregation) aligned parallel to the applied external magnetic field. It is observed
that, under the effects of a magnetic gradient, these structures move faster than would move an
individual colloid in the same conditions, resulting in a dramatic reduction (orders of
magnitude) of the separation time. Thus, the reversible aggregation of colloids becomes a key
aspect when one thinks to enhance the magnetic separation of magnetic colloids. The kinetics of
both separation processes (cooperative and non-cooperative magnetophoresis) has been
characterized theoretically and experimentally as a function of the different properties of the
colloidal dispersion (magnetic response, size of particles, concentration, solvent

viscosity,
etc, ...) and the magnetic separator (magnetic gradient and size of the separator) [5]. Following
the experimental results obtained for different combinations of magnetic colloids and magnetic
fields, the required conditions for the reversible aggregation to occur has been recently
discussed on the base of computer simulations and thermodynamical models [6]. In the much simpler situation in which the separation process is driven by the individual motion
of the colloids, an analytical solution for the kinetics of the magnetophoresis separation process
is provided [7]. The obtained solution is valid under certain restrictive but realistic conditions which
are explicitly discussed here. We also show the utility of the analytical model by comparing our
predictions with experimental results obtained with superparamagnetic particles of different
sizes and magnetizations. We expect that the availability of a simple, analytical model will
allow for a better understanding of the underlying physics of magnetic separation processes and
also allow a rational design of applications.

References:

[1] C. T. Yavuz, A. Prakash, J. T. Mayo, V. L. Colvin, *Chem. Eng. Sci.*, 64, 2510 (2009).
[2] J. L. Corchero, A. Villaverde, *Trends Biotechnol.*, 27, 468 (2009).
[3] G. Friedman and B. Yellen, *Curr. Opin. Colloid Interface Sci.*, 10, 158 (2005).
[4] G. D. Moeser, K. A. Roach, W. H. Green, T. A. Hatton, P. E. Laibinis, *AIChE J.*, 50, 2835
(2004).
[5] G. De Las Cuevas, J. Faraudo and J. Camacho, *J. Phys. Chem. C*, 112, 945-950 (2008).
[6] J. S. Andreu, J. Camacho, J. Faraudo, *Soft Matter*, 7, 2336 (2011).
[7] J. S. Andreu et al. submitted to *Phys. Rev. E*, (2011).

See more of this Session: Magnetic Particle Synthesis and Properties
See more of this Group/Topical: Particle Technology Forum

On the Properties of Uniform Magnetic Microcomposites Assembled Via Microfluidic Spray Drying

Ria Amelia[1], Winston Wu[1], Xiao Dong Chen[1,2], and **Cordelia Selomulya[1]**
[1] Department of Chemical Engineering, Monash University, Clayton VIC 3800 Australia
[2] Department of Chemical and Biochemical Engineering, College of Chemistry and Chemical Engineering, Xiamen University, 361005 Fujian Province, P.R. China

Extended Abstract

A microfluidic jet spray, which produces single trajectory droplets having identical thermal history, was used to obtain monodisperse particles with precise morphology. The method employed a moderate temperature range (≤ 300°C), and was able to handle multi-component precursors to form solid particles in a single step, while post-treatment such as calcination or template removal can be applied as necessary (**Figure 1**). A stable colloidal suspension containing iron chloride, lactose, and silica nanoparticles was processed to produce microcomposites with a platelet-like morphology due to the nanoparticles. The presence of silica nano-particles caused buckling on the particle surface due to slow redistribution of the solutes to support the particle skeleton, thus forming platelet-like particles. Different storage conditions of the as-dried particles influenced the magnetic property of the calcined particles. Storage in a dry environment at room temperature was found to be better than at an elevated temperature (i.e. oven at 40°C) to avoid pre-oxidation of the iron ions on the particles prior to calcination. We also observed the impact of different calcination periods that triggered the formation of iron oxide crystals on the particle surface. Most of the iron crystals were of cubical shapes in chain-like arrangement (**Figure 2**). The phases of these crystals were hematite and magnetite, with distribution depending on the length of the calcination period. This work demonstrated the versatility of the spray drying approach to easily generate complex particles with appropriate post-processing, including calcination under N_2 environment to form magnetite nanoparticles on spray-dried microcomposites, in the synthesis of functional particles.

For more information, please refer to:

Amelia, R., Wu, W. D., Cashion, J., Bao, P., Zheng. R., Chen, X. D., and Selomulya, C. (2011), Microfluidic spray drying as a versatile assembly route of functional particles, Chemical Engineering Science, doi:10.1016/j.ces.2011.07.059 (*in press*).

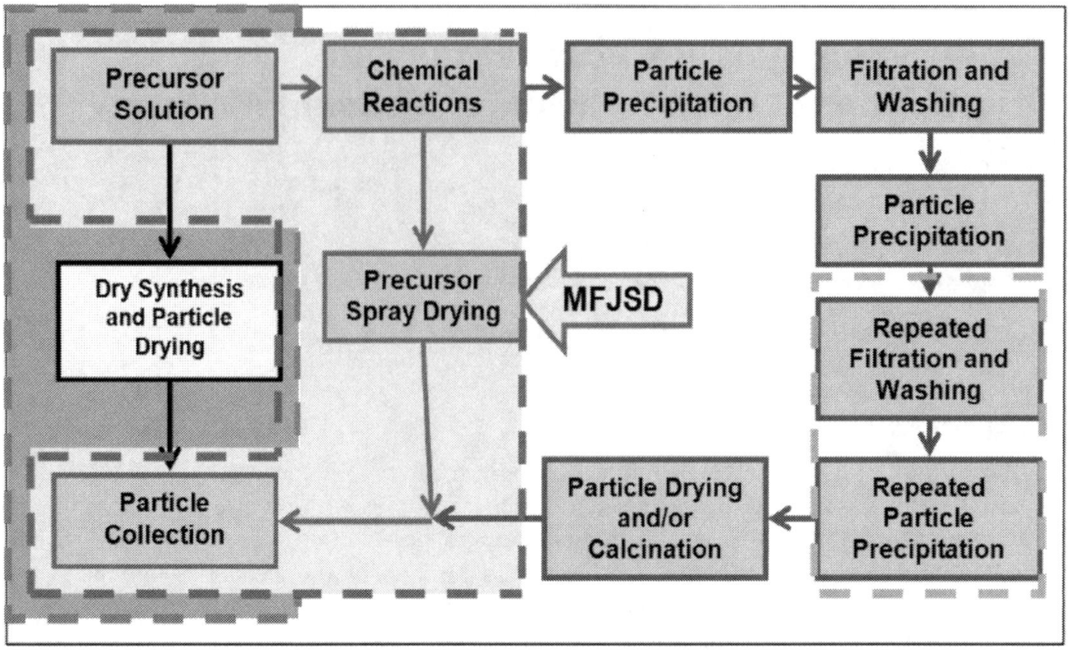

Figure 1 Schematic diagram of the steps involved in the Micro-Fluidic-Jet-Spray-Dryer (MFJSD) process (green) in comparison with wet synthesis route (blue) and dry synthesis route (orange).

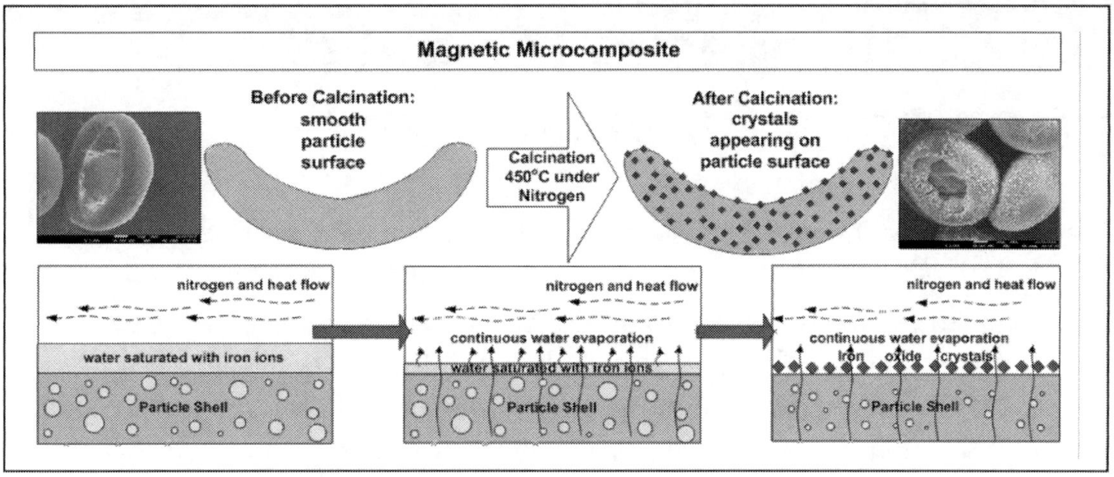

Figure 2 Possible mechanism(s) of iron oxide crystal formation on the spray-dried particle surface upon calcination

(479a) Challenges In the Solids Processing of Biomass

Karl Jacob, Core R&D, Dow Chemical, Midland, MI

It is often assumed for many biomass-based processes that biomass will be available to the plant at low cost and already in a condition which the subsequent processing (e.g. chemical processing or pyrolysis) can handle/tolerate. Unfortunately, this is in most cases an erroneous assumption. Here, we address the myriad of front-end challenges associated with the gathering, long-term storage, size reduction, classification, storage and other solids handling unit operations. The discussion will draw upon the author's experience with a large scale wheat straw handling facility. Numerous opportunities exist for optimizing the collection of biomass and there is inevitably the trade-off between the value of the biomass versus the radius around the plant from which it can be collected. Short term storage of biomass represents a unique challenge as it is prone to elastic wind-up, making it very difficult to remove from hoppers and silos. Subsequent handling within the plant presents its own unique set of challenges – plugging, erosive wear, segregation, process safety hazards, solid-liquid separation, etc. can all be troublesome problems for a biomass-based facility. The advantages and disadvantages of various solids processing strategies are examined and how they impact the operability of a biomass processing plant.

See more of this Session: Solids Processing Challenges In Biomass Conversion
See more of this Group/Topical: Particle Technology Forum

(479b) Nrel's Integrated Biorefinery Research Facility - Pilot Scale Solids Handling Risk Mitigation

Jane C. Fisher[1], Daniel J. Schell[2], Richard T. Elander[2], David A. Sievers[2], Joseph Shekiro[2] and Timothy Johnston[2], (1)National Bioenergy Center, National Renewable Energy Laboratory, Golden, CO, (2)National Renewable Energy Laboratory, Golden, CO

NREL's IntegratedBiorefinery Research Facility Pilot Scale Solids Handling Risk Mitigation

Jane C. Fisher, P.E.,Daniel J. Schell, Richard T. Elander, David A. Sievers, Joseph Shekiro, andTimothy Johnston, National Renewable Energy Laboratory

To achieve commercial-scale production of cellulosic ethanolat a cost that is competitive with gasoline, it is crucial to understand theentire integrated biorefining process and how one stage of the process canimpact the performance of the others.With the addition of the Integrated Biorefinery Research Facility (IBRF)at the National Renewable Energy Laboratory (NREL), the cellulosic biofuels industryhas access to a significantly expanded pilot plant and biochemical conversionprocess research facility. At NREL, thegoal is to improve the cost effectiveness of cellulosic biofuels productionprocesses and thereby to accelerate commercial scale deployment of thesetechnologies. A number of aggressive governmentpolicies are guiding NREL's approach to cellulosic ethanol and other cellulosicbiofuels research and development (R&D), including the 2007 Energy and Independence and Security Act requiring 36 billion gallons ofrenewable fuels by 2022, and President Obama's New Energy for America Plan callingfor 60 billion gallons of advanced biofuels by 2030. The IBRF's new integrated and flexible process R&D piloting capabilities will enable awide variety of biofuels technology developers to reduce risks associated with scalingup to demonstration and full-commercial scales.Many of the risks to be mitigated are related to solids handling, whichpresents a variety of challenging technical issues that the cellulosic biofuelsindustry must resolve to successfully transition the conversion technologies frompilot to demonstration scale and ultimately to commercial scale. The IBRF was completed in two stages andincludes provisions to evaluate a wide variety of sustainable biomassfeedstocks (corn stover, switch grass, sorghum, etc). During completion of the first stage startupand commissioning a number of solids handing issues were encountered and tosome extent resolved. Second stage design modifications were implemented toaddress the solids handling issues experienced during the first stage. The designchanges were primarily associated with storage hoppers and discharge screws,pneumatic conveyance, cyclone separation, weigh belt feed, and dust collection.Design considerations for biomass feedstock handling as well as lessons learnedduring first and second stage facility commissioning and startup will bediscussed.

See more of this Session: Solids Processing Challenges In Biomass Conversion
See more of this Group/Topical: Particle Technology Forum

(479c) Measurement of Unconfined Yield Strength In Small Size Cells

Kerry D. Johanson, Material Flow Solutions Inc., Gainesville, FL

Pharmaceutical companies often create very small quantities of viable drug mixture on a laboratory scale. Extensive analysis is conducted to tease out as much data as possible about the characteristics of the material. Particle size, particle shape, surface area, and other particle scale properties are often the extent of the characterization. While these preliminary properties can help with qualitative product design, they are not generally the primary material behavior measurements the formulator needs to effectively design the product. This information is especially lacking when attempting to infer cohesive flow properties from particle scale flow properties. For example, particle size and surface area are often measured to infer dissolution rates (a bulk scale property); it is much better to be able to measure the property that controls the behavior directly. It would be highly beneficial to have a means of measuring cohesive flow properties directly with very limited material. This paper presents test results from a single axis strength measurement device capable of measuring unconfined yield strengths using a representative sample of 0.1 cc volume of material. This new test technique compacts the material using centrifugal forces and induces failure stresses within the material through controlled application of centrifugal forces. Direct measurement of unconfined yield strength has been accomplished at consolidation stress levels of 0.04 KPa, which is almost one order of magnitude smaller than in typical direct shear cell measurements. This new test technique is compared with standard direct shear measurements using the Schulze, Johanson uni-axial, and Jenike methods. The technique shows good agreement with existing measurement methods. Because of the small volume and lower stress level measurements, this new technique can provide formulators direct measurement of bulk unconfined yield strength before typical process development times, thus shortening the time to market for viable drug mixtures. Consequently, process design can be initiated in conjunction with formulation development.

See more of this Session: Solids Processing Challenges In Biomass Conversion
See more of this Group/Topical: Particle Technology Forum

(479d) Theoretical and Experimental Study of Bulk Strength of Fibrous Biomass Materials

Kerry D. Johanson, Material Flow Solutions Inc., Gainesville, FL

Unconfined yield strength governs the ability of a bulk material to be easily handled in a process. Generally, the greater the strength of a given material, the more difficult it is to handle that material. Strength of simple systems of closely sized spherical particles can be modeled and used as an analytical tool to predict process behavior. However, wet fibrous materials do not follow typical models and rules for nearly spherical materials. This paper explores the relationship between the bulk yield strength and the moisture content, particle aspect ratio, particle size, particle density, and bulk density. Typical biomass materials are of very light density until they are milled. After milling the density increases, creating a more tightly packed product. In general, cohesive flow properties depend on the particle interaction forces. The strength of smaller sized material is generally higher than a corresponding large particle. However, the tendency to exhibit cohesive flow problems is directly proportional to the strength and inversely proportional to the density. Smaller particles produce stronger bulk strength, but larger densities. Thus, it is not easy to predict if a finer or coarser material will flow better in any given process. In addition, interlocking of particles results in large frictional forces between particles that increase bulk strength effects due to particle shape and geometry. This paper presents data of various biomass samples as a function of particle size, particle shape, moisture content, and particle density. These effects are explained using a particle scale theoretical approach and compared to data generated from strength measurements.

See more of this Session: Solids Processing Challenges In Biomass Conversion
See more of this Group/Topical: Particle Technology Forum

(484a) Modeling of Long-Time Flux Decay In Cross-Flow Filtration

Philip P. Schonewill, Richard C. Daniel, Rick W. Shimskey, Justin M. Billing and Reid A. Peterson, Pacific Northwest National Laboratory, Richland, WA

Recent work at Pacific Northwest National Laboratory has been concerned with understanding how cross-flow filter flux decays with time. Understanding the cause(s) of flux decay is important for assessing long-term behavior of this separation technology, especially when used in challenging processing environments e.g. nuclear waste treatment. Fouling in the filter is a complex issue that is affected by many parameters: the chemistry of the solution being filtered, particle size and morphology, operating conditions, and cleaning methods. Based on filtration work performed at PNNL over the past several years, it is clear that a portion of the fouling can be "reversed" by back-pulsing the filter and another portion is irreversible (usually only rectified via cleaning protocols). However, the sheer number of parameters requiring consideration complicates effective modeling of both the long-time behavior of the cross-flow filter and its response to back-pulses.

To combat this, simple models were developed from first principles based on different suspected fouling mechanisms. The models were tested against long-term data (100-hour tests at a constant solids concentration) collected with PNNL's bench-scale filtration system. The flux was found to be well described by a cake mechanism model at intermediate times (starting at approximately 3 hours elapsed time); at longer times (approximately greater than 20 hours of filtration time) the flux behavior transitions to an intermediate mechanism behavior. The decay constant for these models can be predicted using a single dimensionless parameter, the ratio of the transmembrane pressure to the inertial pressure. Additional testing suggested that the filter could be conditioned by back-pulsing. These observations were used to build a comprehensive model to assist in determining a back-pulse strategy for cross-flow filters that will experience long-term decays in the filter flux.

See more of this Session: Theories and Applications of Cross Flow Filtration
See more of this Group/Topical: Separations Division

(484b) Control of Packed-Particle Flow Resistances During Cross-Flow Filtration In Microchannels

Nopphon Weeranoppanant, Levy Amar, **Michael Hill** and Edward F. Leonard,
Chemical Engineering, Columbia University, New York, NY

Cross-flow filtration of fine suspensions through microsieves is a logical, sometimes necessary step in micro-processing. The observed limitation on steady-state filtrate flux accompanying this operation is generally attributed to particle accumulation upon the sieve. We consider two mechanisms: (1) blockage of filter pores by particles in contact with the filter surface, and (2) formation of a packed bed of particles that intrudes upon and is regulated by shear flow within the microchannel. This study examines the conditions under which a packed bed forms on a sieve, and provides a simple model for predicting its thickness when formed by hard non-interacting microspheres. Extensive experimental data are reported that corroborate the model. The good fit of these data lay a firm basis for the semi-quantitative analysis of the behavior of more complex suspensions.

See more of this Session: Theories and Applications of Cross Flow Filtration
See more of this Group/Topical: Separations Division

(484c) Cross-Flow Filtration of Plasma From Blood From a Microchannel

Xiangnan Ye[1], Levy Amar[1], Michael Hill[1], James Jones[2], Jacob Baggerman[3], Cees van Rijn[3] and **Edward F. Leonard**[1], (1)Chemical Engineering, Columbia University, New York, NY, (2)Department of Medicine -- St. Lukes'/Roosevelt Hospital, Columbia University, New York, NY, (3)Aquamarijn Micro Filtration, Zutphen, Netherlands

Steady cross-flow filtration of plasma through microsieves from blood in a microchannel is an important part of several biomedical applications under development, including a new artificial kidney. Erythrocytes, which comprise the largest volume fraction of all particulates in blood, are believed to create the controlling resistance at the sieve surface. Maximum steady-state filtration flux has been observed to be a function of wall shear rate, as predicted by any conventional cross-flow filtration theory, but to show weak dependence on erythrocyte concentration, contrary to theory based on convective diffusion. Maximum filtration flux has also been observed to be a function of sieve pore size and shape (in spite of their very low flow resistance) as well as sieve surface coating.

Recent experiments have correlated macroscopic measurements (filtration rates, transmembrane pressures) with direct observation of erythrocyte behavior at the filtering surface. At low filtration rates (low transmembrane pressures), erythrocytes roll across the filter surface, but at higher filtration rates (higher transmembrane pressures), erythrocytes are observed sticking to the sieve surface. Post-filtration SEM's, even those obtained at very low transmembrane pressures, reveal significant capture and deformation of erythrocytes in the filter pores. Careful balances among sieve design (which can be precisely defined by photolithography), surface treatments of sieves, and flow conditions, allow filtration fluxes at rates exceeding 0.1 cm^3/cm^2-min.

See more of this Session: Theories and Applications of Cross Flow Filtration
See more of this Group/Topical: Separations Division

An electrokinetic model for combined fouling during cross flow filtration of electrolyte solution and charged Colloids

Md Abdullaha Al Mamun, Reeshav Chatterjee, Subir Bhattacharjee,
Department of Mechanical Engineering, University of Alberta, Edmonton, AB, Canada T6G 2G8
Email: subir.b@ualberta.ca
Sirshendu De
Department of Chemical Engineering, Indian Institute of Technology, Kharagpur 721 302, India

Abstract

Colloidal fouling, formation of a cake due to deposition of colloids on to a membrane, is commonly encountered in pressure driven filtration processes, such as nanofiltration and reverse osmosis. The influence of these deposited charged colloids on the ion concentration polarization, ion rejection and the overall permeate flux is the focus of the present work. A common situation encountered during membrane filtration is the combined influence of charged colloids and the associated electrolyte on the flux decline and rejection. In this context, coupled model of concentration polarization (CP), cake layer formation by charged colloidal particles and ion transport through the interstitial spaces of the cake layer has been developed to predict the permeate flux decline and observed salt rejection during cross flow filtration of an electrolyte and colloidal mixture. The suspension structure of the cake layer is considered as a swarm of colloidal particles, represented using the Kuwabara cell model. For this system, the electroosmotic flow is calculated analytically using the Levine-Neale model. The analysis is then combined with the standard film theory of cross flow membrane filtration to predict the flux decline, observed rejection, cake layer thickness and resistance, and osmotic pressure difference. The model illustrates the combined influence of electrical and physical properties of colloids and electrolytes, mass transfer coefficient or crossflow condition and membrane resistance on the separation process. This approach provides a comprehensive insight into the mechanism of cake enhanced concentration polarization for charged colloids.

Keyword: Cake Enhanced Concentration Polarization, Kuwabara Cell Model, Levine-Neale Electrophoretic Model, Film Theory

1 Introduction

Membrane separation processes are increasingly being used in industries as a treatment process for separation of colloids, macromolecules, solutes, and ions. This is often achieved through processes like nanofiltration, whereby a single unit operation is used to remove multiple molecular weight and size species. The major operating costs associated with membranes include creating the pressure driving force for permeation, and mitigation of fouling. Both these operating costs are directly related to the overall energy consumption of membrane filtration. Several methods for fouling prevention and minimization has been suggested over the last few decades. But, most studies on fouling have typically ignored the concept of microscale manipulation of the fouling deposit itself to modify its permeability, or its ability to selectively adsorb ions or small organic contaminants. Therefore study of the fouling mechanism is important to develop a better understanding of the relation between the permeate flux, salt rejection, cake formation and to manipulate the fouled layer to enhance the preformece of the filtration process.

Hoek and Elimelech [1] conducted colloid fouling experiments during cross flow RO and NF, and reported cake enhanced osmotic pressure (CEOP) as the mechanism of flux decline. They proposed a model of CEOP, where hindered back-diffusion of ions within colloidal deposit layers was mentioned as the main mechanism of concentration polarization, resulting in an increase of osmotic pressure and decrease of pemeate flux.

Kim and co-workers [2] presented a detailed model of three component combined fouling to represent flux decline behavior of the system during dead end filtration. The model however introduces several parameters that need to be determined independently, such as, cake layer tortuosity, diffusive tortuosity, and hindered diffusivity of electrolyte through the cake layer. Additionally the model does not incorporate the effect of zeta potential of colloids and electrolyte solution conductivity in the flux decline behavior. However, the model presents a comprehensive mechanism for multi-component fouling from classical filtration concept.

Sim and co-workers [3] investigated the coupled effect of cake enhanced osmotic pressure and colloidal fouling in RO, using cross flow sampler-modified fouling index ultrafiltration (CFS-MFI$_{UF}$). They used CFS-MFI$_{UF}$ measurement to determine the deposition rate of colloidal mass and cake porosity, which then incorporates the CEOP effect in RO fouling. They also considered the hindered diffusivity of ions within the cake layer to model CEOP. Though, the technique provides good estimation of RO fouling profile, the model depends on the CFS-MFI$_{UF}$ or dead-end RO experimental data.

De and Bhattacharjee [4] proposed an electrokinetic model for flux decline during combined filtration of charged colloids and electrolyte. They assumed the cake layer as a bundle of charged cylindrical nano-pores similar to the membrane. Their model considered the electroviscous effect created by the charged nano-pores, which results in additional resistance of the cake layer. Though their model represents an improvement, the actual flow model should consider the flow of electrolytes around the charged colloids in a cake layer to be more realistic.

In this paper, a coupled model of concentration polarization, cake layer formation by charged colloidal particles and ion transport through the interstitial spaces of cake layer has been developed to predict the permeate flux decline and observed rejection due to fouling during cross flow nanofiltration of an electrolyte and colloidal suspension. The model cosiders the cake layer as a swarm of charged spherical particles instead of cylindrical nanopores, which is more realistic, and it can predict the flux decline and observed rejection without considering any correction parameter except the cake tortuosity. The model illustrates the combined influence of colloid mass transfer coefficient, electrical and physical properties of colloids and electrolytes, crossflow geometry and membrane resistance on the cake enhanced concentration polarization of the electrolyte.

2 Mathematical Model

During combined fouling of colloidal particles of radius a, surface potential ψ_c and electrolyte feed concentration of $C_{i,f}$ in a crossflow membrane filtration operation, the colloidal particles form a cake layer on the membrane and the electrolyte solution forms a concentration polarization layer on the cake surface as shown in Fig. 1. The dimensionless permeate flux through the cake layer and membrane can be represented by the following modified Darcy's equation [4]:

$$v_w^* = \frac{v_w}{v_w^0} = \frac{1-\Delta\pi_m^*}{1+R_g^*} \tag{1}$$

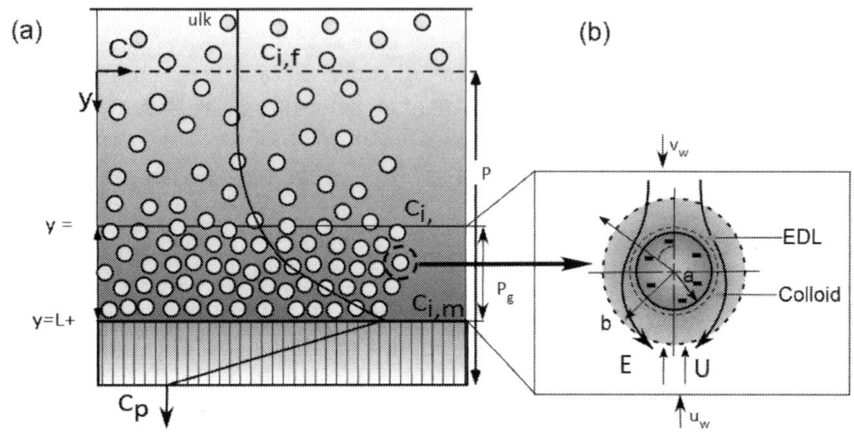

Figure 1: (a) Cake enhanced concentration polarization (b) Electrophoretic flow model around a charged colloid according to the Levine-Neale and Kuwabara cell model

Here, $v_w^o = \Delta P/\mu R_m$ is the pure water flux, $\Delta\pi_m^*$ is the dimensionless osmotic pressure across the membrane and R_g^* is the dimensionless resistance. The cake layer resistance R_g^* has been formulated based on the Kuwabara cell model and Levine-Neale electrophoretic flow model [5] of charged particles in a concentrated colloidal suspension as shown in Fig. 1b. The model considers the combined influence of hydrodynamic and electrokinetic resistance to the transport through the cake layer. According to the Levine-Neale model, electroosmotic flow (U) through the interstices of the cake layer due to induced streaming potential is given by [5]:

$$U = E_\infty\left(\frac{\varepsilon\psi_c}{\mu}\right) g_2(\kappa a, \varphi_p) \tag{2}$$

The expression of induced electric field (E_∞) in Eq. 2 at zero current condition, in accordance with Onsager principle of reciprocity for irreversible phenomena and for thin Debye layer, is [5]:

$$E_\infty = \frac{\Delta P_g}{L}\left(\frac{\varepsilon\psi_c}{\mu\sigma^\infty}\right) \tag{3}$$

Thus, the expression of electroosmotic flux (u_w) from the continuity assumption using Eqs. 2 and 3 is given as follows:

$$u_w = U(1-\varphi_p) = \frac{\Delta P_g}{L\sigma^\infty}\left(\frac{\varepsilon\psi_c}{\mu}\right)^2(1-\varphi_p)g_2(\kappa a,\varphi_p) \tag{4}$$

Therefore, the total flux across the cake layer at no current condition, due to effective pressure across the cake ΔP_g and reverse electroosmotic flow is given as follows:

$$\begin{aligned} v_w &= \frac{\Delta P_g}{\mu r_g L} - u_w \\ \text{or, } v_w &= \frac{\Delta P_g}{\mu r_g L}\left[1 - 9\varphi_p(1-\varphi_p)A_K\frac{\beta g_2(\kappa a,\varphi_p)}{2(\kappa a)^2}\right] \\ \text{or, } v_w^* &= \frac{\Delta P_g^*}{L^*/Mg} = \frac{\Delta P_g^*}{R_g^*} \end{aligned} \tag{5a}$$

$$r_g = \frac{9\varphi_p A_K}{2a^2};\quad M = \frac{2a^2R_m}{9\varphi_p A_K d_e}\quad \text{and}\quad g = \left[1 - 9\varphi_p(1-\varphi_p)A_K\frac{\beta g_2(\kappa a,\varphi_p)}{2(\kappa a)^2}\right] \tag{5b}$$

Here, r_g is the hydrodynamic resistance of the cake layer, $A_K = 1/(1-9\varphi_p^{1/3}/5+\varphi_p-\varphi_p^2/5)$ accounts for the presence of neighbouring particles, $\beta = (\varepsilon\psi_c\kappa)^2/(\mu\sigma^\infty)$ and $L^* = L/d_e$. The function g accounts for the electroosmotic effect in a swarm of colloidal particles, depends on dimensionless Debye length κa, surface potential of colloidal particles ψ_c, solution conductivity σ^∞ and volume fraction of the cake layer, φ_p.

The dimensionless osmotic pressure $\Delta\pi_m^*$, for simple electrolyte (1:1) can be written from the vant Hoff equation:

$$\Delta\pi_m^* = \frac{2R_{ug}T}{\Delta P}(C_{i,m}-C_{i,p}) = \alpha\bar{C}_{i,m}R_r \tag{6}$$

where, $\alpha = 2R_{ug}TC_{i,f}/\Delta P$ and $\bar{C}_{i,m}$ is dimensionless salt concentration at the membrane surface. The expression of $\bar{C}_{i,m}$, derived by integrating the mass balance of electrolytes within the concentration polarization and cake layer with appropriate boundary conditions is:

$$\bar{C}_{i,m} = \frac{C_{i,m}}{C_{i,f}} = \frac{\exp\left[v_w^*(R_i+L^*R)\right]}{R_r+(1-R_r)\exp\left[v_w^*(R_i+RL^*)\right]} \tag{7}$$

Here, $R_i = v_w^o/k_i$, $R = v_w^o d_e/D_i^*$. The observed rejection of the membrane can be written as:

$$R_o = 1 - \frac{C_{i,p}}{C_{i,f}} = 1 - \bar{C}_{i,m}(1-R_r) \tag{8}$$

The dimensionless governing equation for colloidal cake layer growth, obtained by integrating the mass balance equation of colloids in the cake layer with appropriate boundary conditions, is given by [6]:

$$\frac{dL^*}{d\tau} = v_w^*\frac{C_{c,m}^* - \exp(v_w^*R_c)}{1-\exp(v_w^*R_c)} \tag{9}$$

Here, $\tau = v_w^o C_{c,f} t/(d_e \rho_c)$, $R_c = v_w^o/k_c$, $C_{c,m}^* = C_{c,m}/C_{c,f}$. The dimensioless pressure drop with respect to operating pressure can be written as:

$$\Delta P_m{}^* + \Delta\pi_m^* + v_w^* R_g^* = 1 \tag{10}$$

Here, $\Delta P_m{}^*$ is dimensionless pressure drop across the membrane.

3 Numerical Methodology

Equations 1 and 9 are solved simultaneously in conjunction with the Eqs. 6 and 7, using a differential algebric equation solver DDASSL. For a 50 nm particle with surface potential -45 mV and cake volume fraction 0.64 for random-closed-packing, in 0.01 M NaCl solution, the value of g is 0.67. The value of M depends on particle size, cake volume fraction and hydrodynamic radius of crossflow channel and is less than unity for RO or NF. The volume fraction can be as high as 0.72 for hexagonal packing. The viscosity and relative permitvity of water is 0.89 mPa.s and 80.0, respectively. The electrolyte flow turbulence parameter R_i, is considered to be in the order of 1.0 becuase pure water flux and mass transfer coefficient of electrolyte both are of the order 10^{-5}. The corresponding value of R_c is considered to be of the order 10.0, because the diffusivity of colloidal particle is at least one order magnitude less than the electrolytes. For the silica colloidal particles, the bulk concentration is 0.2 kg/m^3, whereas the cake concentration is about 173 kg/m^3 [7]. Typical equivalent diameter of the cross flow channel is 10^{-3} m. For colloid feed concentration 0.2 kg/m^3, with particle density 2317 kg/m^3, τ= 0.01 represents 6.5 hr of operation in real time. The range of dimensionless parameters are listed in table 1.

Table 1: Range of dimensionless parameters used in simulations

Parameter	Range
κa	1 – 100
g_2	0 – 0.9
β	1.0 – 5.0
g	0.05 – 0.95
M	0.005 – 0.5
R_i	0.05 – 1.5
R_c	10.0 – 60.0
R	100 – 1000
α	0.09 – 0.9
$C_{c,m}^*$	800 – 900
R_r	0.8 – 0.9

4 Results and Discussion

Each of the dimensionless parameters listed in Table 1 affect the flux decline and observed rejection during cross flow filtration process. For the present parametric study, the effect of four parameters M,

g, R and R_c are considered and explained. The physical properties of the colloidal particle, cake layer and membrane are incorporated in the membrane and colloidal system interaction parameter M. The value of M is proportional to the square of particle size, i.e., for small particles, the value of M is small resulting in higher resistance since specific resistance is inversely proportional to M. As a result, the permeate flux decline is higher as shown in Fig. 4a. The observed rejection also increases when M is small, this is due to lower convection of salt compared to back diffusion across the cake layer. Therefore, concentration at the membrane surface is less significant for M=0.05 resulting in lower osmotic pressure as shown in Fig. 4a. Therefore, permeate flux and observed rejection can be changed by changing the size of the cake forming particles.

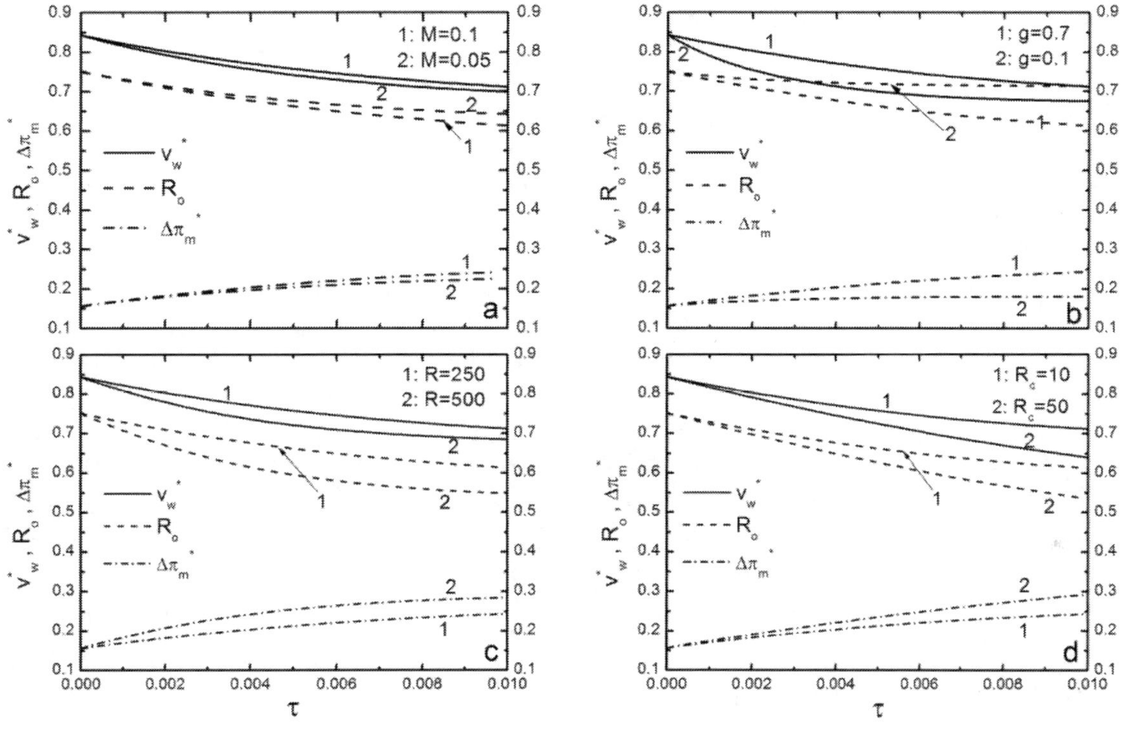

Figure 2: Variation of normalized permeate flux (v_w^*), observed rejection (R_o) and osmotic pressure $\Delta\pi_m^*$. (a) g=0.7, R=250.0, R_i=1.0, R_c=10.0, R_r=0.875, α=0.09 and $C_{c,m}^*$=850.0; (b) M=0.1, R=250.0, R_i=1.0, R_c=10.0, R_r=0.875, α=0.09 and $C_{c,m}^*$=850.0; (c) M=0.1, g=0.7, R_i=1.0, R_c=10.0, R_r=0.875, α=0.09 and $C_{c,m}^*$=850.0; (d) M=0.1, g=0.7, R=250.0, R_i=1.0, R_r=0.875, α=0.09 and $C_{c,m}^*$=850.0

The effect of electroosmotic parameter g, induced due to the presence of charged colloids and electrolyte solution, is shown in Fig. 4b. In absence of surface potential, the value of g goes to unity, implying absence of any induced electroosmotic flow. In presence of charged particles and salt solution the value of g is always less than 1. The parameter g is a strong function of κa and it decreases with increase of κa or with increase of electrolyte feed concentration. The parameter g is also a function of surface potential ψ_c and decreases with increase of ψ_c. Therefore, increase of surface potential of colloidal particles or increase of feed electrolyte concentration causes decrease of g. Since cake resistance is inversely proportional to g, increase of surface potential and feed concentration increases the electroviscous resistance of cake layer. As a result, permeate flux decline is more severe due to higher reverse electroosmotic flow induced from charged colloids. Due to the reverse electroosmotic flow of electrolyte, electrolyte concentration is less at the membrane surface, increasing the observed rejection and decreasing the transmembrane osmotic pressure as shown in Fig. 4b. Therefore, smaller value of g or higher value of κ and ψ_c, decreases the permeate flux and increases the salt rejection siginificantly. Thus, formation of cake layer with nano colloidal particles of high surface potential in concentrated salt solution can enhance the observed rejection of nanofiltration membranes or can enhance concentration polarization on cake surface significantly, which means the cake layer can be modeled as a secondary low salt rejection membrane of nano size pores with high surface potential.

The effect of parameter R on permeate flux and observed rejection is shown in Fig. 4c. The parameter R is defined as the ratio of pure water flux through the membrane to the diffusive flux through the cake layer. Higher value of R indicates that the convective flux is dominant compared to back diffusion through the cake layer which results in enhanced accumulation of electrolyte within the cake layer and increase in transmembrane osmotic pressure. As a result, the permeate flux decline is comparatively higher and observed rejection is lower for large value of R, according to Fig. 4c. Therefore, higher value of R enhances salt concentration at membrane surface and declines the permeate flux and observed rejection.

The parameter R_c is defined as the ratio of pure water flux to colloids mass transfer coefficient. The mass transfer coefficient solely depends on the geometry of cross flow channel and velocity. Since the model is for cross flow filtration lower value of R_c indicates higher mass transfer coefficient or, higher degree of turbulence and shear at the cake surface. Due to higher shear rate at cake surface, the cake layer can not grow much which leads to lower cake layer thickness and resistance, resulting in higher permeate flux as shown in Fig. 4d. Therefore, convective flux of electrolyte is higher for small value of R_c. Though the convective flux is higher, observed rejection is higher and transmembrane osmotic pressure is lower, due to smaller of ratio of convective flux to diffusive flux within the thinner cake layer for small R_c. Thus, higher mass transfer coefficient or small R_c enhances the permeate flux and observed rejection.

5 Concluding Remarks

The electrokinetic model presented in this paper is a simple coupled model for quantifying the combined effect of colloid-electrolyte system during crossflow nanofiltration. Due to reverse electroosmotic flow, the permeate flux and salt transport through the cake layer decrease. The salt concentration at the cake surface increases, which enhances the concentration polarization on cake surface. The cake enhanced concentration polarization will be severe for nano-colloids with higher surface potential. A detailed electrokinetic model of ion transport through the cake layer of nano-colloids and membrane will lead to a better understanding of the combined fouling process for high salt rejection membranes. Consideration of the electrokinetic interaction between the electrolyte and the colloid is important to develop a detailed model of the inter-relations between the permeate flux, salt rejection and cake formation. The study also includes that manipulation of the meso-scale structure of the cake will allow us to alter the porosity of cake layer, extent of adsorption of different species, and alter the salt rejection of the membranes.

Nomenclature

ΔP	Applied operating pressure, Pa
R_m	Membrane resistance, 1/m
R_r	Real rejection of membrane
d_e	Hydraulic diameter of cross flow channel, m
μ	Solvent viscosity, Pa.s
ε	Solvent electrical permitivity
R_{ug}	Universal gas constant, J/mole K
T	Absolute temperature, K
C_f	Solute concentration at feed (i.e., i= ions, c=colloids)
C_m	Solute concentration at membrane surface (i.e., i= ions, c=colloids)
ρ_c	Density of colloids, kg/m^3
L	Cake thickness, m
$D = \frac{k_B T}{6\pi\mu r}$	Brownian diffusion coefficient of solutes (i.e., i= ions, c=colloids), m^2/s
$D_i^* = \frac{(1-\varphi_p)*D_i}{[1-\ln(1-\varphi_p)^2]}$	Hindered diffusivity of ions within the cake layer, m^2/s
$k = 1.86\,\frac{D}{d_e}(\frac{u_c d_e^2}{DL_c})^{\frac{1}{3}}$	Mass transfer coefficient co-relation of solute (i.e., i= ions, c=colloids) for laminar flow in thin channel [8], m/s

References

[1] E.M.V. Hoek and M. Elimelech. Cake-enhanced concentration polarization: A new fouling mechanism for salt-rejecting membranes. *Environmental Science and Technology*, 37:5581–5588, 2003.

[2] A.S. Kim, A.E. Contreras, Q. Li, and R. Yuan. Fundamental mechanisms of three-component combined fouling with experimental verification. *Langmuir*, 25:7815–7827, 2009.

[3] L.N. Sim, Y. Ye, V. Chen, and A.G. Fane. Investigations of the coupled effect of cake-enhanced osmotic pressure and colloidal fouling in ro using crossflow sampler-modified fouling index ultrafiltration. *Desalination*, 273:184–196, 2011.

[4] S. De and S. Bhattacharjee. Flux decline during cross flow membrane filtration of electrolytic solution in presence of charged nano-colloids: A simple electrokinetic model. *Journal of Colloid and Interface Science*, 353:530–536, 2011.

[5] J. Masliyah and S. Bhattacharjee. *Electrokinetic and Colloid Transport Phenomena*. Wiley Interscience, 2006.

[6] S. De and P. K. Bhattacharya. Modeling of ultrafiltration process for a two-component aqueous solution of low and high (gel-forming) molecular weight solutes. *Journal of Membrane Science*, 136:57 – 69, 1997.

[7] S.K. Zaidi and A. Kumar. Experimental analysis of a gel layer in dead-end ultrafiltration of a silica suspension. *Desalination*, 172:107–117, 2005.

[8] S. Kim and E.M.V. Hoek. Modeling concentration polarization in reverse osmosis processes. *Desalination*, 186:111–128, 2005.

(484e) Comparison of Membrane Performance: Analysis with Fouling Models

Mingyan Zhou[1], James E. Kilduff[1] and **Georges Belfort**[2], (1)Department of Civil and Environmental Engineering, Rensselaer Polytechnic Institute, Troy, NY, (2) Department of Chemical and Biological Engineering, Rensselaer Polytechnic Institute, Troy, NY

Membrane research and engineering applications often involve optimizing flux and rejection for a particular application. This process frequently presents the need to compare the performance of membranes having different inherent permeabilities; an important performance criterion is resistance to fouling. To identify the most appropriate (least biased) approaches to compare membrane performance, an analysis using fouling models, with an emphasis on the combined pore blockage and cake filtration model, was performed. We illustrate how the models can be used to provide a general framework for assessing fouling and developing ways to compare membrane performance. Comparisons at constant transmembrane pressure using the same initial flux and the same initial pressure were emphasized. This analysis demonstrates that membrane resistance has a significant effect on flux decline - flux decline is smaller for membranes having greater resistance, especially when the feed solution and membrane-foulant interactions are similar. This effect cannot be normalized regardless of whether flux is plotted as a function of filtration time and volume throughput. A better approach is to compare the effects of fouling by evaluating fouling resistance as a function of volume throughput. Alternatively, comparing membranes based on total resistance provides some advantages because it can be related to permeate production and energy cost. When comparing membranes having different resistance and rejection, the effects of fouling should be compared by evaluating fouling resistance as a function of mass accumulation on the membrane surface. The effects of cake compressibility are also discussed.

See more of this Session: Theories and Applications of Cross Flow Filtration
See more of this Group/Topical: Separations Division

(490a) Scale-up Methodologies When Handling Extremely Dangerous Intermediates and Products

Jerry S. Salan, Shannon Lenahan and Matt Jorgensen, Nalas Engineering Services Inc., Salem, CT

The Department of Defense is currently investigating 'green' primary explosives. As a result, dozens of novel primary explosives have been synthesized and some scaled up. Nalas Engineering Services, Inc. has assisted in the process development of these extremely dangerous materials. Traditional methods for process development were avoided due to safety concerns of isolating large quantities of the intermediates and products. Non-traditional sampling methods, analytical methods and kinetic model development were used to gain process understanding for these materials. In-situ tools included in-situ infrared, focused beam reflectance measurement, and particle vision and microscopy. Off-line tools included atomic absorption, high performance liquid chromatography, karl fischer, differential scanning calorimetry and NMR. The combination of these tools assisted the chemical engineers in solving some very intersting process challenges.

See more of this Session: Advanced On-Line Analytical and Optimization Tools In Pilot Plants
See more of this Group/Topical: Particle Technology Forum

(490b) Automated Pilot Plant System Producing 3,3-Diamino-4,4-Azoxyfurazan

Kimberly Hanson[1], Jerry S. Salan[2], Andrew G. Pearsall[1], Matt Jorgensen[2], Patrick B. Skahan[1] and Tod F. Ricks[1], (1)Chemical Development, Naval Surface Warfare Center, Indian Head, MD, (2)Nalas Engineering Services Inc., Salem, CT

ABSTRACT:

The United States Navy developed a novel continuous process to produce kilogram quanitites of 3,3-diamino-4,4-azoxyfurazan (DAAF) for inclusion in explosive formulations. The process required extremely tight control of pH to maintain the desired impurity profile and resulting thermal stability of the final product. The process required the development of offline analytical techniques to monitor starting materials, intermediates, product and impurities in reaction mixtures and isolated solids. This data was incorporated into a kinetic model which was utilized to design a continuous process to DAAF. The details of the process development, kinetic model and final pilot plant controls will be discussed.

See more of this Session: Advanced On-Line Analytical and Optimization Tools In Pilot Plants
See more of this Group/Topical: Particle Technology Forum

(490c) Process Monitoring of An Energetic Material Pilot Plant

Jerry S. Salan, Nalas Engineering Services Inc., Salem, CT and **Thomas Palathra**, Indian Head Division, Naval Surface Warfare Center, Indian Head, MD

The Naval Surface Warfare Center, Indian Head Division develops processes for energetic material ingredients and their formulations. The energetic material ingredients are developed using similar protocols utilized by pharma and fine chemical companies. However, the Navy is focused on removing operators from the hazardous environment. The Navy has purchased and currently installing novel reactors and chemical reaction monitoring equipment in order to mitigate hazards to operators while maintaining process control.

See more of this Session: Advanced On-Line Analytical and Optimization Tools In Pilot Plants
See more of this Group/Topical: Particle Technology Forum

(490d) Using Process Analytical Technology (PAT) Tools to Support Flow Chemistry Development and Production

Yanqiao Xiang[1], John Lucas[2], John G. Van Alsten[2], Bryan Li[3] and Nandell F. Keene[3], (1)Research Analytical & Technology, Pfizer Inc., Groton, CT, (2) Chemical Research & Development, Pfizer Inc., Groton, CT, (3)Research API, Pfizer Inc., Groton, CT

This work reports applications of process analytical technology (PAT) to flow chemistry development and manufacturing. Case studies demonstrate that the PAT tools provide insight into chemical mechanisms and generate kinetic data to facilitate flow process scale-up. The case studies also present the utility of the PAT tools for on-line monitoring flow production process as it provides real-time data on flow stream dispersion, impurity profile, reaction progress and flow system cleaning verification, proving that PAT are convenient and versatile tools for rapid optimization and synchronized control of continuous flow chemistry processing.

See more of this Session: Advanced On-Line Analytical and Optimization Tools In Pilot Plants
See more of this Group/Topical: Particle Technology Forum

Process Analytics – Intensive Pilot Plants: Learn Early, Save Later

Steven F. Wright, Technology Fellow, Process & Environmental Analytics Group, Eastman Chemical Co.

It is well established that appropriate piloting efforts tend to produce new plants that start up on-time, on budget and running at expected capacity. This is also true for analytical instrumentation used throughout the development and piloting process. Correct inclusion of at-line and on-line analyzers in labs and pilot plants can support rapid, early process knowledge learning while it is still possible to change design features at lower cost, to strengthen models and reduce unpleasant surprises. While good goals, many piloting efforts are hindered because they fail to dedicate sufficient resources to analysis, generally breaking the knowledge link somewhere between the development lab and the full-scale plant.

The author will review some suggested approaches that can be used to identify and justify situations where at-line and on-line analytical will help speed your piloting process, when the normal analytical lab turnaround times just won't suffice for cost, process and/or safety reasons. Examples include highly dynamic reactions, complex samples and intense 24x7 analytical support bills that can wreck the budget while not really giving you the information you need.

Instrumentation is no better than its sampling strategy. The author will review sampling strategies that minimize process disruptions, while insuring representative sampling, especially important for smaller volume units found in bench and small piloting operations.

On-line Analyzers, such as mass spectrometers and Raman spectrometers, while valuable, are expensive to purchase and own. Approaches to instrument selection for maximum flexibility, process sharing, calibration and long-term maintenance in an ever-changing piloting environment will be discussed.

(527a) Formation of Itraconazole-Malic Acid Cocrystals In a CO2-Expanded Liquid

Courtney A. Ober and Ram B. Gupta, Chemical Engineering, Auburn University, Auburn, AL

Itraconazole, an antifungal drug, has poor aqueous solubility. To improve the dissolution rate, cocrystals of itraconazole with l-malic acid were prepared in tetrahydrofuran using supercritical carbon dioxide as an antisolvent. In this novel process, the drug (itraconazole) and cocrystal former (l-malic acid) are dissolved in a liquid solvent (tetrahydrofuran). Cocrystals of the two components are precipitated by pressurization with CO_2. The liquid solvent is then removed and the cocrystals dried with excess supercritical CO_2. These cocrystals were compared to cocrystals produced by using n-heptane, a traditional liquid antisolvent. The cocrystals produced by each method were characterized for crystallinity by X-ray diffraction (XRD), thermal behavior by differential scanning calorimetry (DSC), composition by high pressure liquid chromatography (HPLC), and size and surface morphology by scanning electron microscopy (SEM). Dissolution studies were also conducted for the cocrystals.

The XRD results indicate that cocrystals of itraconazole with l-malic acid can be produced using either supercritical CO_2 or a traditional liquid solvent, such as n-heptane. Both XRD and DSC confirm the presence of an itraconazole-malic acid cocrystal and verify the absence of either of the pure components. HPLC confirmed that the composition of the cocrystals matched that previously reported in the literature. Although the cocrystals produced by CXLCF were slightly larger than those precipitated with n-heptane, their microporous structure allowed for improved dissolution. Through use of XRD, DSC, HPLC, SEM, and dissolution studies, it is shown that the proposed cocrystallization is a viable method for producing pharmaceutical cocrystals of poorly water-soluble drugs to improved dissolution and reduce organic solvent content.

See more of this Session: Engineered Composite Particulate Systems for Pharmaceutical Active Ingredient Delivery
See more of this Group/Topical: Particle Technology Forum

(527b) Preparation of Calcium Stearate Pellets by Extrusion/Spheronization: Effect of Drying Technique

Simone Schrank[1], Aden Hodzic[2], Ben Glasser[3], Andreas Zimmer[1], Johannes G. Khinast[4] and Eva Roblegg[1], (1)IPS/Department of Pharmaceutical Technology, University of Graz, Graz, Austria, (2)Research Center Pharmaceutical Engineering (RCPE), Graz, Austria, (3)Rutgers University, Piscataway, NJ, (4) Institute for Process and Particle Engineering, Graz University of Technology, Graz, Austria

Pharmaceutical pellets intended for oral dosing are frequently prepared via the wet extrusion/spheronization technique. The last "active" step of this process is drying. Although the drying method and/or the drying conditions applied are considered to highly influence the final pellet characteristics, investigations in this field are rare. Some studies are reported, mostly evaluating the moisture removal from pellets containing microcrystalline cellulose (MCC) [e.g.: 1].

The current study focuses on the drying behavior of calcium stearate (CS) pellets containing ibuprofen as model drug. These matrixpellets were shown to be spherical in shape and mechanically stable before [2]. After spheronization three different drying methods were applied; i) desiccation (D) over silica gel at ambient conditions, ii) drying in a fluid bed (FB) apparatus at an inlet air temperature of 40 °C and iii) lyophilization after freezing the pellets in liquid nitrogen (-196 °C). It was ensured via differential scanning calorimetry (DSC) that no interactions between the primary powders occurred during liquid removal. Furthermore, it was proved by X-ray measurements that neither CS nor ibuprofen underwent changes in their crystal stated during preparation.

The drying profile of the pellets dried via desiccation revealed a non-linear shape, where the rate of drying decreased throughout the process. This is mainly attributed to the shrinking of pellets by 13% during liquid removal causing changes in the surface area. CS was shown to swell in contact with the granulation liquid (i.e., 50% ethanol) and is likely to shrink again during drying. The shrinking of pellets during FB drying was less pronounced due to significantly lower drying times (i.e., by 8%). During lyophilization shrinking was largely suppressed since no liquid was involved resulting in minimal capillary flow. The differences in the shrinking behavior correlated with the pellet size distributions. The shape was not influenced by the drying conditions and the aspect ratio was below 1.2 for all batches.

Drug release studies demonstrated that pellets dried via desiccation liberated the lowest and pellets dried via lyophilization released the highest amount of ibuprofen. However, kinetics were best described by the Higuchi approach (i.e., diffusion based on Fick´s law) for all investigated batches. Differences in drug liberation behavior can occur due to variations in the spatial drug distribution throughout the pellets. It was shown before that for high drying rates (i.e., FB drying in the current study) dissolved components are likely to be transported towards the outer shell [3] where they recrystallize. Therefore, the ibuprofen distribution along the pellet cross sections was investigated using Raman mapping. However, ibuprofen was rather homogeneously distributed throughout all pellets despite the different ways of liquid removal. Although ibuprofen is highly soluble in the granulation fluid (i.e., 48 g/l), only small amounts of drug were expected to be dissolved during pellet preparation. Moreover, it is known from literature that ibuprofen causes phase separation of 50% ethanol at 40 °C [4] (i.e., the applied FB drying temperature). Subsequently, the dissolved ibuprofen crystallized in the water phase due to its poor water solubility [5]. Thus, the active pharmaceutical ingredient was not transported towards the external surface. Another reason for the different dissolution rates may be differences in structure and/or morphology. The crushing strength decreased in the order D, FB and L where FB and L showed similar results. This can be explained by the minimal extent of shrinking during L resulting in different morphological characteristics. During FB drying pellets were exposed to mechanical load. Thereby, cracks were likely to be formed which in turn lead to pellet breakage at lower forces. Moreover, it was found from the He-pycnometer measurements that during FB drying closed pores were more pronounced to be formed. Consequently, the variations in drug release characteristics can be explained by drying induced changes in the pellet morphology.

[1] B. Song, S.L. Rough, D.I. Wilson, Int. J. Pharm. 332 (2007), 38-44.

[2] E. Roblegg, S. Ulbing, S. Zeissmann, A. Zimmer, Eur. J. Pharm. Biopharm. 75 (2010), 56-62.

[3] A. Lekhal, B.J. Glasser, J. G. Khinast, Chem. Eng. Sci. 59 (2004), 1063-1077.

[4] J. Manrique, F. Martinez, Lat. Am. J. Pharm. 26 (2007), 344-354.

[5] M. Jbilou, A. Ettabia, A.M. Guyot-Hermann, J.C. Guyot, Drug Dev. Ind.

Pharm. 25 (1999), 297-305.

See more of this Session: Engineered Composite Particulate Systems for Pharmaceutical Active Ingredient Delivery
See more of this Group/Topical: Particle Technology Forum

(527c) Particle Engineering Via Dry Coating: Development of a Novel Material Sparing Technology for Pharmaceutical Powders

Rajesh N. Dave, Chemical Engineering, New Jersey Institute of Technology, Newark, NJ, Lauren Beach, Otto H. York Department of Chemical, Biochemical and Pharmaceutical Engineering, New Jersey Institute of Technology, Newark, NJ, James V. Scicolone, New Jersey Center for Engineered Particulates, New Jersey Institute of Technology, Newark, NJ and Lakxmi Gurumurthy, Department of Chemical, Biological and Pharmaceutical Engineering, New Jersey Center for Engineered Particulates, Newark, NJ

Typically, most Active Pharmaceutical Ingredients (APIs) are extremely cohesive which adversely affects manufacturing processes such as feeding, blending, tabletting and leads to inferior quality of the product. In this work, different pharmaceutical powders (several APIs and excipients) of varying size range are considered for flow and packing density improvement via dry coating based surface modification. The main focus of this presentation is on developing a novel material sparing method that utilizes only small amount of API powders. The idea is to generate information during early API development to assess if the API requires flow improvement and to what extent the flow improvement can be achieved via dry coating. This methodology is validated via multi-faceted characterization, as well as comparison with other batch as well as continuous dry coating methods. Morphology, size and size distribution of these powders are characterized by Scanning Electron Microscopy (SEM) and laser diffraction (Sympatec Helos/Rodos) respectively. Flow characterization is performed using angle of repose (AoR), bulk and tapped density, critical outlet diameter via flodex, Vibration Packed Density (VPD) and shear testing (Freeman FT4). This allows for developing guidelines for the applicability of the novel technology as well as the dry coating approach. The results presented are expected to have an impact on the formulation development of poorly flowing APIs.

See more of this Session: Engineered Composite Particulate Systems for Pharmaceutical Active Ingredient Delivery
See more of this Group/Topical: Particle Technology Forum

(527d) Microstructural Characterization and Dissolution Behavior of Drug/ Semicrystalline Polymer Systems

Qing Zhu[1], Lynne Taylor[2], Michael T. Harris[1] and Hsin-Yun Hsu[1], (1)Chemical Engineering, Purdue University, West Lafayette, IN, (2)Industrial and Physical Pharmacy, Purdue University, West Lafayette, IN

Time-resolved simultaneous small-angle X-ray scattering/wide-angle X-ray scattering (SAXS/WAXS) and second harmonic generation (SHG) microscopy has been employed to study the crystallization kinetics and structural evolution of drug/polymer systems. It was found that the drug formed larger crystallites at higher crystallization temperatures, and the overall crystallization kinetics seems slower. The dissolution study showed that the samples crystallized at higher temperatures shows much slower dissolution rate with different dissolution mechanism, compared to that prepared at lower temperatures, which is consistent with the SHG microscopy results.

See more of this Session: Engineered Composite Particulate Systems for Pharmaceutical Active Ingredient Delivery
See more of this Group/Topical: Particle Technology Forum

(527e) *In Vivo* Cellular Uptake of PEGylated Nanoparticles

Xi Zhan, Kenny K. Tran and Hong Shen, Department of Chemical Engineering, University of Washington, Seattle, WA

Polyethylene glycol (PEG) has been broadly utilized as a strategy in nanoparticle surface functionalization to prolong circulation and reduce nonspecific distribution in drug delivery systems. Modification of nanoparticles with PEG prevents protein adsorption and opsonization to delay non-specific phagocytosis. Previously we have shown PEG-mediated non-specific cellular uptake is cell-type dependent in a range of cell lines. In this study, we examined the *in vivo* uptake of polystyrene nanoparticles with varying PEG density on their surface. Unmodified nanoparticles showed high uptake by different cell types while highly PEGylated particles reduced internalization. Nanoparticles with low and intermediate PEG density showed cell-type dependent internalization. Internalization by phagocytes was highly affected by PEG density while epithelial cells internalized particles independent of PEG density.

See more of this Session: Engineered Composite Particulate Systems for Pharmaceutical Active Ingredient Delivery
See more of this Group/Topical: Particle Technology Forum

On single step assembly of uniform core-shell microparticles with tunable controlled release functionalities

W. **Liu**, W.D. **Wu**, C. **Selomulya**, X. D. **Chen***
Department of Chemical Engineering, Monash University VIC 3800, Australia
Department of Chemical and Biochemical Engineering, Xiamen University, 361005 Fujian Province, P.R. China
E-mail for correspondence: cordelia.selomulya@monash.edu

Core/shell microparticles are of interest for various applications including pharmaceutical engineering, due to the possibility of fine-tuning individual functionalities of the different parts. Previously, core-shell structured microparticles have been produced using concentric nozzles to deliver the core and shell components separately, followed by chemical consolidation of the droplets and further drying steps. To the best of our knowledge, no core-shell structure formation from a homogeneous precursor via spray drying using a single nozzle (orifice) has ever been reported. Furthermore, ordinary lab-scale spray dryers usually produce particles of various characteristics, e.g. size, morphology, structure within the same batch. The lack of reproducibility renders it difficult to correlate the properties of particles formed with the nature of the precursors or other process parameters.

Here we report a new approach for continuous and scalable production of core/shell microparticles, using acrylic nanoparticles (Eudragit® RS) as the shell, and hydrolized tetraethyl orthosilicate (TEOS) forming the silica core upon the single-step assembly. The method is capable of forming monodisperse microparticles from homogeneous precursors without the use of any organic solvents. Precise control of particle uniformity is realized via the use of a micro-fluidic spray drying technique, which has been previously demonstrated to generate monodisperse chitosan microparticles of specific sizes [1] and polymeric microparticles for controlled drug release applications [2, 3].

Evaporation induced self-assembly attained distinct phase separation among the different components during drying and resulted in the core-shell spatial configuration (**Figure 1**), while a precise control over particle uniformity was accomplished via a microfluidic jet spray dryer. Direct control over shell thickness can be achieved from the ratio of the core and shell ingredients in the precursors. A fluorescent compound, Rhodamine B, was used as a highly water-soluble model component to investigate the controlled release properties of these microencapsulates, with the release behaviors shown to be significantly dependent upon their architectures.

References:

1. Liu, W., Wu, W., Selomulya, C., Chen, X. D., 2011, Uniform chitosan microparticles prepared by a novel spray-drying technique, *International Journal of Chemical Engineering*, doi:10.1155/2011/267218, Article ID 267218.
2. Liu, W., Wu, W., Selomulya, C., Chen, X. D., 2011, A single step assembly of uniform microparticles for controlled release applications, *Soft Matter*, 7, 3323-3330.
3. Wu, D. W., Liu, W., Selomulya, C., and Chen, X.D., 2011, Synthesis of uniform silica microencapsulates via spray drying: effects of additives on particle size, morphology and in vitro drug release, (2011), Soft Matter. doi:10.1039/C1SM05879G (*in press*).

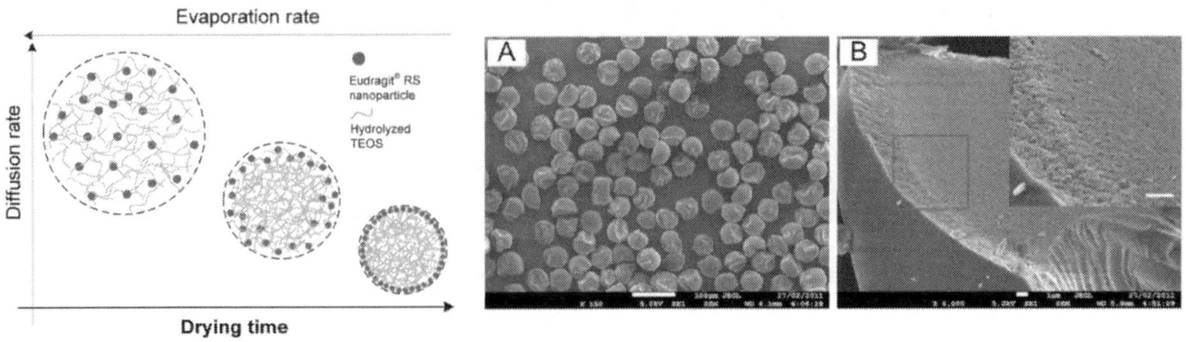

Figure 1 A schematic diagram of particle formation due to evaporation induced self-assembly during spray drying, with FESEM images of (A) monodisperse spray dried microparticles with average diameter of 55 μm, with (B) cross-section displaying (silica) core – (polymeric) shell configuration (inset scale bar: 1μm).

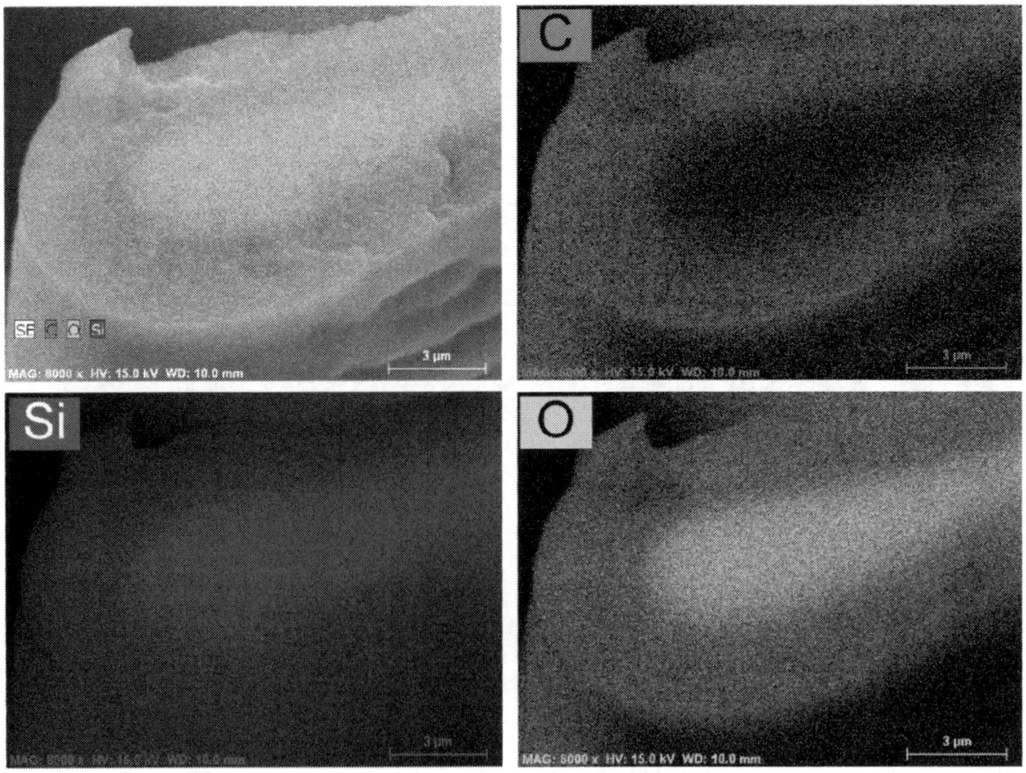

Figure 2 Elemental distribution maps of the cross section of microparticles composed of silica and polymer in 1 : 1 ratio (wt). The elemental "C" originated from Eudragit® RS is mostly present on the outer layer of the particles, indicating the formation of mostly polymeric shell, whereas the elementals "Si" and "O" are concentrated on the inner particles, implying silica as the main core component.

Spray Drying of Uniform Silica Microencapsulates: Implications for Controlled Release

Winston Wu[1], Wenjie Liu[1], Xiao Dong Chen[1,2], and **Cordelia Selomulya**[1]
[1] Department of Chemical Engineering, Monash University, Clayton VIC 3800 Australia
[2] Department of Chemical and Biochemical Engineering, College of Chemistry and Chemical Engineering, Xiamen University, 361005 Fujian Province, P.R. China

Extended Abstract

Spray drying is a scalable route for particle formation with easy product recovery. Conventional spray drying produces broad distributions of particles with various morphologies in a single batch, due to non-uniform distribution of droplets / spray trajectories and different residence times [1]. This renders any interpretation of particle behaviour in relation to their physico-chemical properties virtually impossible. Here, uniform silica microparticles encapsulating vitamin B_{12}, which was homogenously distributed within each particle, were synthesised in a single step via a micro-fluidic-jet-spray-dryer (Figure 1). The dryer utilised a micro-fluidic-aerosol-nozzle for continuous generation of monodisperse droplets from complex precursors [2]. The uniformity of the particles allowed direct correlations between their properties and the release behaviour of vitamin B_{12} to be investigated, without the complications of wide size distribution or non-uniform shapes. The effects of additives in the forms of lactose and Na-alginate to the properties of these microparticles were demonstrated. Spherical particles with relatively smooth surface were obtained with lactose addition, while incorporation of Na-alginate resulted in noticeably increasing surface roughness with higher amount added (Figure 2). Lactose addition also accelerated the release of the encapsulated vitamin B_{12}, due to the relatively fast lactose dissolution that allowed buffer to penetrate into the matrix to facilitate diffusion and silica erosion. On the contrary, Na-alginate slowed down the release considerably by serving as an additional barrier to decelerate silica matrix erosion. Release kinetics data indicated diffusion as the main release mechanism independent of the additives. The release profiles from different compositions of the synthesized particles demonstrated good agreement with computational results, highlighting the reproducibility of particles produced with this technique and the ability to modulate the release behaviour directly from the precursor compositions [3].

References:

1. Woo, M.W., Rogers, S., Lin, S.X.Q., Selomulya, C., Chen, X. D., 2011, Numerical probing of a low velocity concurrent pilot scale spray drying tower for mono-disperse particle production - unusual characteristics and possible improvements, *Chemical Engineering and Processing: Process Intensification*, doi:10.1016/j.cep.2011.02.007 (*in press*).
2. Wu, W., Amelia, R., Hao, N, Selomulya, C., Zhao, D., Chiu, Y., Chen, X. D., 2011, Assembly of Uniform Photoluminescent Microcomposites Using a Novel Micro-Fluidic-Jet-Spray-Dryer, *AIChE Journal*, doi: 10.1002/aic.12489 (*in press*).
3. Wu, W., Liu, W., Selomulya, C., and Chen, X. D., 2011, Synthesis of Uniform Silica Microencapsulates via spray drying: Effects of Additives on Particle Size, Morphology and In Vitro Drug Release, *Soft Matter*, doi:10.1039/C1SM05879G (*in press*).

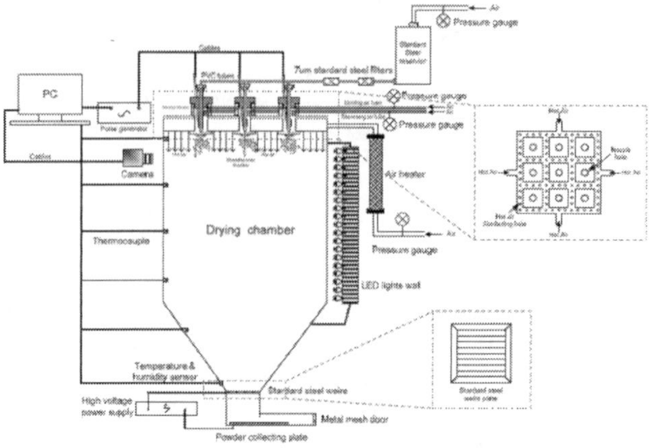

Figure 1 – Schematic of the micro-fluidic-jet-spray-dryer

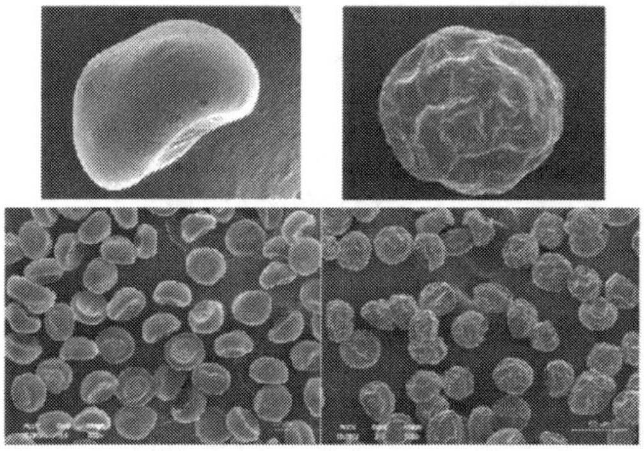

a b

Figure 2 – Uniform silica microencapsulates with
(a) 0.5%w/v Na-alginate; (b) 1.0%w/v Na-alginate

Handling of Fine Particles - a Nuisance or an Opportunity

Jesse Zhu

Departments of Chemical and Biochemical Engineering / Particle Technology Research Centre,
The University of Western Ontario, London, Ontario, N6A 5B9, Canada

More than 70% of chemical processes involve the handling of particles and there are equally if not even more non-chemical processes that require solids handling. The basis for most solids handling is fluidization where particles are suspended in a fluid. However, as the particle sizes become smaller, the inter-particle forces become more significant and their fluidization eventually becomes extremely difficult if not impossible. Those fine particles under 25-30 microns are classified as Geldart Group particles and have been known for strong cohesiveness and agglomeration.

On the other hand, fine particles have many practical applications that give rise to useful advantages over large particles. For example, when liquid coating is replaced by powder coating, the environmental issues related to solvent spray is eliminated, the energy use is also significantly reduced as there is no need for solvent evaporation, and the powdered paint can be nearly 100% recycled. But only very fine paint particles can lead to smooth surface like liquid. For yet another example, inhaling micronized powder medicament through the lung can avoid the inconvenience and the health associated risk when compared to intravenous injection, lead to significant savings on drug usage, and eliminate many side effects compared to oral intake. With the rapid advances in nanotechnology, there are also more and more new applications everyday.

The key to the successful applications of Group C fine particles is its smooth fluidization, which in turn relies on the overcoming of the inter-particle forces, their elimination or significant reduction. The talk will first quickly review some major fluidization research projects carried out in the author's group and then highlight the key work on fine particles carried out in the author's laboratory, that aims to achieve the smooth fluidization and to increase the handlability of fine powders for direct industrial applications. Some experience on successful industrial applications will also be shared.

A short bio:

Jesse Zhu is a Professor and Canada Research Chair in the Department of Chemical and Biochemical Engineering, Faculty of Engineering, at the University of Western Ontario, in London, Ontario, Canada. He is also a Fellow of the Chemical Institute of Canada and a Fellow of Canadian Academy of Engineering.

Jesse Zhu received his B. Eng. in Chemical Engineering from Tsinghua University in Beijing, China in 1982. Between 1984-88 he pursued graduate studies in Chemical Engineering at the University of British Columbia, under the supervision of Prof. John Grace. Upon completion of his Ph.D. in early 1988, he worked as a Research Scientist for two years at the Amsterdam Shell Research Laboratory. In 1990 he returned to Canada and in 1993 he started teaching at the University of Western Ontario in London, Ontario, Canada where he obtained full professorship in 1999. In 1999, he established the Particle Technology Research Centre and serves as its director.

His contributions to fundamental research and technology developments can be reflected by the over 270 referred journal publications, a dozen of granted patents, the over 75 Masters and PhD students graduated under his supervision, and the international scholarly activities such as chairing international conferences and serving on journal editorial boards. Some of the inventions have been commercialized and/or are now in the process of being commercialized. Jesse Zhu also received numerous awards, including two major awards from the Canadian Society for Chemical Engineering. For AIChE, he served as Vice Chair and then Chair for Area 3b –Fluidization and Fluid-Particle Systems, from 2005 to 2009. He also serves as the Conference Chair for the 2011 Annual Canadian Chemical Engineering Conference, to be held in London, Canada in October 2011.

(605b) Shell Thomas Baron Award In Fluid-Particle Systems - Structure, Slip and Rheology of Soft Particle Glasses

Roger T. Bonnecaze, Department of Chemical Engineering, The University of Texas at Austin, Austin, TX

Soft particle glasses form a broad family of materials made of deformable particles, as diverse as microgels, emulsion droplets, star polymers, block copolymer micelles and proteins, which are jammed at volume fractions where they are in contact and interact via soft elastic repulsions. Despite a great variability in origin of particle elasticity, soft glasses have many generic features in common. They behave like weak elastic solids at rest but flow very much like liquids above the yield stress. This unique feature is exploited in industry to process high performance coatings, solid inks, ceramics pastes, textured food and personal care products. These materials also exhibit unusual slip and shear banding depending on whether the surfaces are smooth or rough and the chemical interactions between the particles and surfaces. Here I will show that the slip and bulk rheology are related to the microstructure, which is controlled by an interplay among chemical, elastic and viscous interactions.

For bulk rheology, I present a micromechanical 3-d model that quantitatively predicts the non-linear rheology of soft particle glasses. The shear stress and the normal stress differences depend on both the dynamic pair distribution function and the solvent-mediated elastohydrodynamic interactions among the deformed particles. The predictions, which have no adjustable parameters, are successfully validated with experiments on concentrated emulsions and polyelectrolyte microgel pastes, highlighting the universality of the flow properties of soft glasses. These results provide a framework for designing new soft additives with a desired rheological response.

The slip and shear banding is due to variations of the structure near the surfaces, which surprisingly propagates hundreds of particle diameters away from the surface. This variation and the resulting flow can be described by a non-local fluidity model. Our results establish a link between surface forces, lubrication and yielding in soft glassy or jammed materials and open new routes to manipulate their flow through the surface chemistry of the confining boundaries.

See more of this Session: Particle Technology Awards Lectures
See more of this Group/Topical: Particle Technology Forum

(657a) Resonant Acoustic (Vibratory) Powder Mixing Performance

Juan G. Osorio and Fernando J. Muzzio, Chemical and Biochemical Engineering, Rutgers, The State University of New Jersey, Piscataway, NJ

Powder mixing is one of the most important operations during food, chemical and pharmaceutical processing. Blending performance of various blender designs, including V-blenders, double-cone blenders, tote-blenders and rotary-drum blenders, have been previously studied in depth. The efficiency of the powder mixing process is measured as the speed at which the desired uniformity of a blend is achieved. The blending performance has been found to be dependant on blender type, material properties and even environmental conditions. In these studies, the performance of a new laboratory-scale ResonantAcoustic® Mixer (RAM) is studied. The RAM uses acoustic energy to provide both bulk and micro mixing. Resonant acoustic mixing is known to significantly reduce mixing time making it a good candidate for improving the efficiency of powder mixing processes.

The blend uniformity of lactose and acetaminophen powders blended using the RAM is compared to mixing results obtained using a double-cone tumble blender. Uniformity is determined by the variance in acetaminophen concentration (as measured by near infrared spectroscopy (NIR)) obtained from numerous small samples collected over the duration of mixing. The optimal RAM operating parameters resulted in relative standard deviation (RSD) values for acetaminophen concentration of 0.4 as opposed to 0.8 for the double cone blender. The RAM achieved these minimal RSD results in 30 seconds as opposed to 240 seconds for the double cone blender. In the second part, a fractional factorial design of experiments is performed to characterize and optimize the mixing performance of the RAM. Variations in fill level, mixing time, and mixing intensity were used. Combinations of binary mixtures were used in order to understand the effects of mixing parameters and material properties. Three fill levels (25%, 50%, and 75%), three mixing times (1 min, 2 min, and 4 min), three mixing intensities (25 Gs, 50 Gs, and 75 Gs) were used as mixing parameters. The binary mixtures used were composed of three API grades (micronized and granulated APAP, and caffeine), three API concentrations (3%, 10%, and 30%), and two grades of MCC (Avicel PH102 and Avicel PH200). After each experimental combination, numerous samples were extracted from the 273-ml vessel in order to determine the mixing performance determined by the variance in the acetaminophen concentration (as measured by near infrared spectroscopy). The results indicated that the RSD decreased with increasing acceleration. For the materials and parameters used, the time of mixing was not significant, although higher acceleration results in better mixing. Temperature increased with increasing fill level, time and acceleration.

See more of this Session: Measurement, Monitoring and Characterization Methods for Particulate Systems Research In Powder and Granular Mixing
See more of this Group/Topical: Particle Technology Forum

(657b) Experimental Indicators of Materials Processing Progress In Mechanical Alloying and Reactive Milling

Priya R. Santhanam, Paul Dupiano and Edward L. Dreizin, Otto H. York Department of Chemical, Biological, and Pharmaceutical Engineering, New Jersey Institute of Technology, Newark, NJ

Mechanical milling is widely explored to prepare advanced alloyed and composite materials. However, it is often unclear how to transfer production from small scale laboratory mills to larger scale devices for practical manufacture of the desired materials. Recently, a simplified approach was proposed, in which the rates of energy transfer to the material from milling tools were calculated for three different mills using discrete element modeling (DEM); these rates were correlated with the averaged experimental rates of increase in the yield strength of an oxide reinforced composite as a function of its milling time. The results were encouraging, but two significant limitations remained. First, experimental validations relied on tangible changes in the yield strength of the milled material. Such changes, common for oxide reinforced composites, could be less pronounced or insignificant for other materials. Second, a steady-state refinement was assumed. The latter assumption needs to be revisited if restitution and friction coefficients change during milling, resulting in altered rates of energy transfer from milling tools to the powder. Therefore, one of the goals of this effort is to enable a validation of the DEM calculations for materials, for which measurements of yield strength are impractical. The second goal is to develop real-time indicators of milling progress achieved in an attritor mill, which do not require recovery of the powder samples. The utility of such real-time indicators is two-fold. First, no labor-intensive recovery and examination of the partially milled powders is required. Second, the instantaneous changes in the powder properties, affecting friction and restitution coefficients can be accounted for in modified DEM calculations. The attritor mill is specifically selected for this study as the device most suitable for large scale manufacturing of mechanically milled materials. Correlations between the rate of power consumption, torque, and rpm of the motor and the powder refinement are examined. Yield strength measurements for the recovered samples are correlated with changes in the measured x-ray diffraction patterns.

Investigation of Reaction Mechanism in Ti/C System: Influence of Mechanical Activation

Ya-Cheng Lin[a], Khachatur V. Manukyan[a], Alexander S. Mukasyan[a]

[a]*Department of Chemical and Biomolecular Engineering, University of Notre Dame, Notre Dame, Indiana, 46556 USA*

Abstract

The influence of mechanical activation (MA) on the stoichiometric mixture of titanium and graphite powders at room temperature under argon atmosphere is discussed. Nano-scale TiC (~500 nm) is produced within 9.5 min of high energy ball milling (HEBM). Crystal structures and phases were characterized by X-ray diffractometry (XRD) and energy dispersive X-ray spectroscopy (EDS). Microstructure of the initial and as-milled samples was done by SEM and internal structures characterized by focused ion beam (FIB) technique. Ignition characteristics and activation energy calculations were carried out by the so-called electrothermal explosion (ETE) method [1]. Special attention was focused on the microstructure variations with respect to ignition temperatures at different milling durations. Reaction mechanism regarding chemical route is also proposed.

First, BET specific area analysis showed ~100 (m^2/g) after 2 min of MA (3 m^2/g for initial mixture). At this stage, microstructure images and XRD patterns indicate the complete amorphorization of graphite powders on Ti bulks. Second, preliminary SEM results reveal the fact that Ti/C laminated microstructures are observed as early as 3 min of MA, corresponding to an evident decrease of ignition temperatures from 1950 K to 1800 K. Third, the appearance of nano TiC_x particles (20–50 nm) at the internal as-milled powder microstructure images depicted by FIB also suggests that evolution of fine grains of TiC_x in terms of milling durations (3–7.5 min). Finally, ETA results showed the ignition temperatures further decrease to ~1500 K and stabilize for samples after 7.5 min MA. The results obtained from microstructure evolution with respect to different milling durations (0–9.5 min) are in good agreement with the decrease of ignition characteristics and other metallurgical analyses.

[1] A. S. Shteinberg, Y.-C. Lin, S. F. Son, A. S. Mukasyan. "Kinetics of High Temperature Reaction in Ni-Al System: Influence of Mechanical Activation" *J. Phys. Chem. A*, **2010**, *114* (20), pp 6111–6116.

Outline for Metal Powder Processing Safe Working Practices

Abdollah Kashani
Safety Consulting Engineers, Inc.
Schaumburg, Illinois 60173

Biography

Abdollah Kashani *a Senior Safety Project Engineer has over 30 years experience providing innovative solutions and services for a variety of industrial, pharmaceutical, medical device manufacturing, and governmental customers. He has been providing services for various processes and operations such as material handling, heat generation, material transfer and transportation, hybrid mixtures (liquid-gas-solid) and other operations.*

Involved in design and development of simulated process conditions for material responses. Examples are electrostatic charge generation, electrostatic discharge vapor/gas/solid explosions under elevated temperatures and pressures, identifying factors that influence the process and related hazard and safety issues.

Provided customized services for qualification of materials and equipment for the end user such as ignition protection study of electrical devices, in flammable environments, simulated to understand the cause of fire or explosion related to particular incident investigations.

On an industrial scale, he and his team developed and designed customized solutions such as flame/detonation arrestors, cyclone for high energetic materials, prototype explosive detection system, and conducting process safety/hazard analysis.

His experience also includes material flammability and fire resistance, demilitarization service, open gas/vapor and dust cloud explosion, hydrocarbon spray and pool fire, electrostatic charge/discharge study, propellant and energetic material burning rate combustion, design and setup for testing of hazardous material for DOT Class 1, 2, 3, 4, 5, 8 & 9 per applicable standard methods, and non-destructive evaluations.

He managed the customized service to be efficient and cost effective with quick turn-around-time. Other expertise included limiting oxygen concentration, flammability envelope, and ignition energy, characteristics, assessing the risk runaway reaction of material at variety of conditions and validation for exothermic and kinetic reaction.

He obtained his Associate Degree in Civil Engineering and Bachelor of Science in Medical Technology and Sciences from Iran. He obtained his Master of Science in Bioengineering from the University of Illinois, Chicago, Illinois. He published jointly a number of technical papers.

Abstract

Work related injuries and accidents have been a continuing problem in various industries. Proper training of workers and understanding the various aspects of a process, hazard conditions through customized audio-visual communication tools can be effective and a perfect way of introducing the employer and employees to safety issues and safe working practices.

The human brain remembers visualized information easier and helps workers to ensure a safe work environment. No one and no work place can afford to have accidents.

In this paper a detailed and thorough outline for safe metal processing operations is being presented. The outline presented here can be used to develop a training course or a safety video which will motivate workers toward safe behaviors, and a greater compliance with safe system working practices.

Identification of hazard conditions, implementation of awareness of hazards in the work place, various prevention and protection measures to ensure safe working practices, methodology to determine hazards, powder behavior and its operation as well as adopting suitable means of controlling hazards are discussed herewith.

Customized training and understanding the safety and hazards of a particular operation and process (e.g. metal processing industries) can eliminate or minimize work related injuries and accidents.

Key words: safe work place practice, metal powder process operation, safety training outline, process safety

Introduction

In the metal powder, coal mining, grain handling, plastics, chemicals and wood products industries, dust explosions, fire, and toxicity have been a continuing problem. The most common sources of safety and health problems in manufacturing are the machinery and its operation. Various preventive and protective measures must be taken to ensure safety.

Purpose

This presentation should be used as a communication tool for the promotion of safe working practices. It alerts employers and employees to the hazards of work environment by pointing out the means of identifying, eliminating and controlling hazard conditions. In this paper the focus will be on the **Metal Powder Processing Operation.**

Scope

How dust explosions can occur in the Metal Powder (PM) Producers and Parts Makers operations, how metal working machinery/tools can fail or operate in an unsafe manner and how workplace injuries can occur? What are the steps and processes necessary to cause dust explosions or to generate toxic environments? Safe practice operations must address preventive and protective measures dealing with:

- Generation of a dust cloud in the PM production and PM parts makers
- Generation of a toxic environment in the PM production and PM parts makers
- Generation of conditions necessary for metal dust explosions
- Failure of machinery, or their operations after unsafe conditions are developed

General Hazards Identification

A) Fire

- Extreme fine powder particles when exposed to air could react with air and catch fire.
- A powder layer can be ignited by the following ignition sources:
  - Electrostatic discharge (ESD)
  - Heat
  - Impact
  - Friction
- Metal powder generates high heat flux and can lead to metal powder explosions (and secondary explosions).

B) Explosion

- Fine particles (with high surface area/volume) can be lifted up and create an explosive dust cloud.
- A dust layer can be kicked and lifted up to create an explosive dust cloud.
- Metal powder explosions can be initiated by various ignition sources such as:
  - Pyrophoricity
  - Electrostatic discharge (ESD)
  - Heat
  - Impact
  - Friction
- Dust explosibility characterized by pressure output/rate of pressure rise.
- Hybrid mixture of metal powder and flammable liquid will cause higher pressure output/rate of pressure rise.

C) **Workplace Injuries**

Workplace injuries, which can occur due to lack of safeguarding and safe operation, are mostly related to:

- Improper ventilation
- Improper design or construction
- Improperly or inadequately guarded or unguarded machinery
- Exposed moving parts
- Sharp cutting edges
- Power presses and power hammers
- Furnace and exhaust systems
- Defective tools and/or equipment
- Vibration and noise from machinery
- Physical reaction either voluntary or involuntary

Types of injuries could be:

- Permanent total disability (including death)
- Permanent partial disability
- Temporary total disability

D) **Health and Toxicity**

Health hazards that pertain to fine metal powders/vapors which may enter the body via the following:

- Inhalation
- Skin contact (dermal)
- Ingestion (oral)

Health and Toxicity due to exposure can cause:

- Lung damage (breathing)
- Burns/corrosive burns
- Skin reaction due to contact with metal
- Dust (allergic reaction)
- Interference with normal metabolic function

Process Operation and Hazards

Process Operations:

The main process operations in metal powder and/or parts production are:

A) **Powder/Part Production**

- Atomization
- Chemical reduction
- Spray drying of metal powder
- Milling/grinding
- Pressing operations
- Sintering
- Injection molding
- Parts (machining)secondary operations

B) **Powder/Part Handling**

Major preparation operations for either metal powder and/or parts are:

- Screening
- Cleaning
- Annealing
- Mixing and blending
- Heat treating
- Lubricating

C) **Metal Powder Press**

Mechanically or hydraulically powered machines are:

- Metal sizing press
- Metal coining press
- Compacting press
- P/M briquetting press
- Steel and plate press
- Turning machines
- Boring machines

Hazards:

The hazards associated with any of the above operations are as follows:

A) **Dust Explosion**

- Presence of ignition sources due to:
  - Heat
  - Atomization
  - Grinding and milling
  - Annealing, sintering

- Electrostatic discharge due to:
  - Movement of powder
  - Mixing and blending
  - Screening

- Electric sparks due to:
  - Failure of electrical circuitry
  - Welding operations

- Friction and impact due to:
  - Grinding
  - Milling
  - Parts cleanup
  - Screening

- Thermal effect due to:
  - Annealing
  - Sintering
  - Drying
  - Heat treating

B) **Gas/Vapor Explosion**

- Generation of hydrogen (water reactive metal powder e.g. aluminum)
- Release or leak of hydraulic fluid
- Cleaning Solvent Vapor

C) **Mechanical Hazard**

- Ejection of hard object from presses, grinding, milling, turning, etc.
- Moveable and/or rotating machinery
- Failure of electrical circuitry
- Failure of interlock or interlock bypass
- Inadequate guard, guard removed or barrier guard not provided
- Slippery or rough surfaces
- Impact by moving equipment or parts
- Toppling or overturning of unstable products
- Excessive weight
- Excessive vibration and noise

D) **Toxicity Hazard**

- Generation of toxic dusts
- Generation of release of toxic metal fumes

- Release of toxic vapors/fumes from chemical reactions or processes
- Inadequate oxygen level for breathing
- Direct contact with metal dust

Guidelines to Mitigating Hazards

Hazards can be avoided by proper design, control, procedures and training.

A) **Explosion**

- Minimize powder handling
- Keep the flow of the powder at a low rate
- Use dust collecting/filtration system
- Minimize generation of fine metal powder particles
- Prevent dust accumulation
- Prevent air leak into furnace/equipment
- Removing ignition sources such as ESD, heat, friction, impact, etc.
- Use explosion venting or explosion suppression system
- Housekeeping (maintenance)

B) **Injury Prevention**

- Machines should never be left running unattended unless the machine has been designed to do so.
- Operators should not wear jewelry, loose fitting clothing or long hair, which could be caught in moving parts.
- Use personal protective equipment (PPE), such as respirators, gloves
- Manual adjustment, gauging of work should not be practiced while machine is in operation.
- Metal chips or dusts should not be removed by hand. Use brushes, vacuum equipment or special equipment for removing chips or dust.
- Use only the proper hand tools for each job or machine
- Do not use high-pressure compressed air to blow off dust or dirt. Use a vacuum system instead, whenever possible.
- Machine guards, baffles, or chip guards should be around the machinery to shield the operator.
- Select equipment for installation which has low vibration and noise.
- Equipment which could vibrate should be mounted on firm, solid foundations and secured with bolts.
- Provide sound absorbing enclosures or vibration insulators for noisy or vibrating equipment.
- Safety guards such as, anti-repeat, barrier gate devices, stop control, brake control and spring guards should be used.

Methodology for Example of Dust Explosion Hazard Testing and Analysis

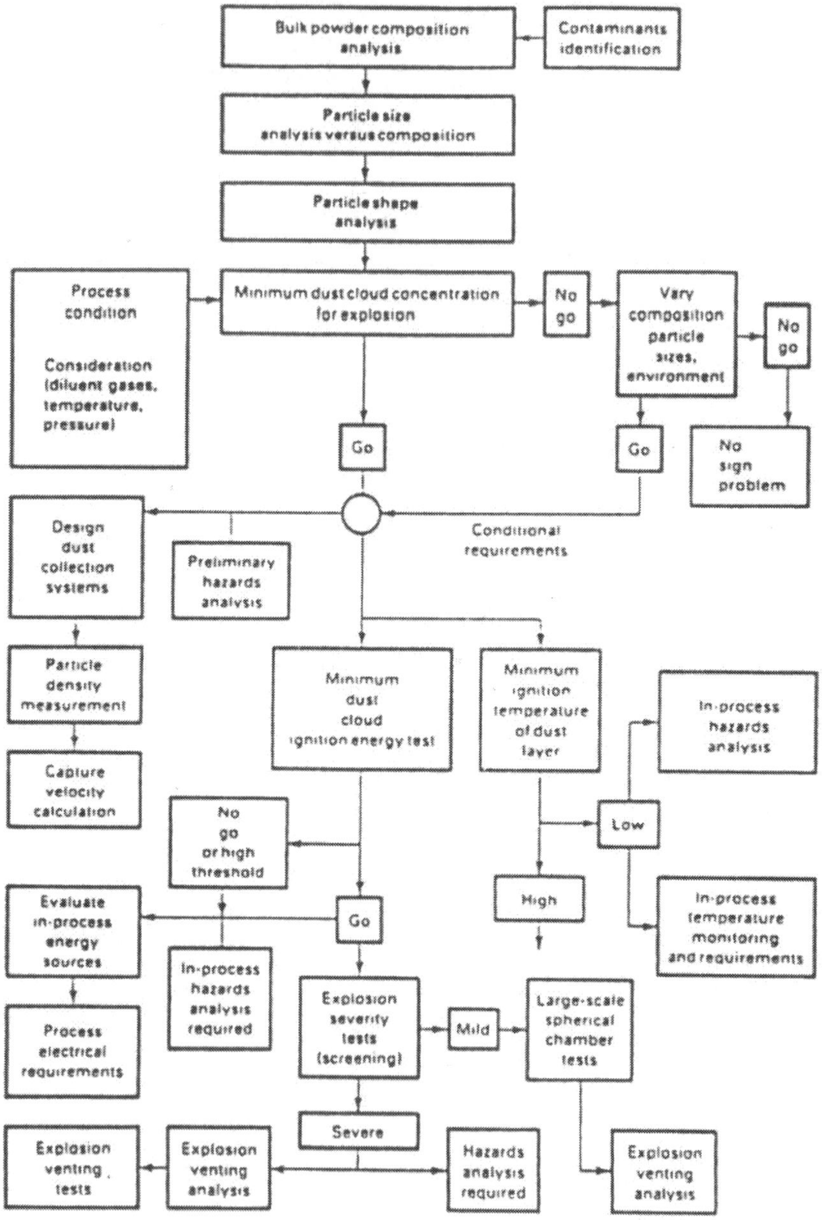

Material dust explosion hazard test methodology.

Table 1

Example of Ignitability and Explosibility of Metal Dust

Dust Type	Minimum Explosible Concentration (gm/m^3)	Output Pressure (barg)	K_{St} (bar·m/sec)	Minimum Ignition Energy (mJ)
Fine Aluminium Type 1	30	11.5	1100	2-3
Fine Aluminium Type 2	60	10.2	100	10-15
Iron Powder, Fresh	500	5.2	50	--
Aluminium/Iron (50:50)	250	9.4	230	--
Fine Silicon Powder	125	10.2	126	3
Zinc Powder DC1	125	7.3	176	2
Zinc Powder DC1	250	6.8	93	--

Conclusion

Each product and design operation may have certain inherent hazards. Proper training through communications will allow employers/employees to eliminate or control hazardous conditions by:

- Identifying hazards that could cause fire, explosion and toxicity so corrective actions and any necessary changes can be implemented, minimizing personnel exposure
- Adopting protective measures means incorporating suitable safety measures in the process design for specific hazards encountered during the operations.
- Contributing to a safe working environment and safe operations means providing accurate safety procedures and standards
- Reporting any hazardous conditions or defective equipment that may exist.
- Provide personnel training on a regular basis.

A financial crisis is more likely to occur when there is an accident, because firms find that their real burden of indebtedness is increasing while there is no increase in the production and in the real value of their assets. The resulting decline in the firm's net worth increases. No one and no work place can afford to have accidents.

A great opportunity to explore a wide variety of smart, safe, sustainable manufacturing solutions tools and programs are available such as:

D) **HAZID** Hazard Identification Study
E) **PHA** Process Hazard Analysis
F) **FMEA** Failure Mode and Effect Analysis
G) **HAZOPs** Hazard and Operability Studies
H) **HFMEA** Health Failure Mode and Effect Analysis
I) **HACCP** Hazard Analysis and Critical Control Points
J) **PHA** Proactive Hazard Analysis
K) **RCA** Root-Cause Analysis
L) **LOPA** Layers of Protection Analysis
M) **TQM** Total Quality Management

These are some solutions which are geared toward improving your safe working practices and your business profitability.

References

Baker, Wilfred E., "Explosions in Air" 1973, University of Texas Press, Published in cooperation with the Southwest Research Institute

Bartknecht, Wolfgang Dr., "Explosions, course prevention protection" 1981, Springer-Verlag Berlin Heidelberg New York 1980

Dahn, C.J., "Explosivity and Pyrophicity of Metal Powders" Metals Handbook, ninth edition. Volume 7 – Powder Metallurgy.

Eckhoff, Rolf K., "Dust Explosions in the Process Industries" third edition 2003, Gulf Professional Publishing

Sax, N. Irving, "Cancer Causing Chemicals" assisted by Elizabeth K. Weisburger, David Schottenfeld, Joanna Haas, Benjamin Feiner, Barry I. Castleman and Richard J. Lewis, Sr., Copyright 1981 by Litton Educational Publishing, Inc. Van Nostrand Reinhold Company

(657e) Experimental and Numerical Study On the Tablet Velocity In Commercial Scale Pan Coaters

Wei Chen, Shih-Ying Chang and San Kiang, Dpst, Bristol Myers Squibb, New Brunswick, NJ

Effective control of coating uniformity in a tablet coating process requires good mixing. Scaling up a pan coating process from lab to commercial scale to maintain the same level of tablet uniformity poses unique challenge. It was found in our previous study that [1] the tablet flow rate, a value calculated from the tablet velocity number density (tablet size) and spray zone width, can be an indicator of the mixing condition. We have also quantified the impact of the tablet velocity and number density on the tablet content uniformity. In this paper, the tablet velocity in commercial pan coaters was investigated experimentally and numerically. This study was carried out in 50kg, 400kg and 600kg pan coaters. Different types of tablets including oblong and round tablets with weights from 200mg to 1500mg were used in the study. A real-time imaging device was developed to measure and analyze the tablet flow behavior within the sprayed area. The current study shows that the smaller round tablets move faster than larger ones in general. However, for the oblong shape tablet the larger tablets move faster than the smaller ones. For all the tested tablet types, the tablet velocities increase with increasing air flow rates. This enhancement by air flow starts to level off at the higher air flow rates. For a given pan size, there appears a pan loading that has the maximum tablet velocities. The existence of such maxima implies an optimal pan loading at which the desired coating uniformity can be consistently obtained. In the parallel study through the numerical simulation, a Discrete Element Method (DEM) was applied to simulate the tablet flow behaviors in a pan coater. The simulation results show a good agreement with the experimentally measured tablet velocities.

See more of this Session: Measurement, Monitoring and Characterization Methods for Particulate Systems Research In Powder and Granular Mixing
See more of this Group/Topical: Particle Technology Forum

(657f) Determination and Modeling of the Limiting Oxygen Concentration of Pesticide Dusts with Highly Corrosive Combustion Products

Hannes Kern and Harald Raupenstrauch, Chair of Thermal Processes, University of Mining Leoben, Leoben, Austria

Within the production of agricultural pesticides various different kinds of dusts occur which can be considered as explosive due to their chemical and physical characteristics. In various steps of the production process hybrid mixtures (e. g. dust in acetone) are present. For most of the considered dusts with high contents of halogens there are no sufficient data available in order to determine the explosion hazardous potentials of the different process steps. In this work safety characteristics, especially the limiting oxygen concentration, of 2,4-Dichlorophenoxyacetic acid- dimethylamine (2,4D- DMA) and different urea herbicides (Chlortoluron, Fluometuron) have been determined experimentally and theoretically.

To characterise the occurring dusts also not only in terms of their chemical composition extensive research on the particle structure and surface as well as on the particle size distribution were carried out. The particles show a crystal structure with no porosities or inclusions. The parameters of the particle size distribution were approximated by a RRSB distribution. All dusts show values around 1 for the scattering parameter n. Besides the particle size distribution also the burning behaviour of the dusts was investigated. For all dusts no autonomous burning could be observed.

After research of the dusts under a microscope and the analysis of the particle size distribution and burning behaviour, experiments in the ISO 20l apparatus of the company FireX were carried out. All observed dusts show significant values of pressure rise and maximum explosion pressure. The maximum explosion overpressures are situated around 8bar and the measured minimum ignition energies lie between 8 and 16mJ.

The measured values for all analyzed dusts where also compared to results of different models for calculation of safety relevant characteristics.

A main goal of the present work was to determine the limiting oxygen concentration of the chosen dusts. In the course of the research the combustion products of the dusts turned out to be highly corrosive, which can be explained by the chemical composition of the dusts and the burning conditions in the ISO 20l apparatus. Some combustion products caused smaller damage on the 20l apparatus, due to corrosion effects. A reliable method to measure the limiting oxygen concentration had to be found as alternative for the 20l apparatus.

By modification of the Hartmann apparatus, which is used for screening purposes, a method was to determine the limiting oxygen concentration (LOC) was investigated. A goal of the chosen method was also to reduce the amount of sample needed for the analysis. The method was first checked by several dusts with well known values of limiting oxygen concentration. All values for the LOC of the investigated dusts, could also be found by this alternative method in consideration of the defined failure limit. The needed sample amount for the analyses is significantly lower compared to the ISO 20l apparatus.

All results of the test dusts were also compared to results of several different models for the determination of the limiting oxygen concentration. The models showed good agreement with the values found in experiment, so that the method could be considered as reliable under the chosen experimental conditions.

As a next step all pesticide dusts where investigated with this alternative method. Compared to all other safety relevant characteristics determined in the course of this work, also the limiting oxygen concentrations of the dusts showed comparable behaviour, regarding their chemical composition. All results of the experiments where compared to results of mathematical models and showed good agreements. The determined limiting oxygen concentrations where also comparable to other known dusts with quite similar chemical composition.

The used method offers the possibility to determine the limiting oxygen concentration of dusts with highly corrosive or hazardous combustions products by limiting the amount of sample needed and simplifying of the cleaning process of the equipment due to the assembly of the used Hartmann apparatus.

See more of this Session: Measurement, Monitoring and Characterization Methods for Particulate Systems Research In Powder and Granular Mixing
See more of this Group/Topical: Particle Technology Forum

(663a) Nanoporous Silicon As a Novel Fuel

Paul Redner[1], Neha Mehta[2], Karl Oyler[2], Gartung Cheng[2], Christopher Csernica[3] and Deepak Kapoor[4], (1)Advanced Materials Branch, US Army, RDECOM-ARDEC, Picatinny Arsenal, NJ, (2)Energetics and Warheads, US Army RDECOM-ARDEC, Picatinny Arsenal, NJ, (3)Pyrotechnics, US Army, RDECOM-ARDEC, Picatinny Arsenal, NJ, (4)Materials Division, US Army RDECOM-ARDEC, Picatinny, NJ

Nanoscale and nanostructured energetic materials are enabling technologies for defense applications. Many current and future requirements cannot be achieved using more conventional materials. A way to overcome the natural barriers to performance, or other undesired material characteristics, is by exploiting the "tunable" nature of these novel materials. Altering the constituents, particle or grain size, surface area and surface characteristics allows one to adjust the energy output, light output, strength of materials, and sensitivity to various stimuli.

Nanoporous silicon is being explored as a novel fuel to replace aluminum and magnesium in energetic formulations for a variety of reasons such as reduction in sensitivity, increased performance, and improved aging characteristics of the material. Additionally, hybrid formulations with silicon and other novel materials will be able to potentially replace legacy materials such lead azide and lead styphnate, thus reducing the environmental impact, and lifecycle cost, of munitions that contain these materials.

See more of this Session: Nanoenergetic Materials
See more of this Group/Topical: Particle Technology Forum

(663b) Aluminum-Based Ternary Reactive Composite Powders

Yasmine Aly, Mirko Schoenitz and Edward L. Dreizin, Otto H. York Department of Chemical, Biological, and Pharmaceutical Engineering, New Jersey Institute of Technology, Newark, NJ

Adding aluminum to propellants, pyrotechnics, and explosives is a common way to boost their energy density. A number of approaches have been investigated that shorten aluminum ignition delay, increase combustion rate, and decrease the tendency of aluminum droplets to agglomerate. Boron as an additive to energetic formulations is also of interest, but due to kinetic barriers in boron combustion, aluminum is the more widely used additive. However, aluminum can potentially be used to enhance boron combustion, making use of the higher volumetric and gravimetric enthalpy of boron. Two composite powders of Al-B-*X*, where *X* is a halogen-containing additive of elemental iodine or PTFE (Teflon), are explored in this work. Both materials are prepared by mechanical milling. For halogen-containing composites, an added objective is the ability to release combustion products that help defeat biological agents. Recently it was shown that composite materials of aluminum and elemental iodine, prepared by mechanical milling have enhanced oxidation rates and generate combustion products substantially improving prompt inactivation of bioaerosols. Biocidal effectiveness of other halogens, such as fluorine or fluorinated combustion products expected in presence of Teflon is also of interest. In addition, Al-Teflon composites are known to be highly reactive. Finally, both Teflon and iodine serve as effective surfactants during mechanical milling, assisting formation of the composites with desired particle sizes and morphologies. In this work, composites containing elemental iodine and Teflon are produced using mechanical milling and characterized. Oxidation, ignition, and combustion of the prepared powders are studied experimentally.

See more of this Session: Nanoenergetic Materials
See more of this Group/Topical: Particle Technology Forum

Effect of Gasifying Agents on Combustion Characteristics of Nanothermites

Zac Doorenbos[1], Ian Walters[2], Wendy Balas-Hummers[3], Michael J. Donadio[3], and Jan Puszynski[*2]

[1]Innovative Materials and Processes, LLC, Rapid City, SD 57702
[2]Chemical and Biological Engineering Department, South Dakota School of Mines and Technology, Rapid City, SD 57701
[3]Armament Research, Development and Engineering Center, Picatinny Arsenal, NJ 07806

ABSTRACT

Metal-metal oxide nanothermite materials have strongly tunable combustion characteristics dependent on the fuel and oxidizer materials and particle size. The ignition and combustion characteristics of binary nanothermite materials with an of fuel 80 nm spherical aluminum and an oxidizers of Bi_2O_3, WO_3, CuO, MoO_3, or Fe_2O_3 have been well characterized over the past decade. This work presents a method for the formation of nanothermite composite materials with highly tunable ignition and combustion characteristics. This contribution focused on the reactivity of mixed nanothermites such as Al-Fe_2O_3, Al-Bi_2O_3, and Al-Fe_2O_3-Bi_2O_3. The ignition and combustion characteristics of the nanothermite composite materials were tuned by: 1) the addition of gasifying agents and 2) composition of the nanothermite composite. The nanothermite composite materials were mixed together using a novel mixing technique involving the utilization of a Resodyn LabRAM mixer will be demonstrated. The effect of slurry concentrations, quantity of binder additives of formability and strength of dried nanothermite composites will be discussed.

INTRODUCTION

Research into nanothermite materials over the past decade has focused on binary reactive systems based on 80 nm spherical aluminum made by NovaCentrix and various submicron and nano-scale oxidizers to tailor the combustion characteristics of the material [1-8]. These energetic materials have been call by many names including metastable interstial composite (MIC), superthermites or nanothermites. While research up to this point in this area has focused on the using nanoscale spherical aluminum as the fuel, in 2009 Innovative Materials and Processes, LLC (IMP) in conjunction with South Dakota School of Mines and Technology (SDSM&T) developed a method for the formation of nanoscale flake aluminum having the same specific surface area and reactive aluminum content as 80 nm spherical aluminum.

This paper presents a new approach for the formation of nanoscale aluminum-metal oxide nanothermite composite materials integrated with nitrocellulose to tune the combustion characteristics of the nanothermite composite material. This work focuses on the use of Fe_2O_3 and Bi_2O_3 as oxidizer materials due to their very different combustion characteristics when combined with nanoscale aluminum and their low solubility in water.

EXPERIMENTAL

Materials

The following materials were used for the formation of the nanothermite composite materials:

- 16 m2/g flake aluminum (Innovative Materials and Processes, LLC)
- 20-60 nm Fe_2O_3 (Nanostructured and Amorphous Materials)

- Submicron Bi_2O_3 (Clark)
- Arabic Gum (Acros)
- Ammonium Dihydrogen Phosphate (ADP)(Acros)
- Distilled H_2O
- Dry nitrocellulose fibers (ATK)

Mixing of Nanothermites

The reactants for the different nanothermite composite materials were mixed in water using a process developed by IMP [9]. The reactants were mixed using a Resodyn LabRAM mixer instead of a using traditional ultrasonic mixing. The Resodyn LabRAM mixer allows for the nanothermite-nitrocellulose composite to be formed using a two-step mixing process. The first step is the dispersion of the nitrocellulose fibers in the water followed by the mixing of the fuel and oxidizer(s) with the nitrocellulose fibers to form the energetic slurry. The entire mixing process for the formation of the energetic slurry takes 20 minutes.

Nanothermite Granule Formation

The nanothermite-nitrocellulose composite materials were formed into granules by dispensing the energetic slurry onto conductive Teflon sheets (0.01" thick purchased from CSHyde) using a pipette. The conductive Teflon sheets were supported on 0.125" thick aluminum. Once the energetic slurry had been dispensed the conductive trays were placed in a drying system and the entire system was grounded. After drying in room temperature air (20-23°C) for 3 hours the granules were stored in a desiccator for a minimum of 12 hours before being tested.

Dynamic Pressure Characterization

Currently there is no reliable quantitative method for the determination of how the fuel and oxidizer particles have been mixed within the nanothermite material. To determine how well the reactants were mixed together dynamic combustion pressure measurements of the nanothermite-nitrocellulose composite materials were conducted. The dynamic combustion pressure measurements were determined using an 8 mL closed combustion vessel. The dynamic combustion pressure was measured using a piezoelectric pressure transducer (Model 617C purchased from Kistler Co.). A schematic of the closed combustion vessel is shown in Figure 1. The 8 mL combustion vessel allows for the dynamic combustion pressure to be measured either perpendicular to on in-line with the combustion front. For all of the results reported in this paper the dynamic combustion pressure was measured in-line to the combustion front.

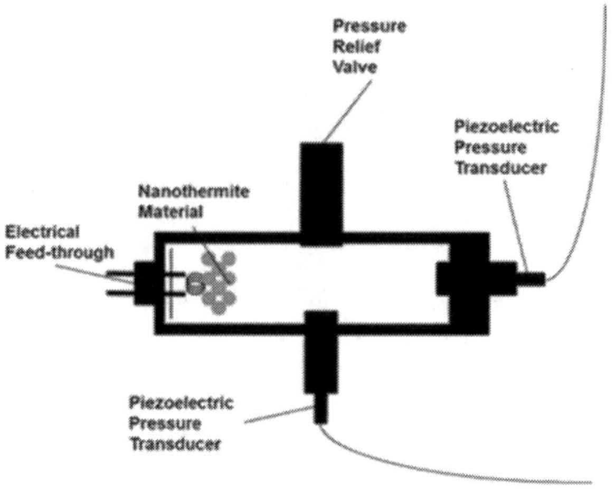

Figure 1: *Schematic of 8 mL closed pressure combustion vessel.*

RESULTS

Nanothermite Composite Granules

The formation of the nanothermite-nitrocellulose composite materials into granules was conducted to reduce the sensitivity of the materials to unwanted ignition specifically electrostatic discharge. A photograph of a selected granule ($2Al-Fe_2O_3$ with 5 wt% nitrocellulose) is shown in Figure 2. The use of nanoscale flake aluminum and the addition of the nitrocellulose to the nanothermite composite materials has shown a qualitative increase in the structural integrity of the nanothermite granules when compared to 80 nm spherical aluminum and no nitrocellulose.

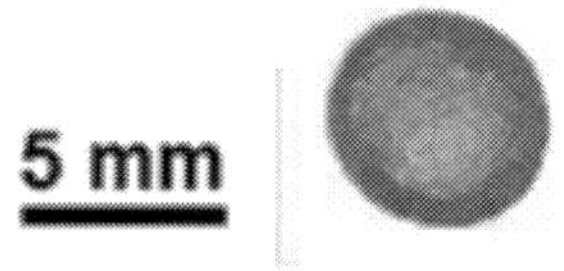

Figure 2: *Photograph of $2Al-Fe_2O_3$ with 5 wt% nitrocellulose nanothermite-nitrocellulose composite granule.*

Dynamic Combustion Pressure

It has been previously reported that the particle morphology of the nanoscale aluminum does not have an effect on the dynamic combustion characteristics of the nanothermite materials [10]. The combustion characteristics of the nanothermite-nitrocellulose composite materials can be tuned by adjusting the ration of the oxidizer materials and/or the mount of nitrocellulose in the energetic composite. Dynamic combustion pressure results showing the effect of that these two parameters have on the combustion characteristics of the nanothermite-nitrocellulose composite materials are shown in Figure 3 [10].

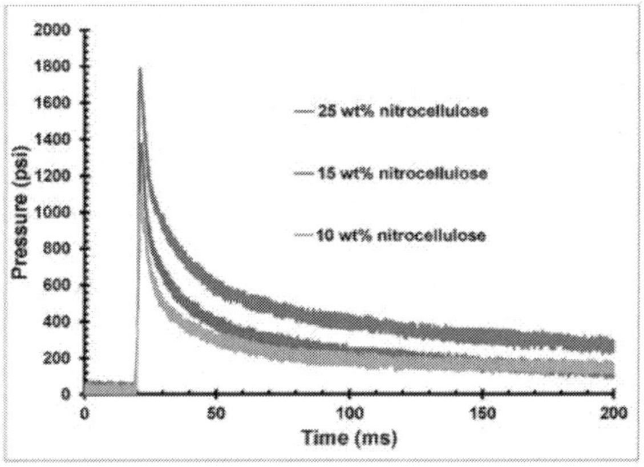

a

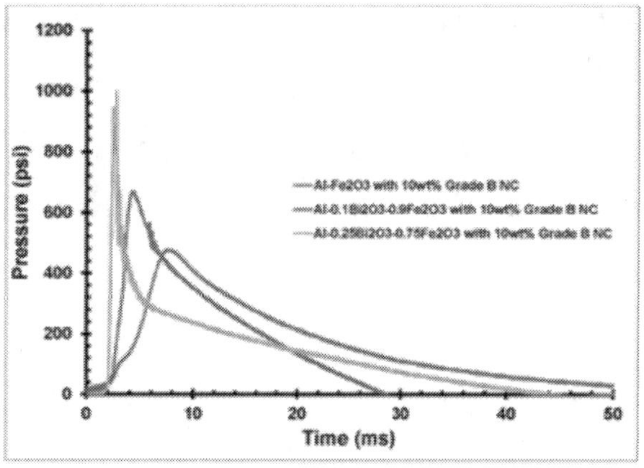

b

Figure 3: *Dynamic combustion pressure profiles for the combustion of 300 mg of nanothermite material in an 8 mL closed pressure cell a) Al-Fe_2O_3 nanothermite with 10, 15 and 25 wt% nitrocellulose, b) nanoscale flake aluminum based nanothermite composite with varying molar ratios of Bi_2O_3 and Fe_2O_3 and 10 wt% nitrocellulose [10].*

CONCLUSIONS

The dynamic combustion results presented indicate that there exists a reliable method for tuning the combustion characteristics of nanothermite materials. The use of the flake aluminum and addition of the nitrocellulose fibers to the nanothermite composite granules increases the structural integrity of the energetic composite granules. The addition of the nitrocellulose can also be used to increase the dynamic combustion pressure of the nanothermite material.

ACKNOWLEDGEMENTS

Funding for this project was provided by the Armament Research, Development and Engineering Center at Picatinny Arsenal, NJ under contract W15QKN-09-C-0154, W15QKN-11-C-0044 and W15QKN-06-D-0006

REFERENCES

1. G.A. Risha, S.F. Son, R.A. Yetter, V. Yang and B.C. Tappan, Proceedings of the Combustion Institute 31, 2029–2036, (2007).
2. L. Meda, G. Marra, L. Galfetti, F. Severini and L. De Luca, Materials Science and Engineering: CVol. 27, (5-8), 1393-1396, (2007).
3. D. G. Piercey and T. M. Klapotke, *Central* European Journal of Energetic Materials, 7(2), 115-129, (2010).
4. N. Muravyev, Yu. Frolov, A. Pivkina, K. Monogarov, D. Ivanov, D. Meerov and O. Ordzhnikidze, Proc. of 36 Inter. Pyro. Seminar, Rotterdam, NL, 43-55, (2010).
5. S.F. Son, R.A. Yetter, Y. Yang, J. Propulsion and Power, *23*(4), 643-644, (2007).
6. K.C. Walter, D.R. Pesiri, D.E. Wilson, J. Propulsion and Power, *23*(4), 645-650, (2007).
7. J.A. Puszynski, C.J. Bulian, J.J. Swiatkiewicz, J. Propulsion and Power, *23*(4), 698-706, (2007).
8. M.L. Pantoya and J.J. Granier, Propellants Explos. Pyrotech. *30*(1), 53-62, (2005).

9. Puszynski, J.A., Bichay, M.M., Swiatkiewicz, J.J., United States Patent 7,670,446 B2, March 2010
10. Doorenbos, Z., Walters, I., Redner, P., Kapoor, D., Balas-Hummers, W., Swiatkiewicz, J.J., Puszynski, J.A., Proc. Of 17th APS SCCM Conference, (Pending), (2011)

(663d) Mechanically Induced Ignition of Thermite Nanocomposite Powders

Rayon Williams, Department of Chemical, Biological and Pharmaceutical Engineering, New Jersey Institute of Technology, Newark, NJ and Edward L. Dreizin, Department of Chemical, Biological, & Pharmaceutical Engineering, New Jersey Institute of Technology, Newark, NJ

Mechanically Induced Ignition ofThermite Nanocomposite Powders

R. A. Williams[1] *andE. L. Dreizin*[1]

[1]*Department of Chemical, Biological, andPharmaceutical Engineering, New Jersey Institute of Technology, Newark, New Jersey 07102, USA*

Thispaper examines the mechanically induced ignition of nanocomposite thermitepowders prepared by Arrested Reactive Milling. Powders are placed on a glass substrate as a thin layer and covered byan aluminum foil. Another powder, typically a nanocomposite thermite with gas-generating oxidizer, such as 2.35Al $+Bi_2O_3$is placed on top of the aluminum foil and is ignited by an electrostaticdischarge (ESD). As a result of itsignition, a pressure wave is generated that compresses the powder placedbeneath the foil and causes its ignition. The experimental diagnostics include optical sensors, high-speed video,in addition to real time monitoring of the ESD current. In separate reference experiments, a pressuresensor is placed between the foil and glass, instead of the powdersbeing ignited. Ignition delays andpressure pulse delays are correlated with each other. Effects of the powder layer thickness, typeof powder used, and discharge settings (voltage, capacitance) are alsoexplored.

See more of this Session: Nanoenergetic Materials
See more of this Group/Topical: Particle Technology Forum

(666a) Particle Engineering Based Framework for Cohesion Reduction of Fine Pharmaceutical Powders Via Dry Coating

Rajesh N. Dave, Chemical Engineering, New Jersey Institute of Technology, Newark, NJ

Flow and handling of fine powders is a topic of great interest to pharmaceutical industry. High cohesion observed in fine pharmaceutical powders lead to problems such as, agglomeration, poor flowability, electrostatic charging and low bulk density. These also lead to downstream problems such as poor content uniformity, marginal improvement in dissolution rate of fine API powders, and need for extra processing steps such as granulation. Some of these flow problems may be mitigated or kept under control via use of flow-aids and glidants. However, dry coating based surface modification is a more rigorous approach due to its ability to achieve precise surface coating of nano-silica and other flow-aides; making it very popular in recent academic research.

In this presentation, we provide results using several devices, both based on batch and scalable continuous operation, for pharmaceutically relevant powders where the surface coating includes not only silica but also several types of "guest" materials. Dry coating devices are categorized in two approaches; the first approach is suitable for powders in 150 down to 10 micron sizes where attrition of powder size is minimal. In the second approach, suitable for even finer particles, including those used in the inhaler application having aerodynamic diameters in the size range of 2 ~ 5 microns, simultaneous micronization and surface modification is considered. Results of the bulk properties for these powders are presented to illustrate the improvement in flow, fluidization, dispersion, lack of agglomeration, bulk density, and electrostatic tendency, all leading to potentially significant cost benefits in industrial processing, handling, storage and transportation. Finally, we propose a 2-D phase-map based on key bulk powder properties to help make manufacturing decisions regarding the formulation strategy for solid pharmaceutical dosages.

See more of this Session: Particle Engineering as Applied to Pharmaceutical Formulations I

(666b) Investigation of Tablet Coating Processes Using Discrete Element Method Simulations

Gregor Toschkoff[1], **Daniele Suzzi**[1], Sarah Just[2], Klaus Knop[2], Peter Kleinebudde[2] and Johannes G. Khinast[3], (1)Research Center Pharmaceutical Engineering (RCPE), Graz, Austria, (2)Institute of Pharmaceutics and Biopharmaceutics, Heinrich Heine University, Dsseldorf, Germany, (3)Institute for Process and Particle Engineering, Graz University of Technology, Graz, Austria

Investigationof tablet coating processes using Discrete Element Method simulations

G. Toschkoff[1], D. Suzzi[1],S. Just[2], K. Knop[2], P. Kleinebudde[2], J. G.Khinast[1,3]

[1] *ResearchCenter Pharmaceutical Engineering GmbH, Graz, Austria*

[2] *Instituteof Pharmaceutics and Biopharmaceutics, Heinrich Heine University, Dsseldorf,Germany*

[3] *Institutefor Process and Particle Engineering, Graz University of Technology, Graz,Austria*

Inthe pharmaceutical industry, drum coating is a widely used unit operation toproduce tablet films of different purposes. In this process, a rotating drumaccounts for the necessary mixing of the tablets, and a coating solution isinjected from above by means of an atomizing nozzle. The appliedcoating layer(s) fulfill different functions, e.g. taste masking and coloring,control of the release of the active pharmaceutical ingredient (API) from thecore of a tablet, application of an additional API, or protection of the tabletcore from environmental influences.

Forall aspects of the coating mentioned above, the uniformity of coating is of greatimportance. This includes both, inter-tablet uniformity and intra-tabletuniformity. In fact, as inhomogeneity in the coating thickness can lead to serioushazards (e.g., significant variations in APIs delivery rate), a single tabletthat fails testing can lead to the rejection of the whole batch.

Althoughdrum coating is a widespread technology in the pharmaceutical industry, processdesign is more often than not based on trial-and-error practices and operatorexperience. In recent years, parallel to an increased effort in experimentalwork [1], numerical simulations of particle motion using the Discrete ElementsMethod (DEM) have proven to be an important tool in the detailed investigationof the tablet coating process, as well as in other particle-basedpharmaceutical processes [2,3].

***Figure 1**: ExamplesofDEM simulations in two tablet coating apparatus that were used in this work.Left: Driam Driaconti cycled continuous coater, right: Bohle BFC5 lab-scalecoater.*

The aim of this work is toanalyze and understand the effects of parameters like tablet form, fill volume orpan rotation speed on the intra-tablet coating variability [3] in differentcoating devices. To this end, DEM simulations are performed to numericallyreproduce the tablet motion inside different coating machines. Materialparameters needed for the simulation are gathered from measurements. For eachgeometry, different tablet shapes, namely bi-convex, oval and/or round, aremodeled by the "glued spheres" approach. Further parameter variations includedifferent fill volumes or different rotational speeds.

Important process attributesthat one wants to know are, e.g., residence time of the tablets under thecoating spray, intra-tablet coating variability, or tablets velocities pattern.While many of these quantities are fastidious or even impossible to get byexperimentation, they are readily extracted from the DEM simulations results.

In the following, twoexamples of the acquired information are given. Figure 2 shows a matrix of themean velocity of the tablets, for different fill levels and different tabletshapes, allowing a direct qualitative and quantitative comparison of thedifferent setups.

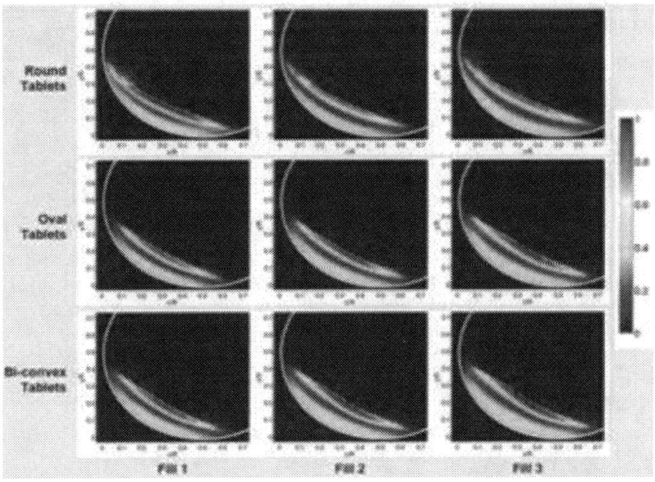

***Figure 2**: Normalizedtime-averaged tablet velocity on the grid for round, oval and bi-convex tabletsat the different coater fill ratios for a vertical slice in the middle of thecoating apparatus. From [4]*

Figure 3 shows the fractional residence time,i.e., the ratio between the time spent by a tablet in the spray zone located atthe top of the surface and the total coating time, for the Driam continuouscoater for a simulation time of 60 seconds. It can be seen that both tabletshape (quantified as length-to-height ratio) and fill level have an influenceon the time that a single tablet spends exposed to the spray. From thisinformation, an expected coating variability and in the end coating processtime can be estimated.

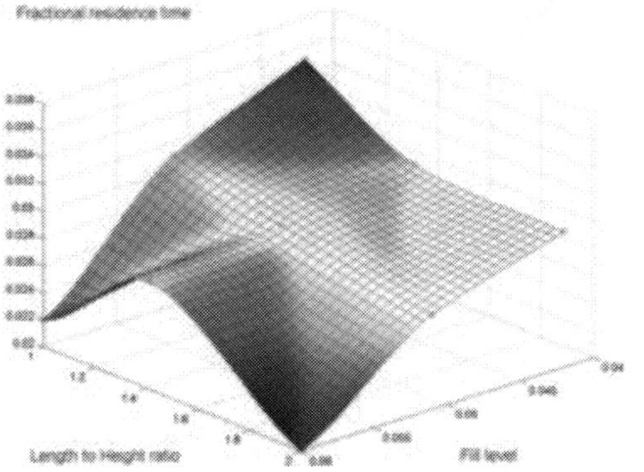

***Figure 3** Averagefractional RT in the spray zone for round, oval and bi-convex tablets at thecoater fill ratios.*

Conclusion

The DEMsimulation has proven to be a valuable tool to gain understanding the dynamicalbehavior of the tablets. The gathered information is essential to e.g. obtain asatisfactory intra-tablet coating homogeneity, which in turn is necessary to minimizethe number of tablet batches that have to be rejected. The outcomes of thiswork aims at demonstrating the utility of numerical simulation in thedevelopment and the design of pharmaceutical tablet coating processes.

References

1. Tobiska,S., Kleinebudde, P., **2003**. *Coating uniformity and coating efficiencyin a Bohle Lab-Coater using oval tablets*. European Journal of Pharmaceuticsand Biopharmaceutics 56, 3-9.

2. Adam,S., Suzzi, D., Radeke, C., Khinast, J.G., **2010**. *An integrated Qualityby Design (QbD) approach towards design space definition of a blending unitoperation by Discrete Element Method (DEM) simulation*. European Journal ofPharmaceutical Sciences, In Press.

3. Ketterhagen,W. R.; am Ende, M. T. & Hancock, B. C., **2009,** *Process modeling inthe pharmaceutical industry using the discrete element method,* Journal ofPharmaceutical Sciences, , *98*, 442-470

4. Suzzi,D.; Toschkoff, G; Radl, S.; Machold, D.; Fraser, S. D.; Glasser B. J.; Khinast,J. G, **2011**, *DEM Simulation of Continuous Tablet Coating: Effects ofTablet Shape and Fill Level on Inter-Tablet Coating Variability*. Submittedto Chemical Engineering Science.

See more of this Session: Particle Engineering as Applied to Pharmaceutical Formulations I
See more of this Group/Topical: Food, Pharmaceutical & Bioengineering Division

(666c) Triboelectric Behavior of Mannitol Powder with Respect to Important Influencing Factors

Stefan Karner, Institute for process and particle engineering, TU Graz, Graz, Austria and Nora Anne Urbanetz, TU Graz, TU Graz, Graz, Austria

Triboelectric behavior of mannitol powder with respect to important influencing factors

Stefan Karner, Nora Anne Urbanetz

Institute for Process and Particle Engineering, Graz University of Technology/ Austria

Since most pharmaceutical powders and especially inhalation powders consist of small particles of insulating material electrostatic charge arises when contact of the particles among themselves or with other surfaces happens in powder handling processes. This triboelectric charge carried by the powder can cause problems like material loss, poor flowability and dosing behavior. The phenomenon of the emergence of charge and its impact on the characteristics of the powder are widely uninvestigated. Several scientists such as Murtomaa et al.[2], Zhu et al.[5], Matsusaka et al.[1], Telko et al.[4] and Rowley et al.[3] examined the emergence of charge in common powder handling processes. The above mentioned problem of uniform dosing is particularly of high interest in dry powder inhalation manufacturing. Therefore the aim of this work is to investigate the triboelectric behavior of the new alternative dry powder inhaler (DPI) carrier material mannitol.

Spray dried mannitol was donated by Roquette Frères (Lestrem, France). It shows the characteristic diameters of d_{10}=80 μm, d_{50}=149 μm and d_{90}=209 μm. Charging experiments were carried out by blending the powder in a tumble blender (T2F Turbula®, Willy A. Bachofen AG - Maschinenfabrik, Switzerland) and measuring the charge in a Faraday cup connected to a high resistance electrometer (Keithley 6517 B, Keithley Instruments, USA). First the dependence of the arising charge on the mixing time was determined. Charge saturation was found after about 20 minutes mixing time. Sign and magnitude of this saturation charge can be influenced by varying the mixing process parameters. Therefore subsequent studies focused on the influences of the particle size, the mixing container size and the fines fraction present in the blend.

In contrast to other studies this work uses statistical tools like design of experiments to investigate the significance of the effects and the interactions of more than one influencing factors. The experimental data revealed strong influences of all investigated factors and also some interesting interactions between them.

In addition to the charging investigations also some discharging tests were performed. This was done by using design of experiments as well to check several influencing factors.

The obtained data are highly useful to improve the understanding of the emergence and impact of electrostatic charge on pharmaceutical powders, especially inhalation powders, and will help to enhance the performance of DPIs.

References

(1) Matsusaka, S.; Masuda, H. Electrostatics of particles. Advanced Powder Technology. 2003;14:143-166

(2) Murtomaa, M.; Savolainen, M.; Christiansen, L.; Rantanen, J.; Laine, E. & Yliruusi, J. Static electrification of powders during spray drying. Journal of Electrostatics. 2004;62:63-72

(3) Rowley, G. Quantifying electrostatic interactions in pharmaceutical solid systems. International Journal of Pharmaceutics. 2001; 227:47-55

(4) Telko, M. J.; Kujanpää, J.; Hickey, A. J. Investigation of triboelectric charging in dry powder inhalers using electrical low pressure impactor (ELPI(TM)). International Journal of Pharmaceutics. 2007;336:352-360

(5) Zhu, K.; Tan, R.B.H.; Chen, F.; Ong, K.H; Heng, P.W. Influence of particle wall adhesion on particle electrification in mixers. International Journal of Pharmaceutics. 2007;328:22-34

See more of this Session: Particle Engineering as Applied to Pharmaceutical Formulations I
See more of this Group/Topical: Food, Pharmaceutical & Bioengineering Division

(666d) Analysis of Non-Linear Tensile Strength Variation In Binary Mixtures Using Multi-Particle Finite Element Method (MPFEM)

Tuhin Sinha, School of Mechanical Engineering, Purdue University, West Lafayette, IN, Jennifer Sinclair Curtis, University of Florida, Gainesville, FL, Bruno C. Hancock, Pfizer Inc., Groton, CT, James D. Litster, Chemical Engineering, Purdue University, West Lafayette, IN and Carl R. Wassgren, Mechanical Engineering, Purdue University, West Lafayette, IN

Pharmaceutical tablets are comprised of a mixture of powders with varying properties. To predict the compaction performance of a drug formulation, it is essential to understand the interaction between the constituents of a mixture that will affect the bulk mechanical properties of the resulting mixture. It has been observed experimentally that mechanical properties of a mixture vary non-linearly with the mass fraction of the constituents, and this non-linear interaction could be either positive or negative. Hence, it is difficult to predict the bulk mechanical properties of mixtures, e.g., tensile strength and Young's Modulus, using simple linear mixing rules. Previous researchers have tried both experiments and analytical models to comprehend the causes of this observed non-linearity but have been unable to provide a proper explanation of it.

The multi-particle finite element method (MPFEM) is used to study the micromechanics of particle interactions and their effect on the macroscopic mechanical properties of binary mixtures. In MPFEM, the particles are modeled in the FEM framework with a given set of material and contact properties. The particles can deform based on the applied boundary conditions and contact interactions. Although the current MPFEM models cannot quantitatively predict the bulk mechanical properties of a mixture,they can give a reasonable explanation behind the experimentally observed non-linear behavior. This talk will focus on analyzing the variation of tensile strength in binary mixtures. Simulations will also be compared to experimental results.

See more of this Session: Particle Engineering as Applied to Pharmaceutical Formulations I

See more of this Group/Topical: Food, Pharmaceutical & Bioengineering Division

(666e) Formulation and Manufacture of Pharmaceuticals by Fluidized Bed Impregnation of APIs Onto Porous Carriers

Plamen I. Grigorov, Benjamin J. Glasser and Fernando J. Muzzio, Chemical and Biochemical Engineering, Rutgers, the State University of New Jersey, Piscataway, NJ

Approximately65% of all prescription drugs are manufactured as solid dosage forms, whichinclude tablets and capsules. In most of these cases the final formulation is aphysical blend consisting of an excipient and an active pharmaceuticalingredient (API), which preferably is homogeneously distributed throughout theexcipient. Mixing of powders almost always poses great challenges toformulators, especially in cases where there is a pronounced difference betweenthe powders' particle size, flowability, surface properties, etc. Thesechallenges often translate into un-homogeneously distributed API. For verypotent drugs the amount of API in the solid dosage form can be as low as 0.1%by weight. This very low API loading poses one of the biggest problems inpharmaceutical product development: the control of blend uniformity. Low API contentvariability in the blend or high blend homogeneity are highly desired andstrictly enforced by the U.S. Food and Drug Administration (FDA). In thecommonly available approaches for blend uniformity control (direct blending,wet/dry granulation) as the API concentration decreases, the variability of theblend increases; this makes it very difficult to meet FDA's requirements forlow drug loadings. Therefore a process or method that is able to tightlycontrol API variability in blends, regardless of drug loadings or physicalproperties of the API, has become very desirable.

In this paper wedescribe a novel method for the formulation and manufacture of pharmaceuticalsthat circumvents all difficulties associated with powder mixing and all negativeconsequences associated with such blends. The method involves fluidized bed(FB) impregnation of APIs onto porous carriers. The main idea behind thismethod is the deposition of the drug material inside the excipient particles.This impregnation is achieved by spraying an API solution onto porous excipientin a fluidized bed, achieving what could be seen as continuous wet impregnationand drying process. Overall quality of resulting impregnated blend depends onprocess parameters such as product drying temperature and spray rate of APIsolution.

Impregnated excipients exhibit one main difference when compared toregular blends. That is, the API is located inside carrier particles whichresults in powder properties that are almost identical to those of the pure excipient, thus eliminating API/excipientsegregation. Combining all this with the facts that FBs exhibit very goodpowder mixing and that each excipient particle is impregnated "one at atime", results in highly homogeneous API distribution. We also discoveredthat consequent milling of the impregnated excipientfurther improves blend uniformity. Included table below shows blend uniformity(as % RSD) for impregnated blends (milled and um-milled) of various loadingsusing acetaminophen (APAP) as the active pharmaceutical ingredient.

Other advantagesof proposed formulation method include: simple and efficient (eliminatesseveral unit operations from drug substance and drug product development), safe(minimizes API powder handling), depends only on excipient'sproperties (independent of API nature).

% APAP Loading	**Blend Uniformity, % RSD**
8.82 % (un-milled)	1.05
8.82 % (milled)	0.54
1.02 % (un-milled)	0.70
1.02 % (milled)	0.56
0.1 % (un-milled)	0.99
0.1 % (milled)	0.42

See more of this Session: Particle Engineering as Applied to Pharmaceutical Formulations I
See more of this Group/Topical: Food, Pharmaceutical & Bioengineering Division

(666f) Attrition and Abrasion Resistance of Particles Coated with Pre-Mixed Polymer Coating Systems

Giacomo Perfetti, Chemical Engineering, NanoStructured Materials, Delft University of Technology, Delft, Netherlands, Frederic Depypere, Laboratory of Food Technology and Engineering, Ghent University, Faculty of Bioscience Engineering, Ghent, Belgium, Sheida Zafari, Nanostructured Materials Group, Faculty of Applied Sciences, Delft University of Technology, Delft, Netherlands and Gabrie Meesters, DSM, NA, Netherlands

Nowadays many particulate products are coated to avoid dust formation and breakage when subjected to forces with different magnitude and direction. In order to distinguish the influence of each force, different testers are needed. In this work Purox-S and Cellets 1000 core particles were coated in a top spray fluid bed coater by three different pre-mixed polymer coating systems (Opadry®, Kollicoat®IR and Instacoat AQUA). The repeated impact tester (RIT) was used to measure their strength towards attrition caused by low magnitude normal forces. Tangential forces may also cause rupture to the particles, so an abrasion tester was used to measure this amount of damage caused to the particles. This work provides information on the resistance of the core particles coated with different mixed polymers against attrition and abrasion.

See more of this Session: Particle Engineering as Applied to Pharmaceutical Formulations I
See more of this Group/Topical: Food, Pharmaceutical & Bioengineering Division

(667a) A Population Balance Model Based Distribution Shaping Control for Tailored Dissolution of Crystalline Products

Zoltan K. Nagy, Chemical Engineering Department, Loughborough University, Loughborough, United Kingdom and Nahid Sanzida, Chemical Engineering, Loughborough University, Loughborough, United Kingdom

A novel methodology for the dynamic optimization of the crystal size distribution (CSD) considering growth, nucleation and dissolution mechanisms for batch cooling crystallization processes is presented. The model based distribution shaping control can be used to design the crystal size distribution of solid products to achieve a desired dissolution profile. By simultaneously optimizing the seed recipe and the supersaturation profile for the crystallization process a tailored dissolution can be achieved. Simulation and experimental studies will be presented to illustrate the potential benefits of the proposed product engineering approach. The population balance model which incorporates both the dissolution and growth mechanism is developed and is solved using a combined quadrature method of moments and method of characteristics approach, which provides a computationally efficient solution of the model, suitable for optimization. The optimization problem is solved to design a fast and a slow dissolution profile, and the operating conditions are evaluated experimentally. The experimental setup includes process analytical technology probes for the monitoring of the crystallization process, including FBRM and ATR-UV/Vis. The approach can have significant potential for application in tailored drug delivery, where the release rate of an active ingredient can be designed by optimizing the dissolution rate. The same approach can have application in case of pesticides for more efficient and environmentally friendly application. Tailoring the particle size distribution of the crystallization product, so that an optimal dissolution rate (heat release rate) is achieved for the case of exothermic reactive processes for which the dissolution is followed by reactions, can improve the safe operation of these systems. Similarly the same strategy can also be applied for reactions, which produce a significant amount of gasses and induce excessive liquid rise, which can lead to the explosion of reactors (two-phase flow can occur in the connection pipes and pressure can build-up).

CONTINUOUS CRYSTALLIZATION FOR DRUG NANOSUSPENSIONS

T. Panagiotou, R. J. Fisher[a], R. Bruno[b]

Microfluidics International, Newton, Massachusetts, USA
(mpanagiotou@idexcorp.com.com)
[a] Dept. of Chemical Engineering, Massachusetts Institute of Technology, Cambridge, MA USA
[b] RP Bruno and Associates, Newton, MA, USA

Keywords: nanoparticles, crystallization, bioavailability.

Presented here is a novel technology (PureNano™) for continuous crystallization and large scale production of drug mirco- or nano- particles with tailored properties, in a highly efficient matter. The technology allows for turbulent mixing of crystallization precursors (solvent, antisolvent, seeds, etc.), the formation of turbulence eddies in the range of 25-50nm, and therefore high spatial uniformity of supersaturation. In addition, PureNano™ allows for precise control of mixing times and intensity, and temperature history. Therefore it allows for precise control of the crystallization process. The energy for crystallization required by this method may be orders of magnitude lower than the energy demands of conventional methods for creating micro- and nano- suspensions, which involve breaking large crystals to smaller entities. In addition, crystallization times may be greatly reduced through minimizing of diffusion limitations.

Background. Estimates suggest that each year roughly half of all newly formulated drugs having potentially high pharmacological value will have little chance to make it beyond the laboratory and into marketplace. Drug formulations exist basically as either emulsions or suspensions, the latter consisting of non- water soluble (hydrophobic) active pharmaceutical ingredients (APIs) suspended in a liquid. Drug suspensions are of great interest because of their high bioavailability which significantly reduces the amount of drug that needs to be delivered as compared to drug emulsions. However, formulation difficulties may still exist since the particle size of the solid API plays a crucial role in 1) the stability of the suspension and 2) the safe delivery of the drug.

Certain crystallization processing technologies claiming to produce stable nano-particles are mainly experimental and may have major shortcomings. As an example, low pressure impinging jet technology[1] is characterized by low flow rates, low jet velocities, and therefore low energy. These jet velocities are over two orders of magnitude lower than velocities used in PureNano™. Low pressures and low energy levels are associated with large particles and inability of the technology to produce stable particles and handle high solid loadings without plugging or fouling. Another example is use of supercritical carbon dioxide spray processing technology. This technology has been proven to be impractical for manufacturing because it is not scalable and is highly energy intensive.
Recent breakthroughs in continuous crystallization technology now enable pharmaceutical and biotechnology companies to overcome issues related to stability and

safe drug delivery. Microfluidics International Corporation is marketing this technology as their PureNano™ Continuous Crystallizer and associated process protocols. Consequently such systems take on an important role in the formulation and the continuous manufacturing of high purity stable nano-suspensions which is not always achievable with conventional particle size reduction methods.The PureNano™ Crystallizer, which incorporates a scalable mixing chamber, is an energy-efficient method. A case study is presented later that demonstrates the success of this system in formulating a nano-suspension cancer drug that conventional particle size reduction methods were unable to achieve.

Principal of Operation/Applications. Applications best suited for the PureNano™ Crystallizer are dependent upon the length of time for crystallization to occur. These "resident" times are critical to the crystallization process and are dependent upon the API involved, which may require residence times that range from a few hundred microseconds to several hundred milliseconds. To accommodate this range of times, multiple system configurations are available.

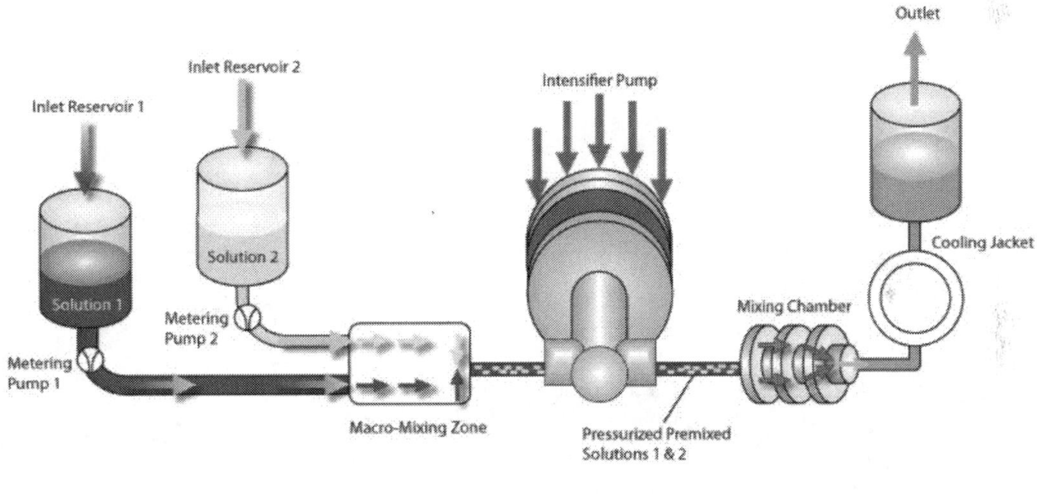

Figure 1. The PureNano Crystallizer configured with coaxial feed

To illustrate the principal of operation we selected the coaxial feed system as depicted in Fig 1. The API is first dissolved in a predetermined solvent, Solution 1, and is placed in one of the two inlet reservoirs. Solution 2, the "anti-solvent," which in many cases may be water, is placed in the second reservoir as shown. This feed configuration allows controlled pre-mixing of the two streams within a "macro-mixing" zone prior to entering the mixing chamber. This pre- mixing occurs for a predetermined period of time within the macro-mixing zone, usually in the order of several milliseconds, creating a small number of product nuclei for "seeding". The feed rates, solution ratios and mixing intensities within the coaxial feed are precisely controlled with metering pumps. This pre-mixed solution is then pressurized and subsequently enters the mixing chamber as a single stream which is internally split into two jets that collide. This forces interact at the nano-scale resulting in a continuous output flow of a stable high purity nano-suspension.

As previously reported by Microfluidics, the PureNano™ Crystallizer has been successful in producing stable nano-suspensions in the laboratory[2,3] where conventional particle size reduction methods were proven to be unsuccessful. In all PureNano™ Crystallizer configurations, post processing may be necessary to prevent crystal growth or to alter crystal habit and/or morphology.

Case Study. Presented here are preliminary results from an ongoing collaborative effort with a well known pharmaceutical company in an effort to successfully formulate a cancer drug identified here as Compound V. To increase bioavailability, it was determined that the drug needed to be delivered intravenously in the form of a nano-dispersion. .

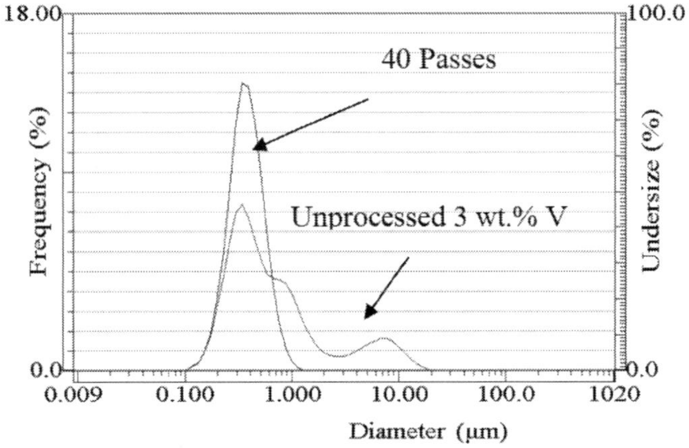

Passes	Particle Size (microns)			
	d10	d50	d95	Mean
Unprocessed	0.220	0.470	7.283	1.401
40	0.227	0.379	0.734	0.410

Figure 2. Results with conventional particle size reduction method

Microfluidics conventional "top down" particle size reduction technology and PureNano were used for the production of the nano-suspension. The required particle size was a median size less than 200 nanometers. Using the top down method, a suspension of the drug in water was prepared and processed using a Model 110EH-30 Microfluidizer processor. After 40 passes through the processor, the mean particle size of the API was reduced from 1.401 microns to 401 nanometers with no further particle size reduction resulting with additional passes (Fig. 2). The resulting suspension was unacceptable; it did not meet the stability requirements.

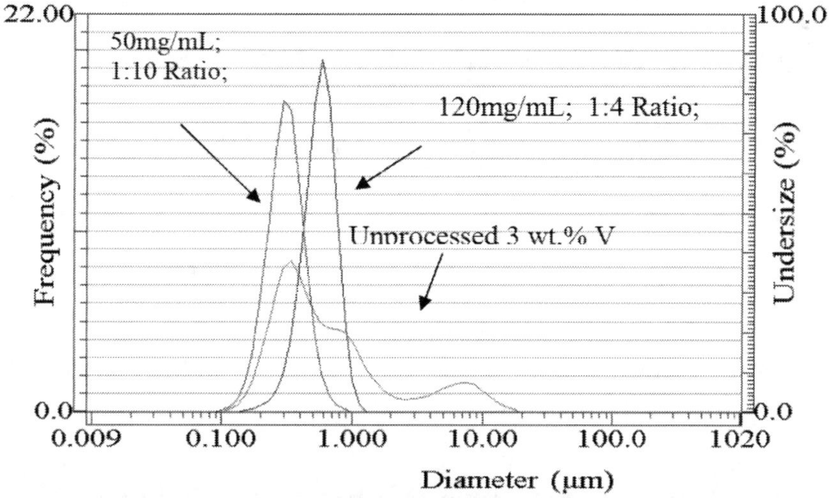

Conc. (mg/mL)	Ratio (Sol:Anti)	Particle Size (microns)			
		d10	d50	d95	Mean
20mg/mL	1:10	0.109	0.177	0.329	0.257
50mg/mL	1:10	0.058	0.102	0.277	0.126
120mg/mL	1:10	0.350	0.529	0.812	0.539
120mg/mL	1:4	0.188	0.287	0.489	0.302

Figure 3. Results with the PureNano Crystallizer

Using a PureNano^{TM} Crystallizer in conjunction with our nano-suspensions development protocol, a continuous crystallization process was developed. Compound V was dissolved in a solvent, polyethylene glycol (PEG), at various concentrations (Fig. 3), in the range of 20-120mg/ml. This stream was mixed with the antisolvent stream (water) at the appropriate proportions using a PureNano^{TM} Crystallizer processor. The ratio of the antisolvent to solvent streams varied from 1:4 to 1:10 producing various degrees of supersaturation. A stable nano-suspension with a median particle size of 102 nanometers was achieved in a single pass. The particle size distribution was well within specifications.

References:

[1] Johnson, B.; Prud'homme, R. Chemical Processing and Micromixing in Confined Impinging Jets. *AICHE J.* **2003**, 49, 2264-2282.

[2] Panagiotou, T.; Fisher, R. Form Nanoparticles via Controlled Crystallization. *CEP*, 2008, 33-39C.

[3] Panagiotou, T.; Mesite, S.; Fisher, R. Production of Norfloxacin Nanosuspensions Using Microfluidics Reaction Technology through Solvent/Antisolvent Crystallization *Ind. Eng. Chem. Res.*, **2009**, 48 (4), 1761-1771.

(667c) High Pressure Homogenization As Particle Size Reduction Technology In the Pharmaceutical Industry: Optimization, Modeling and Isolation Techniques

Plamen Grigorov[1], Christopher P. Long[2] and Azzeddine Lekhal[1], (1)Chemical Process Development & Commercialization, Merck and Co. Inc., Rahway, NJ, (2) College of Engineering, Cornell University, Ithaca, NY

High pressurehomogenization (HPH) is a technology widely used in various industries. In thepharmaceutical industry HPH is used in broad range of applications such as cellrupture, dispersions, emulsions and particle size reduction. Many of today'sactive pharmaceutical ingredients (APIs) exhibit low solubility andpermeability. In order to improve their bioavailability, these APIs are oftenmilled to small sizes. Milling techniques traditionally used in thepharmaceutical industry can be divided into two main categories: wet and dry. Each category has its benefits and drawbacks. Dry milling (e.g. jet milling)can achieve micron size particles with no agglomeration but it is not preferredfor very potent compounds and can introduce amorphous content into the APIs. Wetmilling on the other hand is safe for handling potent compounds but usually cannot achieve very small particles (e.g. rotor-stator mills) or has potential forextraneous matter to be introduced (e.g. media mills). Producing smallparticles by a wet milling technique is also associated with particleaggregation and/or difficult isolation.

High pressurehomogenization can be viewed as a wet micronization technique capable ofachieving micron and sub-micron particle sizes. It is a top-down approach forcontrolling particles size during which slurry of API in a non-solvent media isforced through a very small nozzle under very high pressures. This enormouspressure drop causes cavitation, high shear forces and particle-particlecollisions, which are the main factors for particle breakage. Proper selectionof operation parameters (pressure, nozzle size, flow configuration, number ofpasses) and process parameters (non-solvent media, solid content) can yieldvery narrow particle size distributions (PSD).

In this paper wedescribe our investigational work evaluating HPH as a particle size reductiontechnique and its potential for use in API manufacturing. The work presented consistsof three main parts. In the first part we investigate the effect of process andoperation parameters on particle breakage performance. Very important processdesign considerations for API processing are also presented. In the second partwe propose a mathematical model capable of describing particle breakage and HPHperformance as a function of operation parameters for the specific equipment athand. The proposed model helps with the answer of one very important question:how to scale-up HPH? In the third part we evaluate various particle isolationand drying techniques. These include the following: conventional filtration anddrying, spray drying and freeze-drying. Investigated are the effects theseisolation protocols have on the API's physical attributes such as final PSD, specific surface area (SSA) and bulk density. Finally, a simple decision treeis presented for the selection of these isolation techniques depending on finalAPI formulation method.

See more of this Session: Particle Formation and Crystallization Processes From Liquids, Slurries and Emulsions
See more of this Group/Topical: Particle Technology Forum

(667d) Dynamic Change In Polymer Particle Shape and Size by Simple Emulsion-Diffusion Process

Shaowen Ji, Devesh Srivastava and Ilsoon Lee, Chemical Engineering and Materials Science, Michigan State University, East Lansing, MI

Polymer hollow microspheres have attracted growing attentions in drug delivery, catalyst carriers, and microencapsulation. Here we report a simple preparation of hollow poly(lactic acid) (PLA) microspheres by single emulsion-diffusion method in a turbulent shear flow system. Modified Taylor-Couette mixer was introduced to generate uniform high shear fluid flow, facilitating the formation of hollow structured particles. The critical formation transition of polymer nanospheres to hollow microspheres was studied and controlled. An increasing addition of glycerol into the aqueous phase changed the flow from turbulent inertial to turbulent viscous regime. When the operation temperature was higher than the glass transition temperature of PLA, multiple emulsion droplets were obtained by the inclusion of water droplets during oil drops coalescence. Fast polymer solidification and water entrapping during diffusion process caused the formation of hollow PLA particles. Under harsh conditions, these hollow microspheres collapsed to open structure due to water escaping. Higher mixing speed rate and longer mixing time resulted in more uniform particle size and size distribution. The effects of water to oil ratio, pH of aqueous phase, and electrolyte were investigated for further engineered control of particle shape, size, and size distribution.

See more of this Session: Particle Formation and Crystallization Processes From Liquids, Slurries and Emulsions
See more of this Group/Topical: Particle Technology Forum

(667e) An Investigation of the Applicability of Micro-Calorimetry for the Measurement of Supersaturation During Batch Crystallization

Lotfi Derdour, Drug Product Science and Technology, Bristol-Myers Squibb co., New Brunswick, NJ, Frederic Buono, P R&D, Bristol-Myers Squibb Co, New Brunswick, NJ and Chiajen Lai, Drug Product Science and Technology, Research and Development, Bristol-Myers Squibb Company, New Brunswick, NJ

A new application of reaction micro-calorimetry for early stage crystallization development is described. Using a multi- channel Insight RT-10 reaction calorimeter, critical crystallization information such as detection of primary nucleation can be obtained with minute material quantities (<2 g) . The micro-calorimetry approach was successfully applied to determine the variation of the transient concentration and supersaturation for both cooling and anti-solvent crystallizations. These data can provide information about crystal growth rate at an early stage of the development which permits tailoring robust crystallization processes. Additionally, the ability to analyze up to 8 crystallization processes in parallel, the ease of instrument set-up and sample preparation and the no calibration requirement make micro-calorimetry an ideal tool for crystallization development for early stage projects.

See more of this Session: Particle Formation and Crystallization Processes From Liquids, Slurries and Emulsions
See more of this Group/Topical: Particle Technology Forum

(667f) Ag/Cu Particle Generation by Cosolvent Spray Pyrolysis: Effect of Cosolvent

Kai Zhong[1], George Peabody[1], Amanda Lafleur[1], Howard Glicksman[2] and Sheryl H. Ehrman[1], (1)Department of Chemical and Biomolecular Engineering, University of Maryland, College Park, MD, (2)DuPont Electronic Technologies, Research Triangle Park, NC

A cosolvent spray pyrolysis process was developed for the generation of various single component and multi-component metal particles that require a reducing atmosphere to promote the formation of oxide free metal particles. During the process, a cosolvent was added to a precursor solution, and it decomposed in the furnace. The original work on the cosolvent process utilized ethanol as the cosolvent. To make the process amenable to commercialization, other cosolvents were investigated. Ethylene glycol with a much higher flash point was found to produce oxide free metal particles. We found some differences when ethylene glycol(EG) was used as the cosolvent versus the use of ethanol(ET). In order to understand the effect of the cosolvent on the particle properties as well as to elucidate reaction pathways during the pyrolysis process, single component Cu and two component Cu/Ag particles were generated using either ethylene glycol or ethanol as the cosolvent. Residence times were adjusted from 0.2 s to 3 s by changing the carrier gas flow rate from 2.5 L/min to 10 L/min and furnace temperature from 400 ^{o}C to 1000 ^{o}C. The powder particles were characterized by SEM, XRD, FTIR and He pycnometry. Experimental results showed that $Cu(NO_3)_2$ first decomposes to $Cu_2(OH)_3(NO_3)$ and CuO and then is reduced to form pure Cu particles. Higher density particles were generated with the cosolvent of ET rather than with EG. The use of ET also led to an alternate morphology of product particles. For two component Cu/Ag particle generation, it was observed that pure Ag precipitated out in less than 0.2 s, and promoted the transition from $Cu(NO_3)_2$ to Cu after the evaporation of solvent. Different phase separation behavior was observed for ET and EG, which can be attributed to the different solubility and reactivity of the nitrates in these solvents.

See more of this Session: Particle Formation and Crystallization Processes From Liquids, Slurries and Emulsions

(667g) Experimental Investigation and Population Balance Equation Modeling of Solid Lipid Nanoparticle Aggregation Dynamics

Michael A. Henson[1], Yihui Yang[1] and Alessandro Corona III[2], (1)Department of Chemical Engineering, University of Massachusetts Amherst, Amherst, MA, (2) Fabric and Home Care Innovation Center, Procter & Gamble - Household Care, Cincinnati, OH

Solid lipid nanoparticles (SLNs) have applications in drug delivery and the encapsulation of bioactive, lipophilic compounds such as those required for functional foods. SLNs are commonly obtained by first preparing a lipid emulsion via high pressure homogenization or microfluidization followed by cooling in an ice bath or other low temperature environment such that the lipid emulsion drops crystallize and solid particles are formed. A major obstruction to the industrial use of SLNs is their tendency to aggregate and form large particles and eventually gels when stored at room temperature. The prevailing theory is that aggregation is driven by the lipid crystals undergoing a polymorphic transformation from the thermodynamically unstable a form to stable b form. A large increase in surface area occurs as the spherical a particles are transformed into platelet-like b particles, causing a substantial decrease in surfactant coverage on the hydrophobic surfaces and inducing particle aggregation.

We tested this theory by performing a series of aggregation experiments and by developing a population balance equation (PBE) model for prediction of the aggregation dynamics. Oil-in-water emulsions were prepared by high pressure homogenization with 5 wt% lipid (tristearin) and 1 wt% surfactant (mixture of Tween 60 and Span 60) at 85°C. The lipid emulsions were cooled in an ice bath for 8 hours to obtain SLNs and the resulting solutions were stored at 20°C. Aggregation dynamics were monitored daily by measuring the particle size distributions with light scattering and by examining polymorphic transformation behavior with differential scanning calorimetry. Our experimental results suggest that the polymorphic transformation rather than particle aggregation is the rate determining step of the aggregation process in this system.

The PBE model was formulated to predict the particle volume percent distribution under the assumption that particles collide due to Brownian motion and that only b particles could aggregate. The polymorphic transformation was modeled to be first-order in the concentration of a particles and first-order in the particle surface coverage such that transformation rate was zero if the particle had maximum coverage. An aggregation efficiency was introduced by assuming b particle aggregation was dependent on the surface coverage of each particle, with no aggregation possible when both particles had maximum coverage. The model was able to qualitatively predict our experimental data as well as other SLN aggregation data available in the literature. Our preliminary modeling results further suggest that the polymorphic transformation is the rate determining step of the aggregation process and support the hypothesis that aggregation is driven by the creation of new surface area due to the transformation.

See more of this Session: Particle Formation and Crystallization Processes From Liquids, Slurries and Emulsions
See more of this Group/Topical: Particle Technology Forum

NUMERICAL INVESTIGATION OF HIGH EFFICIENT PRODUCTION OF PROPYLENE IN A MULTI-REGIME FCC RISER ID# 240391

Hui Zhao[1,2], Jieqing Gan[2], Abdallah S. Berrouk[1], Chaohe Yang[2] and Honghong Shan[2]

(1) Chemical Engineering Department, Petroleum Institute, Abu Dhabi, United Arab Emirates
(2) State key laboratory of heavy oil processing, China University of Petroleum. Qingdao, China

Introduction

Fluid catalytic cracking (FCC) is among the most important refinery processes that convert high boiling point petroleum fractions into fuel and light olefins. With the increased demand for high quality products, traditional FCC units have been confronted with multiple economic challenges that limit their role in the transformation of heavier and inferior feedstock to the more valuable light olefins such as ethylene and propylene. In this work, a novel FCC process has been developed and tested on an industrial scale dual-riser FCC unit. This process was based on the TMP process (Two-stage-riser catalytic cracking for Maximizing Propylene yield) (Li at al., 2007). Effectiveness of the new developed multi-regime reactor has been experimentally and numerically proved. Numerical results show clearly the differences in flow field and phase distributions in favor of the novel reactor compared to the traditional ones.

Process Description

The new developed process designed to effectively produce propylene from heavy oil catalytic cracking revolves around the target of high selective conversion of olefin in gasoline (including $C6^{=}$ and $C7^{=}$) into propylene and ethylene (Den Hollander, et al. 2002). Controlling of the competitive relations between thermal pyrolysis and catalytic cracking is the primary concern of this new design. Suppressing of thermal pyrolysis requires good contact between reactant and catalyst, short reaction time for olefin cracking, and also relative low reaction temperature (compared to other heavy oil pyrolysis processes). Fulfillment of these three requirements could hardly be achieved in a traditional FCC riser reactor.

A multi-regime gas-solid reactor is proposed in this work to meet the above requirements. Fig.1a shows a typical TMP process which consists of two riser reactors and a regenerator. The cracking reaction of the hydrocarbon feedstock takes place in the riser while the regenerator reactivates the catalysts by burning the coke deposited on them. The TMP process is based on the two-stage FCC (TSRFCC) technology (Yang et al. , 2005). The first stage of the riser deals with atmospheric residue and C4 mixture gas while the second stage is fed with recycle oil and light gasoline both coming from the first stage. The feed is injected at the bottom of each stage along with a small amount of steam to prompt atomization and reduce coke formation. The feed is subsequently vaporized when contacting with hot catalyst particles flowing from the regenerator. The hydrocarbon vapors go through endothermic catalytic cracking reactions on their way through the riser. Lighter hydrocarbons are produced as main cracking products along with a by-product coke which deposits on the catalyst surface, hence reducing its activity. The detail structure of the combined feeding system is shown in Fig.1b. Lab-scale experiment studies indicate that C4 mixture gas or light gasoline is an ideal component to promote propylene yield (Li, 2007). The preferred reaction condition for C4 mixture gas or light gasoline is high reaction temperature, large catalyst to oil ratio, and suitable resident time. However, for heavy oil, reaction temperature should be kept at low values in order to avoid excess reaction. To ensure suitable reaction conditions for both heavy oil and C4 mixture (sometimes for the light gasoline also), a fuel-stratified injection technique was implemented.

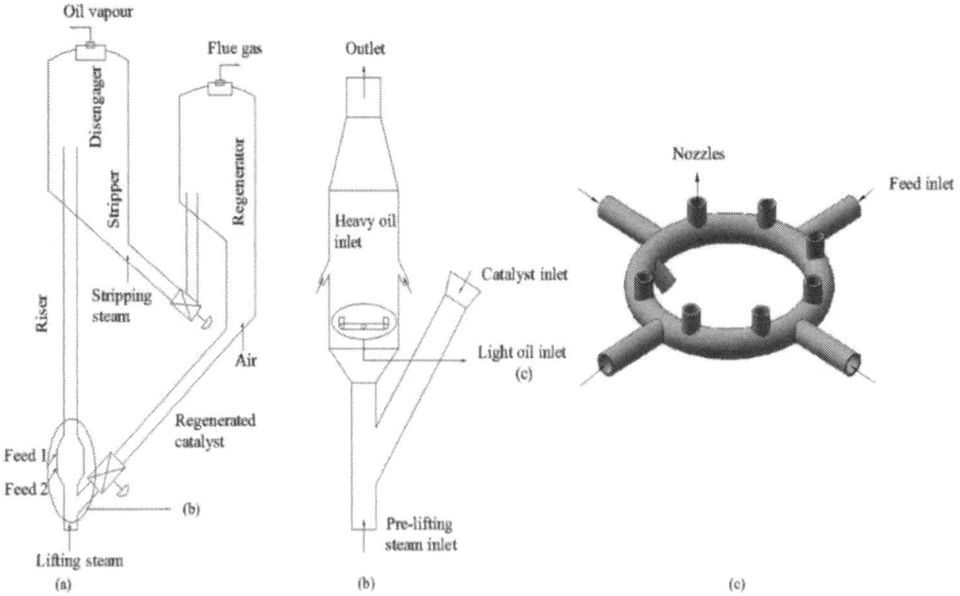

Figure 1. Schematic of the TMP riser reactor unit with stratified injections
(a) The TMP process; (b) Feed mixing section of the TMP riser; (c) Light oil inlet annular pipe

First, the low temperature C4/light gasoline contacts with high temperature catalysts to produce a large amount of $C3^{=}$ and decrease the catalysts temperature. The latter is due to the endothermicity of the cracking reaction and the heat transfer between the gas and solid phases. Subsequently, the low temperature catalysts (compared to the fresh catalysts) contact with heavy oil (AR or recycle oil) giving rise to a low temperature cracking reactions. The latter can stop the cracking reactions of light olefins since the absorption ability of heavy oil is stronger than that of light oil. This ensures short reaction time for the light oil. Moreover, low temperature reactions could also reduce the unwanted secondary reactions and minimize high temperature thermal cracking. To obtain better distribution of light oil, an annular pipe with multi nozzles was adopted (Fig. 1c).

The geometry of the riser reactor has great impact on the gas-solid two-phase flow behavior. For the traditional riser (Iso-diametric riser), typical core-annulus structures and serious back mixing behavior were observed. This unfavors the much-needed contact between reactants and catalysts and could have adverse effects on the yields of the desired products. Therefore, an appropriate change to the riser geometry should be sought to improve the contact efficiency between reactants and catalysts. Preliminary experimental studies show that a diameter-changing riser could enhance contact efficiency between reactants and catalysts at the bottom part of the riser. Thus, a diameter-changing riser was adopted in conjunction with the TMP process. The ratio of the expanding section, D, to the original diameter, d, is taken equal to 2.0.

Mathematical Model

In this work, only the second stage of the TMP riser will be simulated for lack of suitable reaction kinetic models for both atmospheric residue and C4 mixture gas. A three-dimensional Euler–Euler CFD model was adopted. It was coupled to a three-dimension two-phase flow-reaction to simulate the heavy oil catalytic pyrolysis in the second stage of TMP riser reactor. The reaction behavior was described using the 11-lumps kinetic model Guo, 2007).

An Euler-Euler model, closed using the kinetic theory of granular flow, was used to model the hydrodynamics of the gas-solid flow in the riser section of the FCC reactor. One set of mass and momentum conservation equations were solved for each phase. Momentum equations for both phases were linked by an inter-phase exchange term.

To close the governing equations, the solid phase properties should be defined and appropriate closure laws for both phases stress tensors, particulate phase pressure, and momentum inter-phase coefficient are required. Solid phase properties are derived based on the kinetic theory of granular flow (KTGF). The conservation equation for the fluctuating energy of the solid phase, known as granular temperature, can be obtained by solving its transport equation (Gidaspow, 1994)

Numerical Results and Discussion

Fig.2 shows the time-averaged axial solid volume fraction for both the traditional riser and the diameter-changing riser. The solid holdup at the bottom part (the feed inlet zone) of the changing-diameter riser is much higher than that of the traditional riser. At the upper part of this zone, the solid volume fractions of the two geometries are very similar.

Fig.3 displays the solid phase distribution and flow field in the lower part of TMP riser reactor. A dense catalyst bed, which is suitable for gasoline olefin reactions, can be maintained at proper locations in the riser reactor by fine tuning the nozzle configuration and the reactor geometry. In the lower part of the mixing zone, the superficial gas velocity is less than $3 m\cdot s^{-1}$ and it increases to $10\sim15 m\cdot s^{-1}$ in the upper part. This high velocity span in this zone leads to the coexistence of both dense and dilute phases. These results further indicate that the TMP diameter-changing riser exhibits distinct advantage over the traditional one in improving the gas-solid contact efficiency at the bottom part of riser. This favors the conversion of the light gasoline to light olefins. It is worth notice that, the cross-section solid concentration profile is not symmetrical due to the downward oblique nozzle used in the light gasoline injection system. Indeed, the dense bed height in the oblique downward nozzle side is higher than that of the offside part of the system hence the catalyst phase takes on bias flow behavior.

Fig.4 describes the catalyst activity profile contour. Here the catalyst activity is calculated based on the catalyst deactivation function. In the pre-lifting part, few reactions take place, so the catalyst has high reaction activity in this region. As the reactions mainly taken place in the vicinity of both heavy oil and light gasoline feed nozzles, catalyst activity there drops sharply in this region. Therefore, it is of vital importance to improve the feed oil atomization and vaporization efficiency to convert as much feed oil as possible. In addition, similar to the solid flow field, catalyst activity shows obvious bias flow profile due to the downward oblique nozzle. In the oblique downward nozzle side, catalyst activity is higher than that in the offside. Therefore, it is necessary to add another oblique downward nozzle in the offside to form a bilaterally symmetrical flow or let some of the nozzles to oblique downward, which may weaken the bias flow and improve the solid density and activity in the upper region.

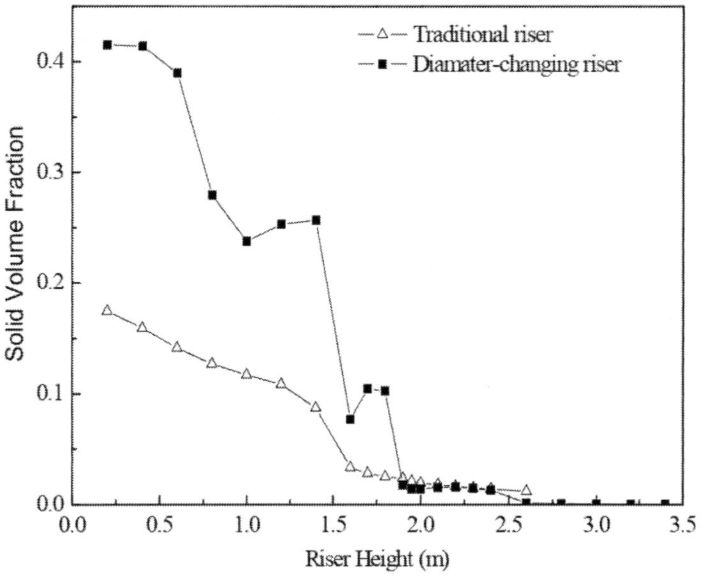

Figure 2. Predicted axial solid volume fraction for two different riser geometries

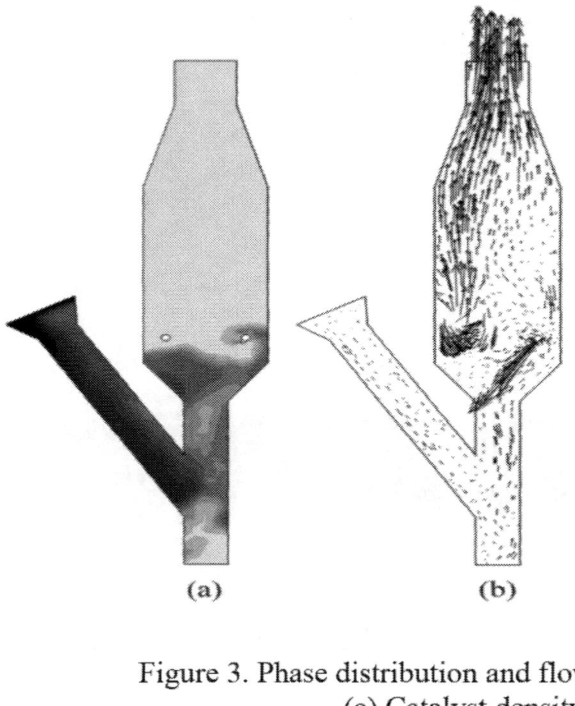

Figure 3. Phase distribution and flow field in the lower part of riser reactor
(a) Catalyst density, (b) Catalyst velocity

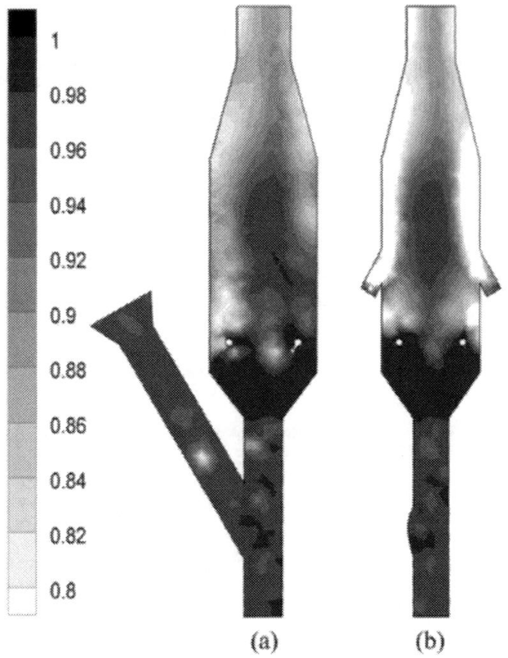

Figure 4. The catalysts activity of the TMP feed injection zone
(a) Catalysts feed cross-section, (b) Heavy oil feed cross-section

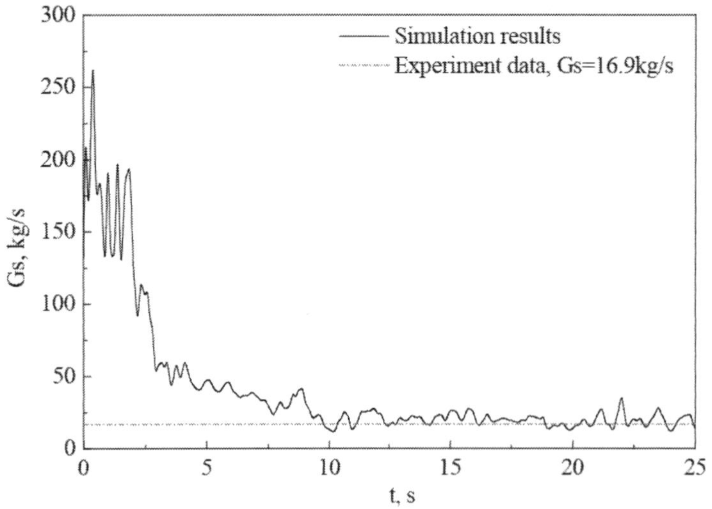

Figure 5. Distribution of solid volume fraction on axis in lab-scale riser reactor

The above simulation results explain the flow and cracking reaction pattern in terms of the hydrodynamics and reaction kinetics. This can play a valuable role in optimizing the geometry of the industrial riser reactor. It is known that investigating hydrodynamic conditions in real industrial reactors is difficult but the overall mass flux of catalyst particles can be measured and used as a factor to evaluate the precision of the CFD simulation. Mass flux of catalyst particles in the industrial reactor was compared with the CFD results in Fig.5. Good agreement between the industrial reactor measurements and numerical findings is observed.

Conclusions

Numerical simulations where the fluid flow is coupled to cracking reactions were conducted to verify the performance of a novel multi-regime gas-solid fluidized reactor for heavy oil catalytic pyrolysis. The numerical findings are in good agreement with test run results of a real industrial unit. This novel process demonstrates its superiority over other catalytic cracking/pyrolysis process. The numerical work provided more detailed insights about the cracking process: *(i)* Catalyst distribution of the diameter-changing riser is better than that of the traditional riser. Various fluidized regimes could be integrated in a single reactor; *(ii)* multi-regime gas solid reactor provides more superiority on the control of fast reaction systems especially for heavy oil catalytic cracking; *(iii)* the cracking reactions mainly take place in the feed mixing section hence optimizing the oil and catalysts contact in this region is of a vital importance.

Acknowledgements

The authors would like to acknowledge the financial support of Takreer (Abu Dhabi Oil Refining Company) through Grant N° 6000.1689.

References

1. Den Hollander M. A., Wissink M., Makkee M., and Moulijn J. A. (2002). "Gasoline conversion: reactivity towards cracking with equilibrated FCC and ZSM-5 catalysts". *Appl. Catal. A,* 2002, 223, 85-102
2. Gidaspow, D., Multiphase Flow and Fluidization: Continuum and Kinetic Theory Descriptions. Academic Press 1994.
3. Guo, J. H. Primary Study of the Lumped Kinetic Model for Heavy Oil Cracking into Propylene by Two-stage-riser Technology[D]. China University of Petroleum, Dongying, China, 2007.
4. Li, C. Y.; Yang, C. H.; Shan, H. H., Maximizing propylene yield by two-stage riser catalytic cracking of heavy oil. Ind. Eng. Chem. Res. 2007, 46, (14), 4914-4920.
5. Li, X. H. Fundamental Studies on the Technology of Fluid Catalytic Cracking with Two-Stage Risers for Maximizing Propylene [D]. China University of Petroleum, Dongying, China, 2007.
6. Yang, C. H.; Shan, H. H.; Zhang, J. F., Two-stage riser FCC technologies. Petroleum Refinery engineering 2005, 35, (3), 28-33

(686b) Particle Population Model for Circulating Fluidized Bed Combustor

P.S.T. Sai, Department of Chemical Engineering, Indian Institute of Technology Madras, Chennai, India

A two-dimensional particle population model has been developed to predict the particle size distribution in a circulating fluidized bed combustor (CFBC) based on the analysis of particle properties and the core-annulus hydrodynamic model. The model incorporates grids in the dense and the dilute regions of the CFBC to consider fuel particle fragmentation, char combustion and particle attrition. The interaction and movement of particles between different grids has also been considered. The simultaneous mass balances in different grids will provide the particle population distribution in the CFBC. The influence of operating conditions and breakage functions on the population distribution of the CFBC has been studied using the model and the process behavior has been analyzed. Optimum feed intervals of the fuel and the bed material have been obtained for the given operating conditions. Also, the influence of kinetic parameters such as rate constant and reaction order has been studied. This information can be used for the deign and control of the industrial circulating fluidized bed combustor.

See more of this Session: Circulating Fluidized Beds
See more of this Group/Topical: Particle Technology Forum

(686c) Hydrodynamic Behaviors of Sand Particles In a Large-Scale Triple Bed Combined Circulating Fluidized Bed

Chihiro Fushimi[1], Yu Nakamura[1], Masanori Ishizuka[1], Guoqing Guan[2], Atsushi Tsutsumi[3], Yoshizo Suzuki[4], Yongpan Cheng[5] and Chi-Hwa Wang[5], (1)Institute of Industrial Science, University of Tokyo, Tokyo, Japan, (2)North Japan Research Institute for Sustainable Energy (NJRISE), Hirosaki University, Hirosaki, Japan, (3)Institute of Industrial Science, The University of Tokyo, Tokyo, Japan, (4)Clean Gas Group, Energy Technology Research Institute, National Institute of Advanced Industrial Science and Technology, Ibaraki 305-8569, Japan, (5)Chemical and Biomolecular Engineering, National University of Singapore, Singapore, Singapore

We have developed a triple-bed combined circulating fluidized bed (TBCFB) as a gasifier for advanced integrated coal gasification combined cycle (A-IGCC) or advanced integrated coal gasification fuel cell combined cycle (A-IGFC) systems to improve of thermal efficiency of coal-fired power plants. In this study, the effect of operating conditions in a gas-sealing bed (GSB) (Φ0.15 m×5.0 m) between a riser (Φ0.1 m×16.6 m) and a bubble fluidized bed (0.75 m×0.27 m×3.4 m) on flow behaviors of silica sand solids in a large-scale TBCFB cold model were investigated in order to achieve high flux and high density solids circulation. The solids mass flux (G_s) monotonically increased with the increase in gas velocity in the GSB. Further increase in GSB gas velocity decreased G_s.

See more of this Session: Circulating Fluidized Beds
See more of this Group/Topical: Particle Technology Forum

An alternative two-fluid model for the hydrodynamics of heterogeneous gas-solid flow in circulating fluidized bed risers

Junwu Wang[1], Quan Zhou[1,2], Yaning Liu[1,2], Wei Ge[1] and Jinghai Li[1]
[1]The EMMS group, State Key Laboratory of Multiphase Complex Systems, Institute of Process Engineering, Chinese Academy of Sciences, Beijing 100190, P. R. China
[2]Graduate School of Chinese Academy of Sciences, Beijing, 100490, P. R. China

Abstract:

An alternative two-fluid model (TFM) is proposed in response to the physical nature of gas-solid flow in risers which is characterized by two-phase structure with particle-rich dense phase and gas-rich dilute phase. The dense and dilute phases are defined as the two interpenetrating continua in the proposed TFM, where the constitutive laws are completely determined by the Energy Minimization Multi-Scale (EMMS) model. It was shown that the experimentally measured axial and radial solid concentration profiles as well as solid circulation flux can be well predicted, indicating its feasibility for describing the hydrodynamics of heterogeneous gas-solid flow in risers.

(686e) Hydrodynamics In a Semi-Dual Fluidized Bed Biomass Gasifier

Son Ich Ngo[1], Young-Il Lim[1], Byung-Ho Song[2], Uen-Do Lee[3], Young-Tai Choi[3] and Jae-Hun Song[4], (1)Integrated Biotechnology, Hankyong National University, Anseong, South Korea, (2)Chemical Engineering, Kunsan National University, Gunsan, South Korea, (3)High temperature processing R&D, Korea Institute of Industrial Technology (KITECH), Cheonan, South Korea, (4)R&D center, SeenTec Co.,Ltd.,, Changwon, South Korea

The bed material circulation rate plays an important role in hydrodynamics and operation ability in dual fluidized-bed (DFB) gasifiers. The high circulation rate results in a high heat demand and a low heat transfer from riser to gasifier, which makes disadvantage in the DFB gasifier. In this study, a semi-dual fluidized bed (sDFB) biomass gasifier is proposed to reduce the solid circulation rate. A cold-rig sDFB (0.8 m × 0.2 m ×3.85 m height) was designed and installed to investigate hydrodynamics of the apparatus and the solids circulation. In order to calculate the amount of direct back-mixing particles through the gasifier–riser interconnection area, an Eulerian-Eulerian two fluids unsteady-state two-dimensional (2D) model incorporating with the granular flow kinetic theory was used. The packing limit of this CFD model was adjusted according to the experimental data for the solid circulation rate and the pressure drops. It was found from the simulation results that the amount of solids is directly back-mixed through the gasifier–riser interconnection area about 5-12% of solids circulation rate.

See more of this Session: Circulating Fluidized Beds
See more of this Group/Topical: Particle Technology Forum

Experimental investigation of silicon grinding with an opposed-air jet mill

Zhimin Xi, Department of Powder Science and Technology, Tel-Tek, Norway
Songxiong Ding, of Powder Science and Technology, Tel-Tek, Norway
Gisle Enstad, Process Department., Telemark University College,

Introduction

An opposed-air jet mill is often used when contamination on powder product has to be minimized; or ultra fine grinding is needed. In a jet mill, streams of particulate solids, driven by high pressurized air, are accelerated and collide headed-on in its grinding chamber, where the particulates are ground. Steams of air and crushed particles are then swept upwards to a classifier. By turning its rotation speed, a classifier allows passage of acceptably fine particles, while returning the coarse particles back into the milling chamber.

In this study, experiments were first carried out to investigate the feasibility to grind particulate silicon with an opposed-air jet mill to meet requirement of final product. At the same time, operational parameters for rotation speed of air classifier and pressure of pressurized air were also sought for optimal performance.

Material and Equipment

Particulate silicon from a processing application was used in this study as provided by the company. The raw material has a particle size distribution as shown in Figure 1.

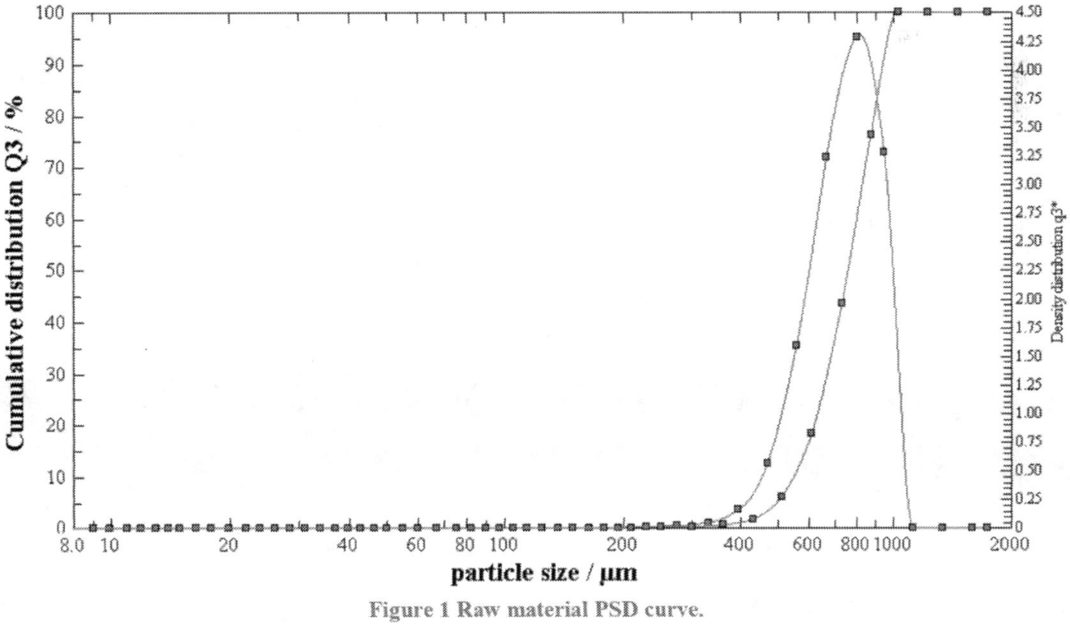

Figure 1 Raw material PSD curve.

A Majac mill was used in experiments as an opposed-air jet mill. It consists of a milling chamber, opposed-air tubes, an air-classifier and a cyclone etc as illustrated in Figure 2. The coarse particles are ground in the milling chamber by high pressure air, whereas the finer particles with required size are selected by the air-classifier as a final product.

In the experiments, the air-classifier was operated with different rotation speeds at 1250, 1500 and 1800 rpm. Four different air pressures were applied to grind silicon particles when the rotational speed of air-classifier is set. The air pressures as adopted in the experiments were 3, 3.5, 4, 5 bars.

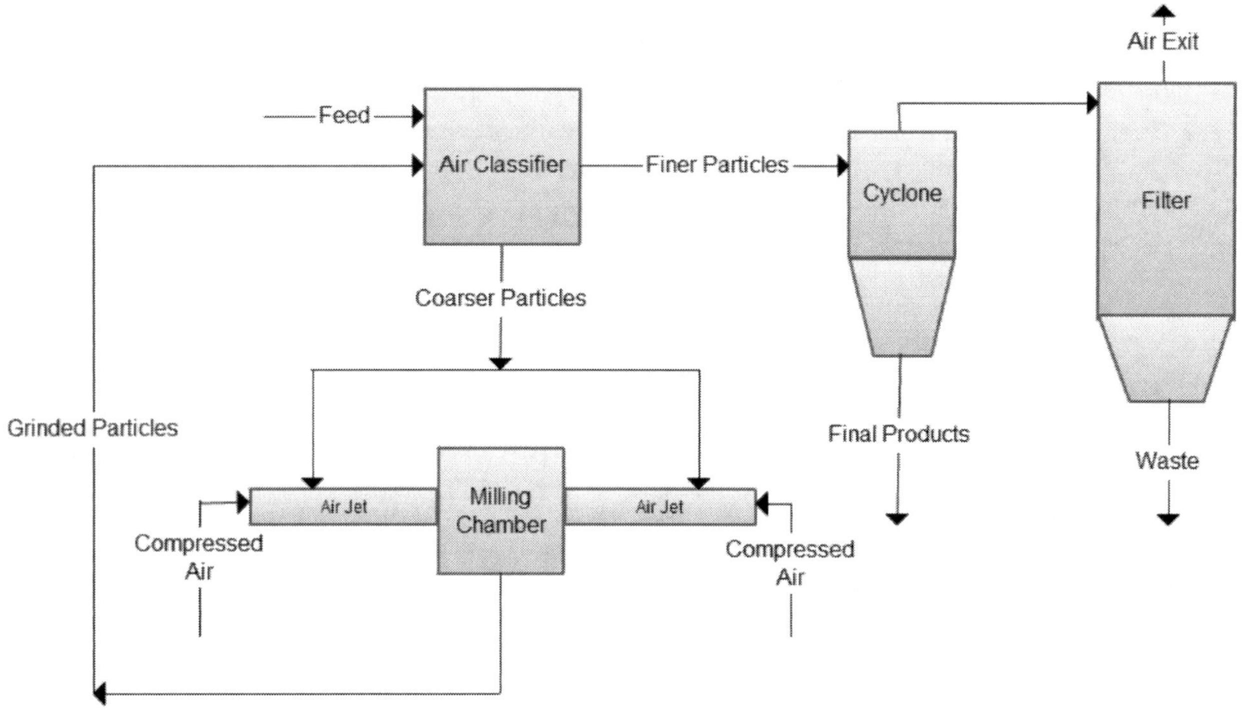

Figure 2 Block diagram of a typical Majac mill.

The product from the air-classifier is measured for Particle Size Distribution (PSD) analysis with a laser diffraction system by Sympatec.

Experiment result and discussion

Experiments were set to be in a batch operation. For each individual operation, an amount of one kg silicon powder was fed with a fixed feeding rate 0.4 kg/min into the Majac mill system to be ground; the grinding lasted 30 minutes.

1. Effect of opposed –air pressure on Grinding

At each of the air classifier rotation speed, the grinding of silicon particle was carried out with various air pressures (pressure increased from 3 bar to 5 bar). Results of PSD analysis for products thus obtained are shown in Figure 3**Error! Reference source not found.**, where no regular pattern was observed in the particle size distribution curves of product. One may also find that an increase of air pressure did not cause significant over-grinding effect.

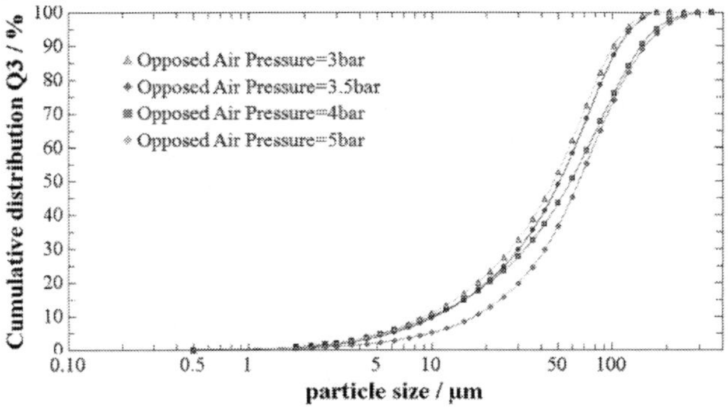

a) Classifier rotation speed = 1250 rpm

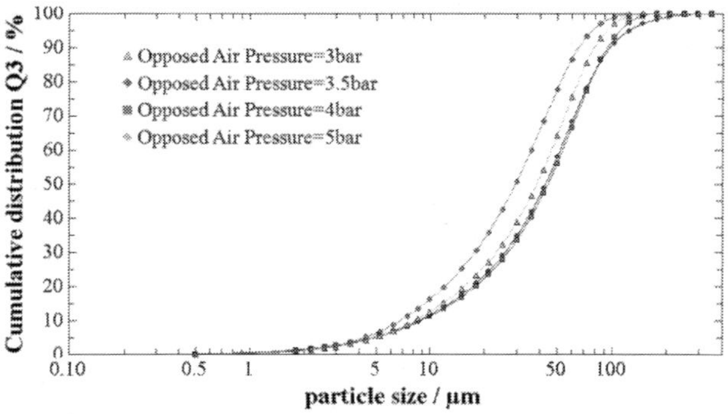

b) Classifier rotation speed = 1500 rpm

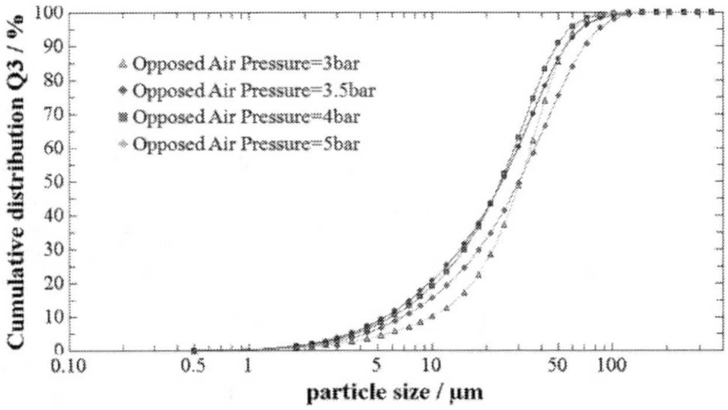

c) Classifier rotation speed = 1800 rpm

Figure 3 Comparison of cumulative size distribution at air classifier rotation speed = 1250rpm (a), 1500 rpm (b), and 1800 rpm (c).

With a higher classifier rotation speed, air pressure in opposed-air jet mill had little effect on the median size of the ground products, whereas at lower classifier rotation speed, air pressure in the opposed-air jet mill had some effects on the median size of ground products, but without clear tendency. To ensure the ground products with a consistent median size, a higher classifier rotation speed is recommended regardless of air pressure in opposed-air jet mill as shown in Figure 4.

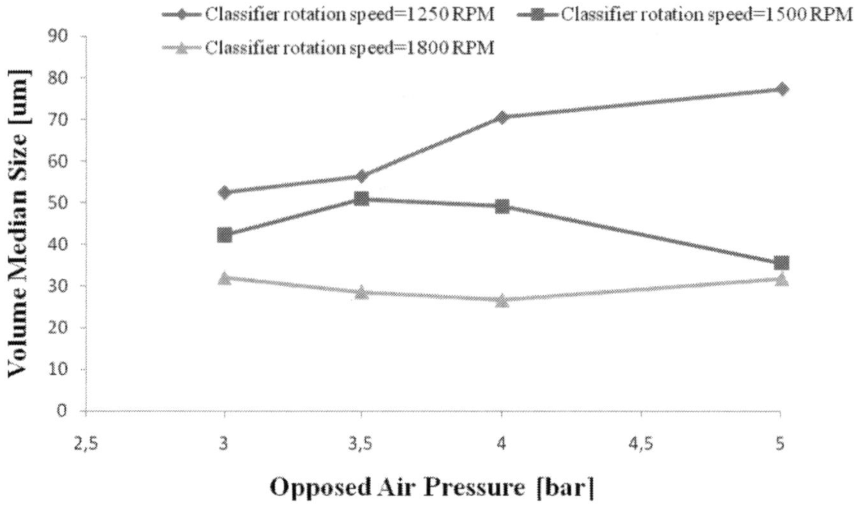

Figure 4 The volume based median size with different opposed air pressure, at each Classifier rotation speed.

2. Effects of Air Classification

In the experiments, the classifier was operated with different rotation speed. They were set as at 1250, 1500 rpm and 1800 rpm. Under each rotation speed of the classifier, several of air pressures in the opposed-air jet mill were introduced to grind the silicon particle (the pressure as used in the present experiment were 3.0, 3.5, 4.0 and 5.0 bars). Results of PSD for the products are given in Figure 5.

One may notice that rotation speed of classifier had a significant effect on PSD in the particle product when the air pressure in the opposed-air jet mill was low as shown in Figure 5a, 5b and 5c (where air pressures were 3.0, 3.5 and 4 bar), the higher the rotation speed in the classifier, the finer are the particle products. This effect was however no longer apparent when the pressure in the jet mill was increased to a higher level. As shown in Figure 5d , at the highest air pressure in the jet mill (5 bar in the present experiment), the PSD of particle product were almost the same when rotation speed in classifier were at 1500 rpm and 1800 rpm, consistent with the results as shown in Figure 5d.

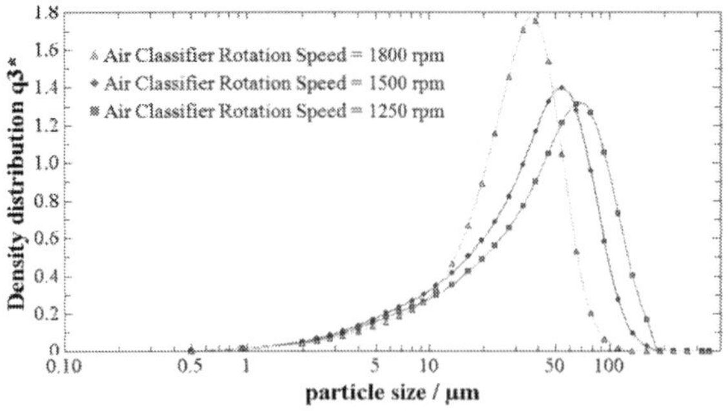

a). Air pressure = 3 bar in jet

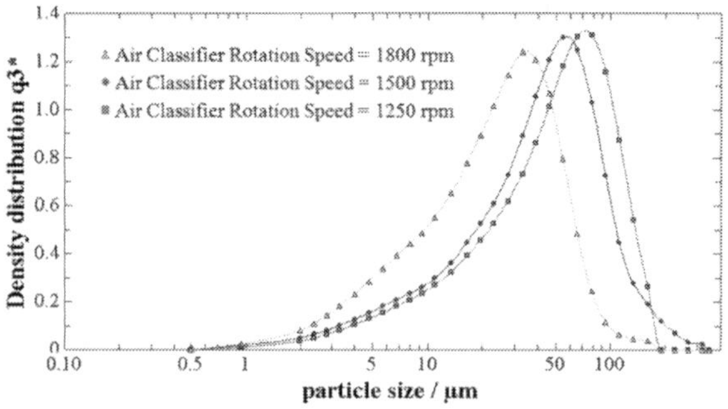

b). Air pressure = 3.5 bar in jet

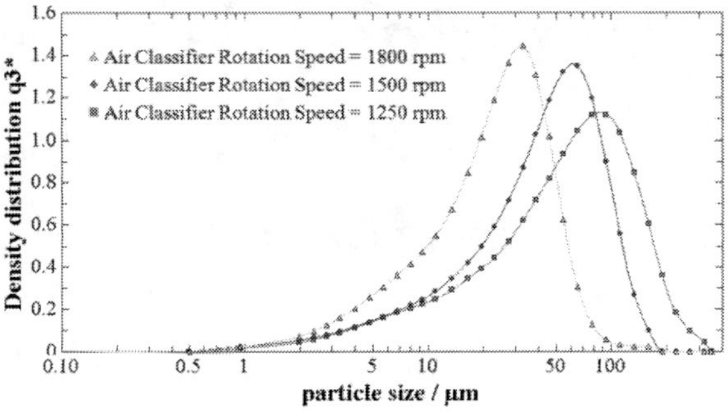

c). Air pressure = 4 bar in jet

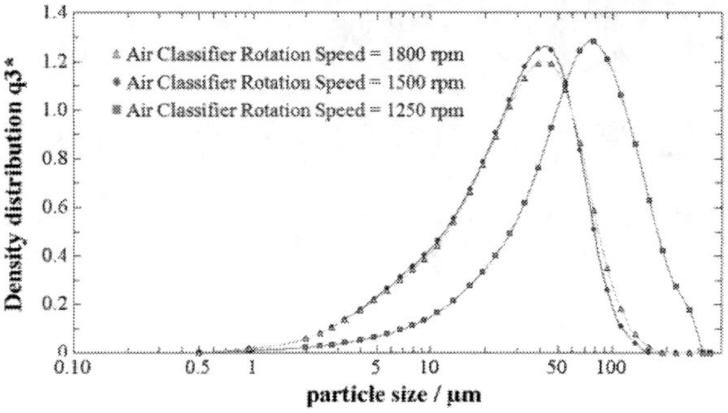

d) Air pressure = 5 bar in jet

Figure 5 Comparison of particle size distribution at opposed-air pressure = 3 bar (a), 3.5 bar (b), 4 bar (c), and 5 bar (d).

(707a) MK 146 Manufacturing Demonstration with the Vacuum Casting System and Twin Screw Extruder

Mark A. Williams, Ordnance Department, Naval Surface Warfare Center, Indian Head Division, Indian Head, MD

The Navy is interested in novel manufacturing processes that will reduce associated manufacturing costs, increase product quality, and increase safety of manufacturing operations.

See more of this Session: Processing and Safety of Energetic Materials
See more of this Group/Topical: Particle Technology Forum

(707b) Effects of High Shear Processing On Single Base Propellant

Jeffrey K. Wyckoff, Duncan Park, Michael Fair and David Thompson, US Army ARDEC, Picatinny Arsenal, NJ

Effects of High Shear Processing on SingleBase Propellant

J. K. Wyckoff, D.Park, M. Fair, D. Thompson

Theperformance of solid gun propellant is historically a function of the initialtemperature of the propellant. At hottemperatures, the maximum pressure reached in the gun and the resulting muzzlevelocity is greater. If the propellantis cold, the opposite is true. Thereforepropellant charges must be designed around the maximum pressure at hot so thegun barrel will not be damaged and the subsequent drop in performance in coldmust be accounted for. This change inperformance over temperature is broadly referred to as the temperaturecoefficient or temperature profile. Thegreater the coefficient is, the greater the difference between hot and coldperformance. Much research has beenperformed to produce flat temperature coefficients. While some new techniques have been developedthat lessen the impact, temperature effects are still noticeable. Some initial results suggest that mixingpropellant with high shear can improve a propellant performance overtemperature by decreasing the variations in burn rate or improving mechanicalproperties. For this study a single basepropellant was chosen. Single basepropellants, in which the bulk of the material is nitrocellulose, are oftenmade with a Beken style mixer that imparts a moderateamount of shear into the material. Resultswill be presented in which different processing techniques using varyingdegrees of shear are used to produce single base propellant. Our investigation included a low-shear, sigmablade mixer and a relatively high-shear, twin screw extruder. A design of experiments approach was adoptedto examine the effects of various processing parameters on the final propertiesof the propellant. Factors such as bladespeed, residence time, and solvent ratio were varied. Closed bomb and mechanical properties testswere performed and compared to baseline propellant manufactured on a Beken style mixer.

See more of this Session: Processing and Safety of Energetic Materials
See more of this Group/Topical: Particle Technology Forum

(707c) Aluminized Explosive Processing Using Twin Screw Mixer/ Extruder Technology

Richard Muscato, Applied Technology, Naval Surface Warfare Center, Indian Head Division, Indian Head, MD

There is a need within the US Navy to develop and demonstrate an alternative manufacturing technique to manufacture an aluminized, pressed explosive for specialized operations. A manufacturing technique has been demonstrated using twin screw mixer/extruder technology that produces an aluminized explosive that exhibits enhanced performance with equal or better sensitivity characteristics. The remote manufacturing process was developed using a 40mm twin screw mixer/ extruder integrated within inline granulator. This manufacturing process is a more safe process that yields product of equal or better quality at a reduced manufacturing cost.

Information shall be presented on the developmental approach, equipment design and installation, manufacturing effort as well as laboratory analysis of the pressed explosive that was manufactured using the 40mm twin screw extruder facility.

The Naval Surface Warfare Center, Indian Head Division has successfully developed a manufacturing method for a pressed, aluminized explosive that has good performance and sensitivity characteristics along with reduced environmental impact.

See more of this Session: Processing and Safety of Energetic Materials
See more of this Group/Topical: Particle Technology Forum

DESENSITIZATION OF ENERGETIC AND HAZARDOUS CHEMICAL MATERIALS

Thaddeus Cooper Speed
and
Abdollah Kashani
Safety Consulting Engineers, Inc.
Schaumburg, Illinois 60173

Biography

Thaddeus C. Speed, physicist and safety professional, is the Technical Director for DEKRA's explosives testing and analysis division, Safety Consulting Engineers, Inc. Mr. Speed has more than 25 years experience in hazards analysis, process and system safety, and in testing, evaluating and characterizing explosives and other energetic materials. He has published numerous papers on electrostatic discharge safety (ESD), specialized explosives testing methods, ignition and initiation phenomena, in-process classification of energetic materials, and similar topics.

Mr. Speed is approved by the U.S. Department of Transportation to examine explosives and recommend classification. Mr. Speed's explosives process hazards and characterization expertise is further enhanced by an in-depth understanding of system safety and the inter-related areas of facility siting, regulatory compliance, process safety management, material characterization and incident analysis.

He has developed various explosives related courses and has trained government officials, plant managers, other safety and engineering professionals and line operators in various aspects of explosives safety, both as general courses and as tailored training for specific issues, companies and operations.

In 1984 he was awarded a Bachelor of Science degree from Brigham, Young University, Provo, Utah in Physics and Astronomy. In 1994 he became a Certified Safety Professional, through the BCSP. In 1999, he was approved by the US Department of Transportation, Office of Pipeline and Hazardous Materials Safety, to examine explosives and recommend classifications (per 49 CFR 173.56).

Abdollah Kashani is Senior Safety Project Engineer has over 30 years experience providing innovative solutions and services for a variety of industrial, pharmaceutical, medical device manufacturing, and governmental customers. He has been providing services for various processes and operations such as material handling, heat generation, material transfer and transportation, hybrid mixtures (liquid-gas-solid) and other operations.

Involved in design and development of simulated process conditions for material responses. Examples are electrostatic charge generation, electrostatic discharge vapor/gas/solid explosions under elevated temperatures and pressures, identifying factors that influence the process and related hazard and safety issues.

Provided customized services for qualification of materials and equipment for the end user such as ignition protection study of electrical devices, in flammable environments, simulated to understand the cause of fire or explosion related to particular incident investigations.

On an industrial scale, he and his team developed and designed customized solutions such as flame/detonation arrestors, cyclone for high energetic materials, prototype explosive detection system, and conducting process safety/hazard analysis.

His experience also includes material flammability and fire resistance, demilitarization service, open gas/vapor and dust cloud explosion, hydrocarbon spray and pool fire, electrostatic charge/discharge study, propellant and energetic material burning rate combustion, design and setup for testing of hazardous material for DOT Class 1, 2, 3, 4, 5, 8 & 9 per applicable standard methods, and non-destructive evaluations.

Other expertise includes limiting oxygen concentration, flammability envelope, and ignition energy, characteristics, assessing the risk runaway reaction of material at variety of conditions and validation for exothermic and kinetic reaction.

He obtained his Associate Degree in Civil Engineering and Bachelor of Science in Medical Technology and Sciences from Iran. He obtained his Master of Science in Bioengineering from the University of Illinois, Chicago, Illinois. He has published jointly a number of technical papers.

Abstract

Many chemicals and mixtures contain a variety of hazardous, reactive materials, including, explosives, energetic materials, catalyst, reactive and toxic compounds. These can represent serious hazards for a specific application, process or other potential uses. Chemical of this type are used in such diverse applications as propellants, explosives and pharmaceuticals. They are also found as precursors and intermediaries in some processes.

These materials can of course be dangerous to transport and store and ways must be developed to make such materials less susceptible to accidental explosions due to thermal and mechanical shock.

Useful techniques to enhance the identification and safe handling of these types of hazardous materials are described in this paper.

Key words: Desensitized Energetic Material, Hazardous Material Identification, Hazardous Material Classification Methods.

Introduction

It is the responsibility of each shipper to identify the substances or articles they ship, provide information on critical physical characteristics and to properly classify all materials that are subject to the Hazardous Materials Regulations (HMR). Employers are similarly responsible for knowing and communicating to their employees the hazards of the process operations and chemicals used at their facilities.

Purpose

How to identify and deal with hazardous and particularly energetic materials in the work place is a constant, on-going challenge in many facilities. Options for desensitization of hazardous, energetic materials are some of the tools available to the process engineer that should not be overlooked.

Outline

- Identifying potentially energetic mixtures and compounds
- Desensitization methodologies to provide reasonable safety, for materials that would otherwise be considered too hazardous to handle in a particular process.
- If you need to transport it, options on how to approach the classification process.
- Overview of the Hazardous Materials Classification process.
- Techniques for reducing the sensitivity of various energetic and hazardous chemical materials to detonation, explosion, self heating, reactivity, fire, friction, heat, etc.
- Guidelines for safely handling energetic and hazardous materials during industrial process; covers disposal and transportation.
- Applicable regulatory methods for hazardous materials and desensitized high energetic materials classification, particularly for those susceptible to accidental explosions due to mechanical shock.

(707e) Application of the Shock Gel Process for the Manufacture of Spherical Propellant Grains

Dilhan Kalyon[1], Halil Gevgilili[2], Semra Senturk[2], Ilknur Kucuk[2], Sarah Longo[3] and John Bolognini[4], (1)Chemical Engineering, Stevens Institute of Technology, Hoboken, NJ, (2)Stevens Institute of Technology, Hoboken, NJ, (3)Rdar-Mee-P, US Army RDECOM-ARDEC, Dover, NJ, (4)Amsrd-Aar-Mee-P, US Army RDECOM-ARDEC, Dover, NJ

"Application of the shock gelprocess for the manufacture of spherical propellant grains"

John Bolognini, SarahLongo, Halil Gevgilili*, Semra Senturk*, Ilknur Kucuk* and

Dilhan Kalyon*

The shock gel process can beutilized for the manufacture of spherical gun propellants. Typically,nitrocellulose syrups which are prepared by the dissolution of nitrocellulose areused in conjunction with gelling baths. Although the process is relativelysimple to operate there are multiple factors that play a role in thedevelopment of the propellant size and shape distributions and the porosity ofthe propellant. In this presentation some of these major factors will beelucidated in conjunction with the results of investigations utilizing bothlive propellants and inert simulants. The issue ofdeterring the grain in situ will also be discussed.

*Stevens Institute of Technology,Highly Filled Materials Institute, Castle Point St. Hoboken, NJ 07030

See more of this Session: Processing and Safety of Energetic Materials
See more of this Group/Topical: Particle Technology Forum

(707f) Linking of Deterrent Concentration Distributions to Ballistic Performance

John Bolognini, Amsrd-Aar-Mee-P, US Army RDECOM-ARDEC, Dover, NJ, Dilhan Kalyon, Chemical Engineering, Stevens Institute of Technology, Hoboken, NJ, Sarah Longo, Rdar-Mee-P, US Army RDECOM-ARDEC, Dover, NJ, Carlton Adam, Rdar-Mee-W, US Army RDECOM-ARDEC, Dover, NJ, Moinuddin Malik, HfMI, Stevens Institute of Technology, Hoboken, NJ and Halil Gevgilili, Stevens Institute of Technology, Hoboken, NJ

"Linking ofdeterrent concentration distributions to ballistic performance"

John Bolognini,Carlton Adam, Sarah Longo, H. Gevgilili*, M. Malik*, D. Kalyon*

A myriad of deterrents areutilized to modify and control the ballistic properties of energetic propellantgrains in conjunction with coating processes. Empirically it is known thatballistic properties of energetic grains depend on the diffusion of thedeterrent into the grains during the coating process and hence on theconcentration distribution of the deterrent in the grain. A mathematical modelof the diffusion process was developed and validated using two separateexperimental methods involving thermo-gravimetric analysis and fluorescence on simulants.The diffusion models are further integrated into ballistic modeling software toallow the solution of the inverse problem of determining what the optimumdeterrent concentration distribution is to be for targeted applications.

*Stevens Institute of Technology,Highly Filled Materials Institute, Castle Point St. Hoboken, NJ 07030

See more of this Session: Processing and Safety of Energetic Materials
See more of this Group/Topical: Particle Technology Forum

(707g) Mathematical Modeling of the Continuous Processing of Energetic Materials In Conjunction with a Pressure Dependent Wall Slip Condition

Dilhan Kalyon, Chemical Engineering, Stevens Institute of Technology, Hoboken, NJ and Moinuddin Malik, HfMI, Stevens Institute of Technology, Hoboken, NJ

Mathematical modeling of the continuous processing of energetic materials in conjunction with a pressure dependent wall slip condition

M. Malik, D. Kalyon

Stevens Institute of Technology

Recent investigations on the development of flow instabilities during the flow and deformation of polymeric melts and concentrated suspensions (1, 2) have revealed that the slip coefficient (3), which links the slip velocity at the wall to the stress field, is itself dependent on the total normal stress (or in some situations the pressure) at the wall. In this presentation the methodology involved in the imposition of the pressure dependent slip coefficient on the coupled twin screw extrusion and die flow will be described. Typical results, focusing on the processing of energetic formulations, will be compared with corresponding modeling results using conventional flow boundary conditions.

1. H. S. Tang and D. M. Kalyon, "Time-dependent tube flow of compressible suspensions subject to pressure dependent wall slip: Ramifications on development of flow instabilities", Journal of Rheology, 52, 5, 1069-1090 (2008).
2. H. Tang and D. M. Kalyon, "Unsteady circular tube flow of compressible polymeric liquids subject to pressure-dependent wall slip", J. Rheology, 52, 2, 507-526 (2008).
3. D. Kalyon, "Apparent Slip and Viscoplasticity of Concentrated Suspensions", J. Rheology, 49, 3, 621-640 (2005).

(733a) Renewable Fuel From Activated Sludge Using Fluidized-Bed Catalytic Cracking (FCC) Process

Emmanuel D. Revellame, William Holmes, Rafael Hernandez and William T. French, Dave C. Swalm School of Chemical Engineering, Mississippi State University, Mississippi State, MS

Annual production of activated sludges in the United States ranges from 7 – 8 million tons. Currently, disposal technologies of these sludges include incineration and landfilling, both of which are heavily regulated and could create adverse environmental impacts. A fraction of the activated sludges can be used as fertilizer once it complies with regulations for Class A biosolids. Activated sludges contain microbial consortia known to produce lipidic materials and many other organic compounds, which could be used to produce biofuels and specialty chemicals. Several studies showed that reasonable yields (2 – 6% weight) of biodiesel could be produced from activated sludge. At these yields however, activated sludge biodiesel is less economical than petroleum diesel. This study describes the production of renewable fuel from activated sludge using Fluidized Catalytic Cracking (FCC). This process could convert most of the compounds present in sludge into fuel, and thus, increase fuel yields and reduce cost.

Several catalysts, sludge to catalyst ratios, sludge moisture content, fluid flow rates and temperatures were tested. Initial results using ZSM-5 at 500°C with air showed considerable production of aromatic compounds, from benzene to methylnaphthalene.

See more of this Session: Conversion of Solid Wastes to Energy and/or Product
See more of this Group/Topical: International Congress on Energy 2011

(733b) Simultaneous Extracellular Polymeric Substance (EPS) and Lipid Production by Activated Sludge Via Fermentation of Glucose

Patrisha J. Pham, Rafael Hernandez, Hien Nguyen, Andro H. Mondala and W. Todd French, Dave C. Swalm School of Chemical Engineering, Mississippi State University, Mississippi State, MS

The use of activated sludge microflora to convert sugars, derived from the hydrolysis of lignocellulose biomass, into lipids for biofuel production is of current research interest in our group. Lipid accumulation in this system has been enhanced by manipulating the carbon-nitrogen (C:N) ratio and initial sugar loading on activated sludge bioreactors with glucose as the main carbon source. Activated sludge is a flocculated suspension composed of microorganisms that can also produce extracellular polymeric substances (EPS) *in situ* during fermentation. Microbial EPS which are secreted largely by bacteria in harsh environments are a matrix rich in polymers including polysaccharides, glycoproteins, proteins, nucleic acids and phospholipids. Thus, studying EPS production along with lipid formation is essential for understanding the technical and economic value of the whole process. In this study, samples were obtained from aerobic batch fermentation inoculated with 20 % (v/v) activated sludge. The total incubation time for the batch fermentation experiment was seven days. EPS from the collected samples were extracted using a Dowex cation exchange resin in phosphate buffered saline solution and precipitated with 95% v/v cold ethanol and dialyzed. Fourier transform-infrared spectroscopy of freeze-dried EPS revealed characteristic IR peaks for polysaccharides and proteins. Total polysaccharide and protein concentrations were estimated in each EPS sample using phenol-sulfuric and modified Lowry methods, respectively. An increase in lipid % (g/g cell dry weight) and a decrease in specific EPS % (g/g cell dry weight) was observed from day 0 to 7 in all treatments. Increasing the carbon loading while maintaining the same C:N ratio increased lipid % and decreased specific EPS %. However for a low C:N ratio, lipid % remained constant while specific EPS % decreased. These results will be useful for manipulation of EPS and lipid production and consequently enable development of valuable strategies for improving the process.

See more of this Session: Conversion of Solid Wastes to Energy and/or Product

(733c) The Hydrolysis of Nylon6 In Ionic Liquid to Recycle Nylon Waste

Xiao-Li Xia and Hai-Jun Wang, Jiangnan University, Wuxi, China

To date, several processes for Nylon6 depolymerization have been put forward. Herein, we first examined several ionic liquids for the hydrolysis of Nylon6. These ionic liquids are Brønsted acidic and they have been used both as a catalyst and environmental solvent for the hydrolytic reaction of Nylon6. The final result showed that using 1-butyl-3-methylimidazolium chloride ([Bmim]Cl) as solvent and catalyst for hydrolysis of Nylon6 ,the high conversion of Nylon6 and yield of Aminocaproic acid were obtained. In order to confirm the product, FT-IR and ^{1}H NMR were performed. After the experimental, theoretically, we also used high level ab initio methods with the 6-31G(d) basis sets to calculate the interaction between [Bmim]Cl and Nylon6. Based on our calculated interaction, we conclude that hydrogen bonds exist between [Bmim]Cl and Nylon6 which explain that Nylon6 dissolve in [Bmim]Cl.

See more of this Session: Conversion of Solid Wastes to Energy and/or Product
See more of this Group/Topical: International Congress on Energy 2011

(733d) Process Development for Efficient Utilization of Waste Plastic to Liquid Product for Rural Development In Nigeria: Gasification Option

Bamikole Amigun, Environmental Biotechnology and Conservation Department, National Biotechnology Development Agency (NABDA), Abuja, Nigeria, Christie Onyia, Environmental Biotechnology and Conservation Department, National Biotechnology Development Agency, Abuja, Nigeria and Bamidele Ogbe Solomon, Director General's Office and consultancy, National Biotechnology Development Agency (NABDA), Abuja, Nigeria

Waste production is a very important problem in all countries especially in developed countries and Nigeria is no exception. Waste plastics create a very serious environmental challenge because of their huge quantities and their disposal problems, due to their non-biodegradable attributes (or the biodegradation process is very slow). Plastic waste poses a serious environmental problem that is not addressed by recourse to landfill. Open dumping has been the predominant solid waste disposal option in Nigeria. Their destruction by incineration poses serious air pollution problems due to the release of airborne particles and carbon dioxide into the atmosphere. Only a limited amount of plastics can be recycled, while most plastics could be used, for example in energy production. The management of plastic waste using gasification in fluidized bed technology is a technically feasible option. Apart from eliminating the waste, a chemical recycling is produced during this process, obtaining a gas (syngas) which can be used in different applications such as ethanol through bacteria conversion of the syngas intermediate. Ethanol can be further converted to ethanol gel fuel to substitute the use of paraffin and traditional firewood in rural communities. This is doubly environmentally friendly since it will reduce the volume of plastic waste being disposed of in landfill whilst producing green fuel without generating green house gases. This application could give some prospect of self-reliant energy supplies at local levels, with potential economic, ecological, social, and security benefits; thereby improving the socio-economic and political wellbeing of the rural people.

See more of this Session: Conversion of Solid Wastes to Energy and/or Product
See more of this Group/Topical: International Congress on Energy 2011

(733e) An Anaerobic Bacterial Treatment of Corn Stover to Prepare a Biorefining Feedstock

James MacLellan, Biosystems and Agricultural Engineering, Michigan State University, East Lansing, MI

Anaerobic Bacterial Treatment of Corn Stover to Prepare a Biorefining Feedstock

James MacLellan, Zhengbo Yue, Robert Kraemer, Wei Liao*

Department of Biosystems and Agricultural Engineering, Michigan State University

East Lansing, Michigan 48823

Abstract

The objective of this study is to apply an anaerobic bacterial consortium to uptake the hemicellulose in corn stover to prepare a relatively cellulose-enriched feedstock for biofuel production. Previous research demonstrated that submerged anaerobic bacterial culture is capable of converting hemicellulose in raw lignocelluloses to methane gas. The in-depth study is needed to optimize the culture, enhance the hemicellulose utilization efficiency, and improve the quality of treated cellulose as a biorefining feedstock. An anaerobic bacterial consortium from animal manure will be screened and enriched as the seed for this study. Two feedstocks of raw corn stover and steam-treated corn stover will be used to compare the effects of structure on the anaerobic culture of hemicellulose uptake. A completely Randomized Design (CRD) on seed-to-feedstock ratio, culture temperature, and culture time will be carried out. Fiber composition, methane production, and microbial consortia will also be analyzed. An optimal condition can then be established based on the quality and quantity of cellulose in the treated corn stover. The ethanol production on the treated corn stover will be fulfilled in order to elucidate the effects of anaerobic bacterial treatment on biofuels production.

*: Corresponding author

Dr. Wei Liao

Department of Biosystems & Agricultural Engineering

Michigan State University

202 A.W. Farrall Hall

East Lansing, MI 48824-1323

Tel: 517-432-7205

Fax: 517-432-2892

E-mail: liaow@msu.edu

See more of this Session: Conversion of Solid Wastes to Energy and/or Product
See more of this Group/Topical: International Congress on Energy 2011

(733f) Towards Developing Anaerobic Digestion Based Biorefinery - Research On Food Wastes As Feedstock

Shulin Chen[1], Liang Yu[2], Jingwei Ma[2] and Craig Frear[2], (1)Department of Biological Systems Engineering, Washington State University, Pullman, WA, (2) Washington State University, Pullman, WA

Anaerobic digestion (AD) is a proven technology for waste treatment. However, as the organic material becomes more valuable and being treated as resources, traditional AD process is no longer adequate. The Bioprocessing and Bioproduct Engineering Laboratory at Washington State University has been devoting its effort in developing the second generation of AD technology with the concept of biorefinery, featured by producing various products from improved AD processes. Key to an AD biorefinary is innovative design of the digestors. This presentation reports the recent results from a study aimed at evaluating a two stage-digester designed for treating food wastes operated within a solid concentration range of 15-20%. The digester had a unique two-stage configuration that resulted in high biogas production rate than the existing technologies based on the laboratory data. The experimental observations also led to a simple mixing design that has low capital and operational costs. The information provides some new insights in AD research and development related to municipal solid waste applications.

See more of this Session: Conversion of Solid Wastes to Energy and/or Product
See more of this Group/Topical: International Congress on Energy 2011

(740a) Pneumatic Conveying and Packaging of Flame-Made Nanopowders

Karsten Wegner[1], Vera Tschedanoff[1], Max Eggersdorfer[1], Christian Amann[2] and Steven Multer[2], (1)Particle Technology Laboratory, Swiss Federal Institute of Technology (ETH Zurich), Zurich, Switzerland, (2)Hecht Technologie GmbH, D-85276 Pfaffenhofen, Germany

After a decade of intense research on nanoparticle development, a plethora of new materials has been synthesized at the laboratory scale, as is apparent for instance from a recent review on flame aerosol technology (Teoh et al., 2010). Today's challenge is the translation of these achievements into an industrial production environment which poses questions on continuous nanoparticle manufacture, safe handling and packaging.

Here, a fully automated pilot plant for nanoparticle manufacture by flame spray pyrolysis (FSP, Bickmore et al., 1996; Mädler et al., 2002) is presented that allows continuous production of nanopowders at rates up to 500g/h. Nanoparticles are collected with bag-house filters and are discharged to a pneumatic conveying system that transports the dry powder to a packaging site.

Bulk nanopowder flow characteristics and requirements for the design and operation of the pneumatic conveying system are studied by the example of two differently-sized zirconia nanopowders. A filling head with a continuous polymer inliner is investigated for contamination-free nanopowder packaging. Release of nanoparticles into the workspace during production, conveying and packaging is monitored with a scanning mobility particle sizer (SMPS) at different locations in the pilot plant in order to develop suitable standard operation procedures and to guide equipment design. Finally, results on cleaning of the nanopowder handling equipment for product change or maintenance are reported along with nanoparticle release levels.

References:

Bickmore, C.R., Waldner, K.F., Treadwell, D.R. and Laine, R.M. (1996), *J. Am. Ceram. Soc.* **79**, 1419-1423.

Mädler, L., Kammler, H.K., Mueller, R. and Pratsinis, S.E. (2002), *J. Aerosol Sci.* **33**, 369-389.

Teoh, W.Y., Amal, R. and Mädler, L. (2010), *Nanoscale* **2**, 1324-1347.

See more of this Session: Handling and Processing of Nanoparticles
See more of this Group/Topical: Particle Technology Forum

(740b) Investigating the Effect of Precursor Composition On the Size Distribution of Droplets Generated by An Ultrasonic Droplet Generator

George Peabody[1], Kai Zhong[1], Howard Glicksman[2], Sheryl Ehrman[1] and Andre Marshall[3], (1)Department of Chemical and Biomolecular Engineering, University of Maryland, College Park, MD, (2)DuPont Electronic Technologies, Research Triangle Park, NC, (3)Department of Fire Protection Engineering, University of Maryland, College Park, MD

Ultrasonic atomization processes serve as an important source of monodispersed highly tunable micro and nano droplets for use in a variety of industries. Specifically, common applications include drug delivery, micro and nano materials processing and environmental emulation. The nature of the processes developed provides several distinct advantages over other systems designed to similarly generate droplets. The ultrasonic generators can be run in parallel to generate large amounts of aerosol. To fully appreciate this intrinsic value and allow for future applications we have begun developing an understanding of the physics behind the droplet formation process. We have examined the correlation between the surface tension and viscosity of the precursor solutions and carrier gas flow rates on the size distribution of the generated droplets. Our investigation utilized a generator that operates at a frequency of 1.7 MHz, when operating the generator forms a cone with precursor solution, on which the droplets with diameter around 10 microns are formed, probably due to the capillary theory. Observations from a *Malvern/INSITEC* system were used to characterize the aerosol droplet size distribution, which revealed an additional, larger size distribution of the droplets under certain design conditions. By manipulating the precursor salt and cosolvent volume percent, one can generate a second mode at 50 microns, and also a lower atomization rate, when a viscosity of at least 3mm/s was measured. The second mode was confirmed through the use of Shadowgraphy: a high-speed camera and laser illumination source. The images taken indicate that the larger mode was formed via a different process than the current capillary theory indicating the physics of the system are not fully appreciated in the literature.

See more of this Session: Handling and Processing of Nanoparticles

(740c) Integrated Rapid Purification of Polymeric Nanoparticles for Drug Delivery In Microfluidic Channels

Minsoung Rhee[1], Jong-Min Lim[1], Robert Langer[2], Omid C. Farokhzad[3] and Rohit Karnik[4], (1)Mechanical Engineering, MIT, Cambridge, MA, (2)Department of Chemical Engineering, Massachusetts Institute of Technology, Cambridge, MA, (3)Department of Anesthesiology, Harvard Medical School, Boston, MA, (4) MIT, Cambridge, MA

Integrated Rapid Purification of Polymeric Nanoparticles for Drug Delivery in Microfluidic Channels

Minsoung Rhee[1,2], Jong-min Lim[1], Robert S. Langer[1], Omid C. Farokhzad[1,2], Rohit Karnik[1]

[1]Massachusetts Institute of Technology, USA.

[2]Brigham and Women's Hospital, Harvard Medical School, USA.

We report here the development of an integrated microfluidic system that performs polymeric NP synthesis for drug delivery followed by rapid in-line purification in a continuous manner. Polymeric nanoparticles (NPs) comprised of poly(lactide-*co*-glycolide)-*b*-polyethyleneglycol (PLGA-PEG) block copolymers have attracted much attention as targeted nanocarriers for delivery of chemotherapeutic agents [1] due to their promising physicochemical characteristics such as easy incorporation of various targeting agents, enhanced stability, tunable drug release, and the ability to carry a payload of thousands of drug molecules per NP [2]. To optimize such characteristics that are not independent of each other, systematic synthesis of such targeted NPs and screening have become very important since precise control of the physicochemical parameters of NPs is required to achieve optimal therapeutic efficacy [3]. Regardless of whether the NPs are synthesized in the microfluidic system [4-5], NP synthesis procedures involve a large quantity of organic solvents and potentially toxic or non-biocompatible materials like unencapsulated drug molecules, large polymer aggregates, and stabilizing agents that substantially limit direct use of synthesized NPs without purification for the following in vitro and in vivo screening. Most of conventional NP purification method typically relying on centrifugation are highly labor-intensive or require extended processing time to achieve desirable concentration and purification of NPs. Besides, separate batch-to-batch procedures, which can significantly retard the overall preparation process, often becomes a bottleneck for screening and optimization of NPs. Therefore, the development of efficient NP purification system has been a common issue in the field of NP synthesis.

We have designed a microfluidic technology that performs counter-current flow dilution where prepared NPs solution with impurities and washing deionized water are oppositely directed in the microchannel network. This counter-current flow maintains maximum concentration gradients of impurities that in turn facilitate diffusive dilution by washing flow. Since the organic solvent and non-encapsulated drug molecules have higher diffusivities than that of polymeric NPs, the impurities are washed out and NPs can be purified and concentrated with the specially designed microfluidic system where convection between two opposite flows is minimized by imposing high hydrodynamic resistance in connecting bridges. The device operates with input polymer precursors along with drug molecules at programmed flow rates by syringe pumps, and drug-encapsulated NPs assembled in the channel will flow into the purification microchannel network where organic solvent and non-encapsulated drug will be rapidly removed. Such integration of purification functionality eliminates the need for manual collection, delivery, and filtration and thus accelerates the NP preparation and screening process by orders of magnitude. Reduction in processing time also results in faster stabilization of NPs, minimizing the possibility of premature release of encapsulated drug molecules or NP aggregation.

To ensure sufficient time for diffusive dilution, the system may run at a low flow rate, which potentially reduces the throughput. Parallelization of microfluidic purification channels is thus pursued to achieve higher purification capacity. Using the parallel channel architecture constructed by using the multilayer microfluidic device fabrication technique, By simulations and experiments, we propose and examine a microfluidic purification system to achieve 99.9% v/v removal of organic solvent and 99% w/w removal of free drug molecules. We believe that this approach will lead to a platform technology for fully automated and high throughput preparation of NPs that can be directly used for screening without further processing steps. Consequently, development of an integrated microfluidic system for single-step NP preparation will have a substantial impact on current NP research in our lab and other labs by enabling high throughput and increased stability of purified NPs.

REFERENCES:

[1] "Formulation of functionalized PLGA-PEG nanoparticles for in vivo targeted drug delivery," J. Cheng *et al.*, *Biomaterials,* **28**, 869 (2007).

[2] "Targeted nanoparticles for cancer therapy," F. Gu *et al.*, *Nano Today,* **2**, 14 (2007).

[3] "Translational Nanomedicine: Status. Assessment and Opportunities," J. S. Murday, R. W. Siegel, J. Stein, J. F. Wright, *Nanomedicine,* **5**, 251 (2009).

[4] "Microfluidic Platform for Controlled Synthesis of Polymeric Nanoparticles," R. Karnik et al., Nano Lettlers, **8**, 2906 (2008).

[5] "Synthesis of Size-Tunable Polymeric Nanoparticles Enabled by 3D Hydrodynamic Flow Focusing in Single-Layer Microchannels", M. Rhee et al., *Advanced Materials*, **23**, H78 (2011).

See more of this Session: Handling and Processing of Nanoparticles
See more of this Group/Topical: Particle Technology Forum

(760a) Development of Accurate Intermolecular Potentials for Mono and Dinitro Aromatic Comounds

Navendu Bhatnagar, Chemical Engineering, Wayne State University, Detroit, MI, Ganesh Kamath, Computer Science and Mathematics Division, Oak Ridge National Laboratory, Oak Ridge, TN and Jeffrey J. Potoff, Chemcial Engineering and Materials Science, Wayne State University, Detroit, MI

Over the past two decades, there has been considerable work in the development of insensitive munitions (IM), which exhibit low shock sensitivity and high thermal stability, due to the increased safety and environmental concerns associated with the traditionally used explosives. Contamination of the environment (ground water, soil and sediment) by explosives due to military activities such as weapon production and handling, waste discharge, testing and training of weapons and demilitarization has become a multi-billion dollar problem. Therefore, development of methodologies that can be used to predict the environmental fate of a particular compound before its deployment or even pre-synthesis is of great importance, and could lead to significant long-term cost savings. Molecular simulation has emerged as an attractive tool to study and predict the physicochemical properties of complex species for which little or no experimental data exist. In the case of energetic materials, the use of molecular simulation provides a means to screen candidate molecules for environmental impact prior to synthesis. Such a priori estimates are expected to lead to significant cost savings in the development of green munitions.

In this work, molecular models are developed for mono- and di- nitro aromatic compounds, including 2.4-dinitroanisole (DNAN), N-methyl-p-nitroaniline (MNA), 1,3-dinitrobenzene (13DNB), 1,4-dinitrobenzene (14DNB), 2-nitroanisole (2NAN) and 4-nitroanisole (4NAN), based on the Transferable Potentials for Phase Equilibria (TraPPE) [1,2,3]. These models are used in molecular dynamics to predict physicochemical properties, such as log K_{ow}, K_H, crystal density, melting point, vapor pressure and hypothetical vapor-liquid coexistence curves. Molecular dynamics simulations are performed with NAMD using united atom TraPPE force field and SPC/E water model. Absolute values of the 1-octanol/ water and air water partition coefficients were calculated directly using the adaptive biasing force method [4].

1. Martin, M.G.; Siepmann, J.I. *J. Phys. Chem. B* (1998), 102, 2569.

2. Chen, B.; Potoff, J.J.; Siepmann, J.I. J. Phys. Chem. B (2001), 105, 3093.

3. Wick, C.D.; Stubbs, J.M.; Rai, N.; Siepmann, J.I. J. Phys. Chem. B (2005), 109, 18974

4. Rodriguez-Gomez, D.; Darve, E.; Pohorille, A. *J. Chem. Phys.* (2004), 120, 3563.

See more of this Session: Thermophysical Properties of Energetic Materials
See more of this Group/Topical: Particle Technology Forum

Physical properties of munitions compounds -- Chinalake-20 (CL-20), Diamino-Dinitroethylene (DADE), Cyclotetramethylenetetranitramine (HMX), and Cyclotrimethylenetrinitramine (RDX)

Thermophysical Properties of Energetic Materials
Section: 03E03

Siri Chakka[1], Carmen Costales-Nieves[1], Veera M. Boddu[1], and Reddy Damavarapu[2],

(1) Environmental Processes Branch, US Army ERDC-CERL, Champaign, IL,
(2) Energetics and Warheads Division, U.S. Army Armament Research, Development and Engineering Center, Picatinny Arsenal, NJ

Physicochemical properties of chemical compounds over a series of temperatures are very essential for understanding their environmental fate and distribution, biotransformation, and for developing potential wastewater treatment processes. In this study, aqueous solubility (S_w) and octanol-water partition coefficient (K_{ow}) were measured for four insensitive munitions compounds, Chinalake-20 (CL-20), Diamino-Dinitroethylene (DADE), Cyclotetramethylenetetranitramine (HMX), and Cyclotrimethylenetrinitramine (RDX) from 288K to 313K. Differential Scanning Calorimetry (DSC) was used to obtain melting and boiling points. The DSC data were correlated using the standard Van't Hoff equation in order to determine the influence of temperature on the solubility of the munitions compounds. The enthalpy and entropy of phase transfer were also derived from the experimental data. The experimental results obtained were compared with available literature values and predicted values using EPI suite and the SPARC models. Details of the experimental results for the four munitions compounds and model predictions will be presented.

A manuscript being submitted to a peer reviewed journal.

(760c) Influence of Dry and Wet High Energy Ball Milling On Reactivity In Ni-Al Gasless Heterogeneous System

Khachatur V. Manukyan[1], Aaron Mason[2], Robert V. Reeves[2], Steven F. Son[2] and Alexander S. Mukasyan[1], (1)Chemical and Biomolecular Engineering, University of Notre Dame, Notre Dame, IN, (2)School of Mechanical Engineering, School of Aeronautics & Astronautics, Purdue University, West Lafayette, IN

The influence of short-term (up to 20 min) high energy ball milling (HEBM) on the reactivity of a heterogeneous Ni-Al system has been investigated. More specifically, we studied the effects of the dry and wet HEBM on the self-ignition temperature and kinetics of chemical reactions in the stoichiometric 1:1 mol Ni-Al mixtures. It was shown that HEBM under dry inert conditions leads to formation of coarse (0.1–2 mm) Ni/Al composite particles with defined layered metal microstructure. The electron microscopy investigation revealed that the thickness of metal (Ni, Al) layers in such composite particles varies from 0.1 to 1 μm. However, XRD analysis showed that no intermetallic phases were formed during the dry HEBM. In turn, the thermal analysis revealed that 20 min of dry HEBM leads to the significant decrease of the exothermic reaction onset temperature from ~660 °C, i.e. eutectics in Ni-Al system, to ~400 °C. For more enhancement of the reactivity, the complex approach has been used, i.e. the dry HEBM process was "arrested" just before the reaction initiated in the milling vial, followed by additional wet HEBM in hexane for 10min. Such approach decreases the size of Ni/Al composite particles to 30–40 μm, as well as leads to their microstructural transformation. Thermal analysis revealed that self-ignition temperature for such media is extremely low and equals to 300 °C. Apparent activation energies of different HEBM-formed reaction medium were also measured by using conventional differential thermal analysis, as well as unique, so-called, electro thermal explosion method. Both methods showed the significant influence of HEBM on the effective activation energy for the considered reactions. The mechanisms of the dry and wet HEBM's influence on the chemical reactivity of the considered systems are also discussed.

See more of this Session: Thermophysical Properties of Energetic Materials
See more of this Group/Topical: Particle Technology Forum

Vapor Pressure of RDX - Experimental and Correlation

Michael L. Reinig, Tushar K. Ghosh and Dabir S. Viswanath, Nuclear Science and Engineering Institute, University of Missouri, Columbia, MO

This presentation discusses the vapor pressure of 1,3,5-Trinitroperhydro-1,3,5-triazine (RDX) available in the literature and current experimental data. Experimental data have been collected using two techniques via Knudsen method and the Gas Chromatography-Retention Time (GC-RT) method. The two techniques - Knudsen technique at low temperatures (320 -350 K) and GC-RT technique at high temperatures (400 - 495 K) were adopted to collect consistent, accurate vapor pressure data. The Knudsen effusion method, which is based on the kinetic theory of gases, measures vapor pressures of test substance by molecular effusion through a small orifice which encloses the sample cell. The vapor pressure, P (Pa), is related to the diameter of the hole by the expression:

$$P = \frac{g}{Aft}\sqrt{\frac{2\pi kT}{m}} = \frac{g}{Aft}\sqrt{\frac{2\pi RT}{M}}$$

where k is the Boltzmann's constant, T is the temperature in K, A is the area of the orifice, t is the time in seconds, g is the mass loss, R is the universal gas constant, m is the molecular mass and M is the molar mass. The effective area of the hole is A·f where f is the Clausing factor and depends on the ratio of hole length to hole radius.

The GC-Retention time is based on the relation between isothermal vapor pressure and the retention time for a homologous series at different temperatures expressed as:

$$\mathrm{Log}_{10} P = A - B \cdot t_M$$

where P is the vapor pressure (Pa), t_M is the retention time (min), and A and B are constants. The current work has expanded the vapor pressure and temperature data for RDX, a well known explosive material. The data should be useful in building sensors for detecting RDX.

The vapor pressure versus temperature data was fit to Antoine's equation resulting in the constants A = 34.88, B = 25418.7 and C = 310.88. The maximum deviation was 13.1% and an average absolute deviation of 5.0% in the temperature range of 328.15 – 493.15 K. The maximum uncertainty in the vapor pressure data based on an error analysis is 1.01% in the case of the Knudsen method and uncertainty values for the GC-Retention Time method are less than at 10%.

(760e) Combustion Dynamics of Aluminum Particle In Air

Bing Wang, Salil Mohan, Carlo Badiola and Edward L Dreizin, Chemical Engineering, New Jersey Institute of Technology, Newark, NJ

Experiments on combustion of single spherical Al particles with the sizes in the range of 2 – 20 μm were processed to determine the Al burn rate as a function of particle size, temperature, and preceding burn history. In experiments, Al particles were ignited by a CO_2 laser beam in air at room temperature.

Temperature history of Al particles was measured. This temperature history was used in a heat transfer model formulated to determine the rate of heat release necessary to balance convection and radiation heat losses from the burning particle. The rate of heat release was related to the rate of oxygen diffusion from air towards burning particle, and respective Stefan flow was included into the heat transfer model. The calculation considered three different combustion modes. In the first case, the oxide layer grows uniformly on the top of the Al particle. The second case assumes the pure vapor-phase combustion so that no oxide is present on the particle surface. Finally, the third case considers partial removal of the produced oxide, so that a fraction of the particle surface is covered by an oxide cap. The size of the growing oxide cap was adjusted to match the experimental data for different particle sizes. Additional experiments were performed to burn Al particles in air and collect and analyze combustion products to provide a reference point for the model development.

See more of this Session: Thermophysical Properties of Energetic Materials
See more of this Group/Topical: Particle Technology Forum

Degradation of Insensitive Munitions Compound Diaminodinitroethylene (DADE) Using Activated Nanoparticles of MgO and ZnO

Kelsey N. Wiles (Speaker), Veera M. Boddu*, Carmen Costales-Nieves,
Environmental Processes Branch, US ARMY ERDC-CERL, Champaign, IL

Anthony Bednar
Environmental Laboratory, US ARMY ERDC-EL, Vicksburg, MS

Abstract

Degradation mechanisms of an Insensitive Munitions (IM) compound, Diaminodinitroethylene (DADE, also known as FOX-7), were studied, in a laboratory environment, on the surface of catalytic nanoparticles mixed oxides of magnesium oxide (MgO) and zinc oxide (ZnO). A mixture of nanoparticles MgO and ZnO was synthesized following a sol-gel approach. These nanoparticles were characterized using Scanning Electron Microscopy (SEM), Fourier Transform Infrared Spectroscopy (FTIR), and X-Ray Diffraction (XRD) techniques. DADE was dissolved in a solvent and placed in a batch reactor containing the nanoparticles. Oxygen and hydrogen gases were bubbled through to facilitate oxidation/reduction reactions. Samples that were taken before and after these reactions were analyzed using proton Nuclear Magnetic Resonance Spectroscopy (^{1}H NMR), Liquid Chromatography (LC), and Mass Spectrometry (MS) in order to determine the reaction products. Degradation of DADE was at its maximum under reduction conditions, and also evidence of acetic acid production was observed.

1. Introduction

During the past decade, the Department of Defense (DoD) led an initiative requiring all new munitions to withstand accidental handling, high temperatures, mechanical impact, and shock. Thus, Insensitive High Explosives (IHEs) are being developed. They are considered insensitive based primarily on their insensitivity to accidental explosions or impact. DADE is one of the most common IHEs. It was first synthesized in 1988, and it poses a higher activation energy (56 kcal/mol) than other commonly used explosive compounds such as RDX (E_a = 40 kcal/mol) and HMX (E_a = 35 kcal/mol) (Karlsson et al., 2002). The DADE molecular structure is shown in Figure 1.

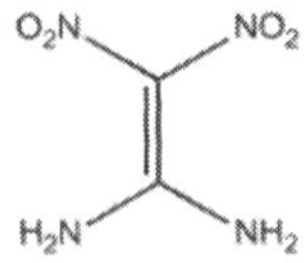

Figure 1: 1,1-diamino-2,2-dinitroethylene (DADE)

**corresponding author:* Veera.Boddu@usace.army.mil

When observing the degradation mechanisms and products of a similar compound such as nitroethylene, the degradation products are observed to be $H_2C=O$, H-C≡N-O, HCN, and formic acid (Shamov et al., 2001). When the temperature is increased above 200°C, DADE degrades rather than melting (Bellamy, 2007). This degradation has been studied by mass spectrometry, and it has been determined that one path of degradation may be H_2O condensation, while another possible path is NO_2 elimination (Karlsson et al., 2007). They also claim that breaking of the C-NO_2 bond may occur as well. According to one source (Gao et al., 2006), the end products would be HCN, NH_3, CO_2, NO, and HN=C=O. Other sources (Politzer et al., 1998 and Gindulyte et al., 1999) stated that the end products would be CO, N_2, and H_2O.

The main objective of this work was to investigate the degradation mechanisms of DADE on the surface of catalytic nanoparticles of mixed oxides of MgO and ZnO. The MgO and ZnO nanomaterials were synthesized using the sol-gel approach, and their bandgap energy was effectively manipulated by varying the oxides compositions. When subjected to reduction/oxidation agents such as hydrogen gas (H_2) and oxygen gas (O_2), these materials are expected to facilitate electron transfer mechanisms promoting surface reduction-oxidation reactions. Pure ZnO has a bandgap energy of 3.3 eV, and pure MgO has a bandgap energy of 7.8eV (Fujita et al., 2005). By combining the oxides in different ratios, nanoparticles can be manufactured to give a spectrum of varying bandgap energies (Zhang et al., 2005). Nanoparticles are used because they have a high surface area to volume ratio, with a high density of ions on their surface. We hypothesized that proper selection of chemical composition of nanomaterials will result in bandgap energies that will facilitate oxidation/reduction reactions to transform IM compounds into environmentally benign compounds.

2. Experimental Procedure

2.1 Materials

The chemicals used for the synthesis of nanomaterials were obtained from commercial sources. For magnesium oxide synthesis, materials used were: pure magnesium metal ribbons, 99% pure optima grade methanol, nitrogen gas (99.998% purity), laboratory deionized water and $C_6H_3(CH_3)(NO_2)_2$ (1-methyl-2,6-dinitrobenzene or DNT). For zinc oxide synthesis, materials used were: anhydrous zinc chloride powder, 99.9% HPLC grade isopropanol, and a 5.0N standardized solution of ammonium hydroxide. DADE was obtained from US Army Armament Research, Development and Engineering Center/Research, Development, and Engineering Command (ARDEC/RDECOM) facilities (Picatinny Arsenal, NJ 07806-5000).

2.2 Preparation of Nanomaterials

Both nanomaterials were prepared using sol-gel method, following different procedures. The chemical processes for preparation of the MgO particles can be summarized by the following chemical reactions (Boddu et al., 2008):

$$Mg + 2CH_3OH \rightarrow Mg(OCH_3)_2 + H_2$$
$$Mg(OCH_3)_2 + H_2O \rightarrow Mg(OH)_2 + 2CH_3OH$$
$$Mg(OH)_2 \rightarrow MgO + H_2O$$

Similarly, ZnO particles were synthesized starting with anhydrous zinc chloride and ammonium hydroxide described in the literature (Vaezi, et al., 2007). The synthesis involved several steps. The overall chemical reaction can be summarized as:

$$(NH_4)_2ZnO_2 + H_2O \rightarrow ZnO + 2\ NH_4OH$$

In the preparation of the mixed oxides nanoparticles, the oxides of magnesium and zinc were added together in isopropanol at selected compositions; washed with isopropanol; sonicated to speed mixing, and the solids were then separated using centrifugation. Then the mixture was subjected to thermal treatment in a pressurized vessel at very high temperatures (~500 °C) in an inert nitrogen atmosphere, in order to remove completely any residual organics and to crystallize the powders. After the thermal treatment, mixtures were re-heated at high temperatures ~400°C in an open air atmosphere to remove their gray or yellowish appearance (MgO and ZnO, respectively) and any other residuals. Characterization details of the nanomaterials are not presented here. Refer to (Boddu et al., 2010).

2.3 Reactivity Studies

DADE degradation was studied by a reduction/oxidation reaction with hydrogen or oxygen catalyzed with different compositions of zinc and magnesium oxide nanoparticles. A mass of 0.100 g of nanoparticle was measured into a weigh boat. A volume of 50 mL of the ~5 ppm stock solution of DADE in H_2O was measured into a graduated cylinder. This solution was used to quantitatively transfer the nanopowder into the three neck flask before being added to the flask itself. A condenser was attached to one of the necks of the three neck flask, with the cooling system set to 4°C. Gas tubing was attached to another neck of the three neck flask so that catalytic oxidation/reduction gas could be pumped into the system to ensure gentle bubbles were present in the solution. A stir bar was placed in the flask to continuously stir the reaction mixture. The gas was turned on to start the reaction and left on for a total of 24 hours. The third neck of the three-neck flask was stoppered to facilitate substance transfer in and out of the system. Four samples of 1 mL each were taken at the beginning of the reaction. After 24 hours, the gas tank was turned off and the reaction was stopped. The mixture was transferred to a graduated cylinder, using washes and a rotary evaporator if necessary, and the volume of the end solution was brought to 50 mL. Four samples of 1 mL each were also taken at the end of the reaction. Samples were submitted to HPLC and FTIR for analyses.

3. Results and Discussion

3.1 Degradation Studies of DADE on MgO and ZnO Nanostructures

Many different methods were used to analyze the results of the reactions and compare them to the solutions prior to the reactions.

i. High Performance Liquid Chromatography (HPLC) Analysis of DADE in H_2O

HPLC analyses were performed and the results and experimental parameters are shown in Table 1. The peak produced by DADE eluted at around 5.5 min in this mobile phase, and the area underneath this peak was measured to determine the concentration of DADE in each individual sample.

Table 1: HPLC Results for DADE Degradation Experiments

Composition of Nanomaterials		Redox agent	Concentration before reaction (ppm)	Concentration after reaction (ppm)	% Difference
% of ZnO	% of MgO				
100	0	Hydrogen	4.851	4.538	6.438
0	100		5.419	3.587	33.807
40	60		7.165	3.592	49.859
60	40		7.430	3.688	50.357
100	0	Oxygen	3.974	2.848	28.323
0	100		3.345	2.644	20.939
40	60		4.390	2.808	36.041
60	40		5.204	3.413	34.423

The HPLC analyses that were performed for the explosive DADE have shown degradation of DADE varying from 6-50%, as shown in Table 1. HPLC analyses are shows around 50% degradation of FOX 7.

ii. Nuclear Magnetic Resonance Spectroscopy (^1H NMR) Analysis of DADE

^1H NMR spectra for DADE in DMSO is shown in Figure 2. Degradation of DADE was at its maximum under reduction conditions, and also evidence of acetic acid production was observed.

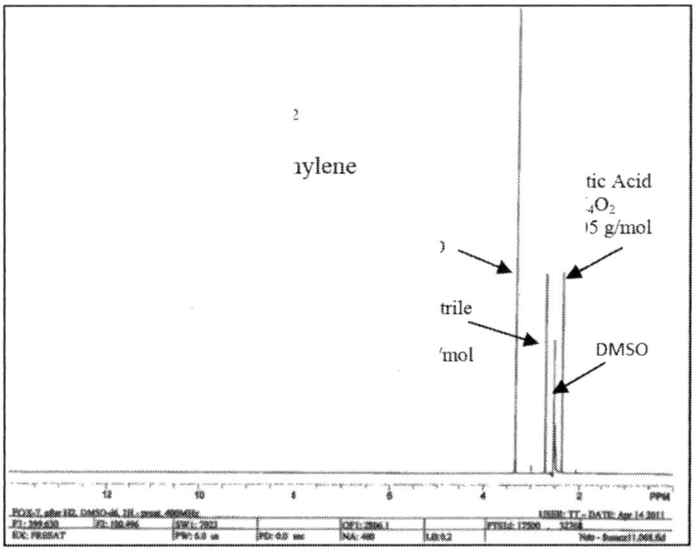

Figure 2: H NMR Spectrum of DADE in DMSO after reduction reaction with DMSO

4. Conclusions

The Magnesium Oxide and Zinc Oxide nanomaterials were synthesized using the sol gel approach. Surface morphology was observed showing nodular and needle structures. The bandgap energy values were obtained by measuring photoluminescence of the nanomaterials. Preliminary results show that these nanomaterials of the mixed oxides can catalytically degrade DADE and other insensitive munitions compounds at room temperature. Preliminary results show that the degradation products from DADE include acetic acid and acetonitrile.

5. References

[1] Bellamy, A.J. DADE *Struct. Bonding* 2007, *125*, 1–33.

[2] Boddu, V., Viswanath, D.S., Maloney, S.W. Synthesis and characterization of coralline magnesium oxide nanoparticles. *Journal of American Ceramic Society*. 2008. 1-3.

[3] Boddu, V., Viswanath, D.S., Cataneda, Y., Costales-Nieves, C., Kirgan, R., Bednar, A., Qasim, M. Designer nanomaterials: bandgap energy Manipulation of mixed metal oxides of magnesium and zinc oxide through sol-gel synthesis. *27th Army Science Conference, Orlando, FL.* December 2010.

[4] Database of NMR Spectra. http://www.nmrdb.org/predictor (accessed April 7, 2011)

[5] Fujita, S., Tanaka, H., Fujita, S. MBE growth of wide bandgap wurtzite MgZnO quasi-alloys with MgO/ZnO superlattices for deep ultraviolet optical functions. *Journal of Crystal Growth.* 2005, *278,* 264-267.

[6] Gao, H.-X., Zhao, F.-Q., Hu, R.-Z., Pan, Q., Wang, B.-Z., Yang, X.-W., Gao, Y., Gao, S.-L., Shi, Q.-Z. Thermochemical properties, thermal behavior and decomposition mechanism of 1,1-diamino-2,2-dinitroethylene (DADE). *Chinese Journal of Chemistry.* 2006. 24(2). 177-181

[7] Gindulyte, A., Massa, L., Huang, L., Karle, J. Proposed mechanism of 1,1-diamino-dinitroethylene decomposition: A density functional theory study. *J. Phys. Chem. A.* 1999, 103, 11045-11051.

[8] Karlsson, S., Östmark, H., Eldsäter, C., Carlsson, T., Bergman, H., Wallin, S., Pettersson, A. Detonation and Sensitivity Properties of DADE and Formulations Containing DADE, *12th Symposium (International) on Detonation*, San Diego, California, 11-16 August, 2002.

[9] Politzer, P., Concha, M.C., Grice, M.E., Murray, J.S., Lane, P., Habibollazadeh, D. Computational investigation of the structures and relative stabilities of amino/nitro derivatives of ethylene. *Journal of Molecular Structure.* 452 (1998) 75-83.

[10] Shamov, A.G., Khrapkovskii, G.M. A theoretical study of the gas-phase pyrolysis of nitroethylene. *Mendeleev Commun.,* 2001, 11(4), 163-164.

[11] Vaezi, M.R., & Sadrnezhaad, S.K. Nanopowder synthesis of zinc oxide via solochemical processing. *Material and Design.* 2007. *28,* 515-519.

[12] Zhang X., Ma H., Wang Q., Ma J., Zong F., Xiao H., Ji F., Hou S. Structural and optical properties of MgxZn1-xO thin films deposited by magnetron sputtering (2005) *Physica B: Condensed Matter,* 364 (1-4), pp. 157-161.

AUTHOR INDEX

AUTHOR INDEX

AUTHOR INDEX

AUTHOR INDEX

AUTHOR INDEX

AUTHOR INDEX

AUTHOR INDEX